ABOUT THE AUTHOR

The author operates "ED'S ENGINES", a small engine building shop in Santa Fe, New Me[illegible] drag races a big block Chevy Nova. He started writing about big block Chevys in 1995 and his first book CHEVROLET BIG BLOCK CHEVROLET PARTS INTERCHANGE MANUAL was published by CARTECH® in April of 1996. HOW TO BUILD MAX PERFORMANCE CHEVY RAT MOTORS is his second book about big block Chevy engines. Ed now spends most of his time writing, taking pictures, drag racing and working on computer models of Chevy engines. A third book on interchanges for the small block Chevrolet engine was published by CARTECH® in early '97.

ACKNOWLEDGMENTS

I was thinking the other day that this all started because I was trying to get an updated list of Chevrolet big block casting numbers together for my own use. One thing led to another. Dave Arnold the publisher of CARTECH and S/A DESIGN books thought, like me, that updated and new information on Chevrolet big blocks was needed and useful. He gave me the chance to write two books on the subject.

John Baechtel, my editor, as usual worked long hours to edit the text and do the photo layouts. He also provided some of the sidebars and photos. During this time John still managed to run his own engine dyno testing facility, WESTECH PERFORMANCE GROUP, edit a number of books by other authors, write his own magazine articles and books and still find enough time for his family.

David Hakim of Federal-Mogul Corporation also provided photos and technical data about their products for big block Chevys. Ken Casey, Jr. at Burt Chevrolet/GM Performance Parts in Englewood, Colorado helped me get some more photos of GM Performance Parts.

Todd Ryden at Autotronic Controls in El Paso, Texas, the makers of "MSD" ignitions, kindly provided photos. So did Mark Campbell, of Street & Performance. Lori Brotherton, of Brodix Cylinder Heads also provided photos and technical information on their big block castings. The guys at Super Shops, in Albuquerque let me pull parts off of their shelves and take photos. Thanks folks!

Santa Fe Auto Machine let me make take photos of motors they were working on at their shop.

Andrew Cordova and Larry Millington, both of Albuquerque, each helped me get photographs of the motors in their big block Chevelle drag racers and lent their support and encouragement.

COVER ENGINE

Our cover engine is a 520 cubic inch big block equipped with a Bowtie block, Callies crankshaft, Oliver rods, JE pistons, Brodix aluminum cylinder heads and intake manifold, Carburetor Shop Dominator, Hooker Competition headers, Jesel belt drive and Competition Cams camshaft and valvetrain. This single four barrel combination delivered a reliable 1076 horsepower. Cover photo by John Baechtel

By Ed Staffel

Photo courtesy of Racing Head Service

HOW TO BUILD MAX PERFORMANCE
CHEVY RAT MOTORS

By ED STAFFEL

CarTech®

39966 Grand Avenue
North Branch, MN 55056
651-277-1200 or 800-551-4754 • FAX: 651-277-1203
www.cartechbooks.com

OVERSEAS DISTRIBUTION BY:

Brooklands Books Ltd.
P.O. BOX 146, Cobham, Surrey, KT11 1LG, England
Telephone 01932 865051 • FAX 01932 868803
www.brooklands-books.com

Brooklands Books Ltd.
3/37-39 Green Street, Banksmeadow, NSW 2019, Australia
Telephone 2 9695 7055 • FAX 2 9695 7355

COVER PHOTO BY
John Baechtel

EDITED BY
John Baechtel

PRODUCED BY
John Baechtel

ISBN-13 978-1-884089-20-6

CONTENTS

S-A DESIGN

How To Build Max Performance Chevy Rat Motors

Planning

Photo courtesy of Nostalgia Supercars

You have already taken the first step towards planning your next motor, you have chosen a big block Chevrolet. It's the best choice in large cubic inch engines for a variety of reasons. They are cheaper to build than those made by the other guys. They have been around since 1965, so there is a long list of GM and after-market performance parts and support. Parts interchange between motors easily and with few restricting guidelines. The individual parts are very strong and the power potential is truly awesome. You can build from 396 to almost 800 cubic inches of earth shaking motivation. Its your choice and that makes it uniquely American. In this age of small, fuel-sipping, four and six cylinder sewing machines that are made to power look-alike moving boxes with weird sounding names, the big block Chevrolet V-8 seems to be something from another planet. Younger folks who have grown up with late model "High Output" engines that have 130, 160 or even 200 horsepower, have difficulty comprehending what it is like to drive something with 450 horsepower and 500 ft./lbs. of torque. With a few easy tweaks, 500+ horsepower and enough torque to make you beg for mercy are on tap and there is more if you want it. Interested? Sit down. Let's talk.

One of the most important things you need to do before building any engine is to ask yourself what is it you want to do with the engine. Forget what your friends have done. Don't try to do what your favorite racing hero has

Photo by Ken New

By choosing to build a Chevy rat motor, you have already given yourself a tremendous advantage. You can now benefit from the vast bank of engine building knowledge acquired by thousands of engine builders who have made the big block Chevy one of the most popular racing engines in history.

done. What do you want to do? How much can you realistically afford? Can you afford to maintain it once the motor has been built? Try to think ahead. If you build a high performance or race motor and if it later breaks or parts wear out, can you afford to fix it? Do you want to run the engine on the street with pump gasoline and low octane levels or do you want to go racing and use a higher octane race fuel? Can you build and maintain the engine and the rest of the vehicle or are you going to have to pay someone else to do it?

If you are going to go racing, is it going to be a hobby? Are you the type of person who is looking for some fun and the personal satisfaction of building and then racing your own car or are you the type of individual who has to win and won't be satisfied until you do? Are you doing this alone or with the help of family and/or friends? Do you have anyone who can help sponsor you and defray some of the expenses? Can you meet your obligations and responsibilities to a sponsor?

Be honest with yourself. Are you good at starting projects but not so good at finishing them? Are you patient or easily frustrated? Are you flexible and willing to learn? Do you quickly blame others for problems or do you spend your time thinking of solutions for problems? Do you have a goal?

Competition of any kind is something that brings out the best in people and sometimes the worst. Racing or competing in a car show reflects the lives of human beings and machines that people build. You can exhaust tremendous amounts of energy, time or money and still lose in the first round of eliminations. Your hopes and dreams can succeed beyond your expectations or be shattered by something as simple as a loose plug wire. So maybe the first question isn't what are you going to build, the first question is who are you?

Still want to do this? Good! I'll stop philosophizing. Let's figure out what you want to build and how we are going to do it.

This particular book has to do with the big block Chevrolet family of engines, but this section can be used with any motor planning. If the car will

RAT MOTOR Performance Tips & Tricks

TORQUE & HORSEPOWER

There is something that I want you to understand.

Torque is an expression of physical force acting on a lever over time and distance. Horsepower is not a force, it is a method of measurement that is arrived at by comparing the force of torque and in our case, its relationship to the rpms an engine turns. The classic formula for measuring torque and rpm and then expressing this measurement as horsepower is: HP = (RPM x TORQUE) divided by 5252. The number 5252 is a constant in the formula that relates to time and distance and so horsepower and torque measurements will always be equal at 5252 rpms. Look at the plot curves on a dyno sheet and the two plot lines showing horsepower and torque will always cross each other at 5252 rpms and be equal at that point. When an engine is on the dyno, the dyno actually determines how much torque is being made and then a calculation is done to determine horsepower at that given rpm.

It happens that when an engine reaches its peak torque and then starts to fall off to a lower torque figure, horsepower can continue to climb upwards because the rpms can continue to go up. On a motor this multiplying effect of the rpms can still act positively on the horsepower side of the equation depending on how quickly or the rate at which the torque curve falls lower after peak torque.

So what, you say? You already know all of this. My point is when you build the motor, what you are trying most to attain is torque and large amounts of torque over the broadest, flattest range you can achieve in your circumstances. If you can do this, horsepower will take care of itself. Remember that most motors do not spend all of their time at peak torque or peak horsepower. It is possible to keep a motor close to peak power thru the proper selection of transmission and rear gear ratios or if you have a boat. At the low end and middle of the engine's operating rpm range, keeping cylinder pressures as high as possible will accelerate the vehicle quickly. At the high end of the engine's operating range if you can keep the torque from dropping off too quickly after torque peak and help the engine to hang on longer in terms of torque production, horsepower will continue to climb even though torque is declining. This is generally due to better breathing heads, valves sizes, induction systems and longer duration camshafts that operate well at higher rpms.

Let me give you an example. We have two engines and at peak torque they are both making 600 ft./lbs. at 4500 rpm. At this point both motors are putting out 514 horsepower. As the rpm continue to climb and torque begins to fall lower, we find that at 5500 rpm motor "A" is now putting out 580 ft./lbs and is making 607 horsepower. The other "B" has dropped lower and is making 520 ft./lbs., which means at 5500 rpm it is making 544 horsepower.

We then take both motors to 6500 rpm. Torque is still falling off and motor "A" is now making 540 ft./lbs which at 6500 rpm comes out to 668 hp. Motor "B" has fallen to 475 ft./lbs. and is making 588 hp. At 7500 rpm, torque has fallen further still. The "A" engine is making 480 ft./lbs. and 685 hp. "B" is down to 400 ft./lbs. and horsepower has fallen to 571 hp and is headed down not up.

	MOTOR "A"		MOTOR "B"	
RPM	TORQUE/HP		TORQUE/HP	
4500	600	514	600	514
5500	580	607	520	544
6500	540	668	475	588
7500	480	685	400	571

At 7500 rpm motor "A" is making 114 more horsepower than motor "B", but it didn't do this because of horsepower. Motor "A" did this because the torque curve dropped off at a slower rate than motor "B". At all rpm points above 4500 rpm "A" had more torque than motor "B".

It's TORQUE that you should be looking for when you build an engine. Horsepower is an interesting afterthought.

RAT MOTOR Performance Tips & Tricks

ENERGY LOSSES

When the fuel and air mixture is burned in an engine, heat and energy are produced. This heat and energy is what does the work in an engine. This sounds deceptively simple and like most "simple" things it is extremely complex. When you make changes to an engine to improve its performance, you are really making changes that increase the amount of heat and energy the motor produces. Getting more air/fuel mixture into and out of the motor allows more heat and energy to be made. But there is a catch. Some of the energy winds up being removed by the cooling system as heat. A portion of the available energy goes into the lubrication system as heat generated by friction losses. A big percentage leaves the motor by way of the exhaust system. Only a small percentage, 20- 30%, of the remaining energy goes to mechanical work that pushes the piston down and actually turns the rear wheels and moves the vehicle. Some extremely heat efficient race motors may hit 35% usable power, but this is difficult to do.

If you consider it, anything you can do to keep heat and energy from being lost to the exhaust, cooling system or to frictional losses, will leave energy to do work. The more energy and heat that can be trapped in the cylinder to work on the piston, the more efficient the motor becomes. However, you still need to keep the motor cool enough to avoid overheating and exhaust still must be removed from the engine in a way that promotes good cylinder breathing. Frictional losses need to be minimized. Its an extremely difficult and delicate balancing act that makes the difference when seeking maximum usable power. These are the real secrets performance engine builders try to discover, understand and then take advantage of.

Think about it.

be driven on the street using pump gasoline some of the parameters have already been decided for you. The static compression on the motor will have to be 9.5-1 or less to avoid detonation when using low octane pump gasoline. There are some parts combinations which will allow you to have a higher static compression ratio and still use pump unleaded gas but generally speaking, keep this static compression limit in mind. Cam, piston, intake manifold, carb and cylinder head selection will be dictated by the need to have a "streetable" combination. On the street, engines should develop as much power as possible in low to mid range rpms where the motor will spend most of its time. You will also need mufflers and an exhaust pipe system of some sort. Vehicles required to have environmental emissions controls will need to maintain those controls in order to pass visual equipment and tailpipe emissions tests. Manufacturers are making many aftermarket parts which will enhance the performance of street driven, late model vehicles, and still pass "50-state" emissions requirements and laws. Treaded street tires will be required.

Most streetable big block Chevy engine combinations should use large oval port heads. The best open chamber cast iron variety and the new aluminum large oval port heads available in late 1995, make a lot of low and mid range torque because these style intake ports keep air flow velocities high at these rpms. Throttle response is excellent. These heads can also be used in bracket racers with an engine rpm limit of about 6500. Two-bolt blocks and cast cranks will be more than sufficient in these motors. Stock 3/8" rods will also work at these rpm limits, although I prefer to use premium aftermarket 3/8" rod bolts for extra security. A split plenum intake manifold with a vacuum secondary carb or some of the new electronic fuel injection systems, will also operate well in the 0-6500 rpm range on the street. Nitrous oxide can enhance the power of a low compression street operated setup. You will also need to pick a camshaft that will operate with the lower compression ratios dictated by the use of pump gasoline. A short duration, high lift cam will be streetable and tractable. Don't forget the rear end gear ratio and suitable tire sizes. Building a performance motor may necessitate the use of a lower rear gear so that the engine can operate in its best power band. Some of the stock ratios from the factory were intended to keep rpms low at highway speeds in order to improve mileage. I'm talking about the 2:70 to 3:08 economy gears commonly found. Moving up to a 3:73 or 4:11 rear gear on the street will improve performance, but it might come at the expense of fuel mileage. To offset this many people are using the newer overdrive transmissions to keep the performance but still get decent mileage when cruising down the highway.

If you are going to use square port heads, on a stock size engine, you should realize that because the intake ports are so large, flow velocities do not get high until you are in the higher rpm ranges above 6000-6500 rpms. Then these heads surpass the flows found in large oval port heads. In other words if you are going to operate your motor above 6500 rpms then move to the rectangular port heads. However, if the motor goes over 6500 you will need to move to the stronger 7/16" style rods to have a margin of safety and strength. A four bolt block and forged crank will strengthen the bottom end. At rpms over 6500 an open plenum intake and double pumper carb will add top end power.

If you are going to race the vehicle, is it going to be used both on the street and the strip? Or is the car only going to be used at the race track or off road? Try to keep everything you plan to build operating in the same power range. By this I mean that if you plan to use large oval port heads for instance, you don't want to pick a camshaft that operates best at 7800, because the oval port heads won't move enough air to feed the motor at 7800. Likewise don't use a stock connecting rod with a stock 3/8" rod bolt if you are planning to rev the motor to 7500 rpm. The stock production 3/8" rod won't take that kind of abuse for long. Make some decisions about how fast you are going to spin the motor and then make your parts selections accordingly.

A full time race car does not require some of the things a street ride needs, but a race car needs some things that a street car doesn't. A race car doesn't need a heater or A/C, but since it is going to go faster (hopefully) than your grandmother's grocery getter, you have to upgrade the safety equipment to protect yourself. Many race sanctioning bodies, such as the National Hot Rod Association and others, have very strict safety rules. The rules are there to protect not just you but also the other racers and the spectating public. The faster you go, the more safety equipment rules apply. No one wants to be in a wreck or a fire, but if it happens everyone wants you to get out of the car uninjured so you can get started repairing the damage. If you are hurt your friends will have to clean up the mess. That's not fair. If you get hurt, you can't work and if you can't work how can you fix the car? See it's a vicious cycle. You're better off not getting hurt.

This brings us to the budget section. How fast can you afford to go is an old question in automobile performance. If you are on a small budget, then try to build a motor that is low maintenance and very reliable. Trade some peak power for consistency and reliability. Seek this balance throughout the entire car and its various systems. It's no good if you build a killer big block and then try to save some money on the transmission or rear end. The motor may turn out fine and be a real torque monster, but if the trannie is tired or just not equal to the challenge from the motor, which part do you think will wind up on the pavement when the light goes green? The weakest part in the driveline or suspension doesn't get better or stronger just because you built a powerful engine. The weakest parts will remain weak and you will quickly find them if the powertrain is not in balance. So in a way when you think about what you can afford and how to spend your budget, the costs of going faster, include not only what the engine will cost, but what other non-engine related costs will be incurred because you want to go quicker. Safety upgrades, transmission, rear-end, suspension, tires and vehicle weight reductions all add to the costs. Will you need a trailer to pull the race car, boat or show car? Will you need a truck to pull the trailer and car? If you own a truck already, is it set up to tow a trailer and car? Does the truck have enough power to tow a vehicle and trailer?

You don't have to be rich to go racing and I don't want you to feel that you can't afford to try. On the con-

RAT MOTOR Performance Tips & Tricks

BIG ENGINES

By "Big Engine" I generally mean something with 475 cubic inches or larger. These engines sizes require a larger bore and/or stroke not commonly found with factory combinations with the exception of the 502. Their appeal lies in the fact that you can achieve higher power levels at lower rpms, but if you can use parts that will keep them together at even higher rpms, the power can be tremendous. Let's assume that you have a 454 based combination now and you want to move up to a larger combination.

When you build a large motor, you have the same concerns that you have with any motor. What can you afford? What do the rules allow you to use? What parts do you already have and can use on a larger motor? Are the parts strong enough to withstand the increased power? The block is a basic concern. How big can you physically go with it? Certainly a four-bolt block with siamesed cylinder barrels is going to be a requirement. The rotating assembly is another concern. You can easily build a 502 or 509 cubic inch motor with a factory steel 4" stroke crank if your block can take a bore of 4.470". You will need a 502 factory block, a "Bowtie" or an aftermarket block to get this bore size. Forged piston costs remain pretty much the same even at these bore sizes. Connecting rod costs depend on what you want to use. Aluminum rods are cheaper, but eventually will need replacement sooner than a steel rod and aluminum rods aren't suitable in a long duration motor. Aftermarket rods with increased center to center lengths are priced about the same. If you increase the stroke length more than 4", then the costs go up because of the cost of a new aftermarket crank. The cylinder heads will also be required to feed more cubic inches, so more than likely, the production heads that you used on a smaller engine won't be enough, so that means a set of square port heads with even larger port volumes and bigger valve sizes.

For a motor in the 500-510 cubic inch range, you are going to need, at a minimum, a different 4-bolt block that can be bored from 4.470" to 4.600", a 3.76" or 4" stroke forged crank and a set of heads that can supply the increased airflow demands of this engine size.

An alternative would be to use a longer stroke crank in a production block, keep the bore to 4.310" and try to fit a 4.250" stroke crank in to get 496 cubic inches. The cost difference between a good aftermarket, long stroke crank and a "Bowtie" block are not that much, but if you go with a production block you are then limited in any bore or stroke increases at a later time. Some of the truck tall deck cases can take a larger bore, but the stroke limitations are still there.

If you desire even more cubic inches, (above 515, .060 over 502), practically speaking, it will require that you increase the stroke of the motor, which means an aftermarket crank and increased cost, in addition to a block that can provide the necessary bore sizes and crank stroke clearance. Using a "Bowtie" or aftermarket block will allow you to make bore and stroke increases at a later time.

RAT MOTOR Performance Tips & Tricks

COMPUTER PROGRAMS

Face it, computers are everywhere, including the automotive and racing world. Computers are managing engine systems in every daily driver now made. In the race world, they are being used to build simulated engines, test cams, create fuel curves for electronic fuel injection, do chassis setups, 4-link adjustments, collect performance data, predict 1/4 ET's and MPH, manage clutch, ignition and fuel systems and make predictions based on weather conditions and changes.

If you have access to an IBM style computer, I want to make a suggestion. Get yourself a copy of an engine building program, cam selection program or desktop dyno. There are a number of fine programs on the market some which are relatively inexpensive and some very sophisticated and higher priced. They all have very useful utility programs which can figure out cubic inches, proper rear gear ratios, compression ratios, horsepower and torque figures and a great deal more. Many have complete lists of all cams made by a particular cam maker and will help you select the best one for your engine application. Some list every piston made by TRW and Speed-Pro and you can select which one you want to use and it will figure out the compression ratio for that particular piston in your particular motor. Engine dyno programs, like Motion Software's "Desktop Dyno", (available from GM/Performance Parts or Mr. Gasket), let you build a motor from various parts and then "run" that engine to see what kind of power it makes. Some other programs like Racing Systems Analysis (RSA) "Quarter, Jr." and "Quarter Pro" allow you to take the simulated engine and run it down the quarter-mile to see what the engine would do. RSA also has programs for weather stations "Density", engine simulation "Engine Jr." and "Engine Pro" and race clutch adjustment "Clutch Pro".

Now no computer simulation program is 100% accurate, however many of these programs come very close to actual results found on engines that have been dyno tested. The more factors you can put into the computer model, the more accurate the results. The best attribute of computer modeling is that it is much less costly than purchasing dyno time or buying your own dyno and you can quickly change various factors in an engine and then view the results on your computer screen. To me, the ability to compare the trends and differences between different parts is most useful. Whatever you spend on these programs, they will more than pay for themselves on the first engine you build. They will help you build it right the first time.

trary, anyone who wants to can do it. This is America after all, everyone has a God-given right to go as fast as they or their sponsors can afford. I just want you to think about it a little before you make decisions that unexpectedly can turn out to be wrong or financially difficult to achieve.

One way to stay in your budget is to make the most of what you have now. All too often, folks think that the answer to more performance is something they can buy, take out of the box and bolt on. Sometimes this is true, but it depends on so many other factors. I'm always particularly impressed by drag racers in the Stock Eliminator classes. The existing class rules force them to use what they have and for the most part do not allow for radical modifications and changes. They spend their time and efforts looking at what they can do to make their respective combinations run better and they look at the total picture. The results they achieve are incredible. In late 1995, Bobby DeArmond of Gallup, NM, driving a 1969 427 Camaro held the A/SA class 1/4 mile NHRA national record with a posting of 10.46 ET at 126.09 mph. Have you paid attention, not only to the engine, but engine tuneup, suspension, tires, gear ratios, track conditions and driving consistency? Many of these Stock Eliminator car owners have been working on and racing the same car and engine combination for years. Over that time they have tried everything imaginable, within the rules, to improve performance and consistency. They have discovered what works for them and what doesn't.

Engine costs for a high performance big block Chevy start out fairly reasonable. They are a bit higher initially than the costs for a small block Chevy, but then again a base level big block gives you more to begin with than a base level small block. As the power levels go up, the price of power goes up exponentially. Getting 100 more horsepower out of a motor with 450 horsepower may cost you "x", but getting 100 more horsepower out of say a 750 horsepower engine could cost you "3 times x" or "4 times x". In the search for more power you may need to switch from a $250.00 solid lifter cam and valvetrain to a $1,000.00 roller cam and related valvetrain parts. A change from a $150.00 wet sump oil pan to a $2,000.00 dry sump oil system will find more power if it is set up right, but what would happen if you spent that same $2,000.00 (or less) on some other aspect of the engine?

Weight savings should also be a consideration. Many people overlook the performance gains from weight reduction. Money and time spent on engine performance increases can alternatively be spent on decreasing vehicle weight which will result in performance gains. If you increase engine power and reduce vehicle weight, performance really takes off. Here's an example. If we had a vehicle that weighed 4000 lbs. with the driver and a load of fuel and the driver ran a 1/4 mile ET of 10.777 and a mile per hour of 127, how much horsepower would it take to do that? What would then happen if the weight of the vehicle was reduced? How much power would be needed then to run the same ET and MPH?

WEIGHT	HP	ET	MPH
4000	720.5	10.777	127
3500	630.5	10.777	127
3000	540.4	10.777	127
2500	450.3	10.777	127
2000	360.2	10.777	127

As you can see it takes less power to run the same predicted ET and MPH as the weight is decreased. Now what would happen if the same 720.5 hp engine, that was in the 4000 lbs. car, was in the 3000 vehicle? The predicted ET and MPH would be 8.08 ET at 169.4 MPH. Now these prediction calculations, which can be found in the ENGINE MATH chapter, don't take into account the aerodynamics of the vehicle and things like weather and track conditions. They do give you an idea of what trends you can expect.

Another important benefit of weight reduction is that drivetrain parts tend to last longer. Weight reduction also has an exponential cost. In other words it costs more per pound lost, the more pounds you remove. It doesn't cost much to remove unneeded seats, heater cores, air-conditioning, undercoating or exhaust pipes and mufflers. It just takes time and your unpaid labor. However, fiberglass body parts or switching to a chrome moly tube chassis and rack and pinion front end cost money. The overall reduction in weight is happening, but now its costing more. Aluminum heads, radiator, water pump, intake manifold, and wheels have their related costs, but titanium or carbon fiber component costs make aluminum cheap by comparison. Superchargers and turbo systems make power, but their costs make weight reduction a cheaper alternative. If you have a choice between a part made of cast iron and the same part made of a lighter material, such as aluminum, use the lighter part. Likewise if you have a part and you can lighten it without compromising its strength, durability or usefulness, then lighten the part. One other nice thing about aluminum is that welding on aluminum is much easier to do when compared to weld repairs on cast iron.

RAT MOTOR Performance Tips & Tricks

TYPICAL COMBINATION WITH CALCULATED PERFORMANCE

475 SHORT DECK

BLOCK	**Mark IV Bowtie short deck bored to 4.350" (+.100"), zero piston deck to block deck, 4-bolt main caps w/studs**
CRANK	**Forged 5140 or 4340 steel 4" stroke, w/Fluidampr balancer**
RODS	**Aftermarket steel or aluminum 6.535" (+.400") center to center length, 7/16" bolts, Rod ratio 1.63-1**
PISTONS	**JE Pistons #-111598 (+.100") 50cc dome forged piston, w/pin height of 1.245", piston weight - 552 grams**
HEADS	**Ported aluminum square port castings, 2.25"/1.88" valves, 118cc open chamber, Crane #-99876 double valve springs, installed height 1.968" gives 160 lbs. closed and 490 lbs. open spring pressures, Roller Rocker arms, 1.72-1 rocker arm ratio, Head stud kit**
PUSHRODS	**7/16" diameter, make sure you check for correct valvetrain geometry and pushrod length**
HEAD GASKET	**Composition, .041" thick, with this combination gives 13.49-1 static compression ratio**
CAMSHAFT	**Crane solid roller (PN-138181, Grind #-TR-268/4167-2S-8) .712"/.712" lift, duration @ .050" 268/272, Seat to seat timing 46-82-84-48, @.050" cam timing 29-59-67-25, hot lash .022"/.022", LSA 108, Intake centerline 105.**
INTAKE	**Dart, Edelbrock Victor, or Weiand Team G, open plenum, aluminum hi-rise, square port, Dominator carb bolt pattern for use w/short deck block, match intake manifold port opening to the intake port of the head.**
CARB	**Holley Dominator, List #-9377, 1150 cfm, race gas**
HEADERS	**2" diameter primaries, 30" primary length, 4" diameter collector**

Calculated Results

RPM	HP	TQ
2000	175	459
2500	242	508
3000	305	534
3500	381	571
4000	470	616
4500	564	658
5000	645	677
5500	714	682*
6000	768	672
6500	804	649
7000	805	603
7500	809*	566
8000	782	513

This motor could be used in a lighter vehicle and spun to 7500 rpms. The piston average speed is 5000 fpm at 7500 rpm. Valve spring pressures aren't outrageous so valvetrain reliability won't be so problematic. A short deck, four bolt, production block could be substituted for the "Bowtie" block, but it would not be as strong and the "Bowtie" case would allow for a number of subsequent bore jobs and later stroke increases if you wanted to make them. Torque converter stall speed would have to be in the 5000-5500 rprange where max torque is made.

Don't forget to have the rotating assembly balanced, check the open valve to piston deck clearance, and check the rocker arm geometry.

Make your finances do the most for you. Chrome parts look great, but don't add any more power. An expensive, well made, aftermarket ignition that will work accurately at 10,000 rpms is great, but may not be a wise decision if your motor will only be spinning to 6,500 rpms. Instead of buying those gorgeous wheels for $750.00 maybe for the same amount of money you can make sure the engine is properly machined, or buy a more suitable cam or buy the safety equipment you know you will need.

Next is **THE RULES.** If you plan on racing a boat, or an off road or drag racing vehicle or the event is sanctioned by a racing association, you need to become familiar with the rules governing the association. Most rules have safety as their major concern. But there are also other rules concerning types of vehicles, allowed engine modifications, vehicle minimum class weights and much more. You must be aware of what you can and cannot do and what you must do. This will have a direct effect on the costs involved. Most racing associations have classes or levels which enable the beginner or budget racer to compete effectively and without disadvantage. However, if you decide to move up in class or move to a faster level of competition, expenses will increase.

Get a copy of the rule book for whichever racing association is sanctioning a race event you plan to attend. You need to know the rules before you get started on a race project, not after the motor and vehicle are built.

Decide on what your goals are and plan and work toward them. Talk to people who have similar ideas. Find out what they did. What mistakes did they make and try to learn from them. If your goal is to build a drag car to run in the Super Gas class or H/Stock Automatic or in the bracket classes, find someone in those classes who is willing to talk to you about it. Find out what works and what doesn't. Make your own assessments. Sift thru the hype and use your brain and common sense.

Which brings us to the most important part of planning: information. No plan can be made without it. Accurate information increases the chances of success. Without accurate information, no successful plan is feasible. These days, like everyone else, we are passing into the Information Age. Even information regarding high performance engines is expanding and more importantly, becoming available to anyone who cares to look for it. Books, magazines, computer programs and computer bulletin boards, the computer Internet, World Wide Web and video sources are available and contain information for the novice and the expert. Data, which a few years ago, was only available to the initiated few is becoming available to anyone. I agree that the information deluge is sometimes too great, the numerous options provided can make decision making difficult. However, you do have choices and the same computers which make all of the information available can also help you make decisions. There are a number of computer programs on the market which enable engine builders to make informed choices and build computer performance models. The cost of these programs range from inexpensive ($40.00) to very sophisticated and expensive ($500-1,000.00). However, the cost of the programs is cheap when you consider that they allow you to build a specific engine on a computer and then run that engine to see the type of power it makes or run that engine down the quarter mile in a specific vehicle and predict performance and then change the components and make comparisons or spot trends.

Also growing is the use of data recorders in a race vehicle. I know, most of the systems are relatively expensive, but if you can afford them and if the class rules allow you to use them, vehicle on-board sensors are available to keep track of exhaust temperatures, clutch and driveshaft rpms, tire spin, engine parameters and much more. This information can be downloaded and analyzed and gives you more data to find out what happened and what to base your future decisions on.

Software is out there that will do a large number of engine or vehicle related calculations, such as compression ratios, rear gear and tire size, cam selection, effects of weather conditions on elapsed times and throttle stop adjustments, engine dynamometer simulations, piston selection, valve spring requirements, maximum piston velocities, fuel curve mapping, correct exhaust header lengths and diameters, minimum carb size, cubic inches, port flows, volumetric efficiency, proper transmission gear selections and much more. With some of the dyno simulator programs you can build a model motor and then, for instance, make a specific cam change to see if there is a power increase. You can change the valve size or carburetor and the program will help you find a better or faster combination. Some of the dyno simulation programs are more sophisticated than others and allow for more engine parameters and factors to be utilized. The more factors, the more accurate the predictions. However, most of the programs allow you to quickly make changes and then make power comparisons or spot trends. You can try ten different cam shafts, for instance, in only a few minutes if you have the cam timing data. In this book and it may be the first engine building book to do this, I have used the "Desktop Dyno" program from Motion Software (available thru GM/Performance Parts, Item #PCDS) and two cam programs, one from Crane Cams called "Cam Choice" and one from Wolverine Blue Racer called "Cam Select" to produce data. All three of these programs are relatively inexpensive (about $50.00) and if you have access to a 386, 486 or Pentium chip computer with a math co-processor, you can do the same thing.

Now before you scoff at the idea of using computer simulated dyno data as opposed to actually building a "dyno mule" engine and swapping parts around to get actual dyno results, let me say this. Ten or twenty years ago, you couldn't use a computer to do some of this stuff. Computers of today's power and size were not in the hands of average people like you and me. Computer programs related to engine building, race prediction, etc. weren't commonly available. Now they are and as each year goes by these computer programs become more sophisticated and less expensive. From my point of view, I can't afford a $15,000-30,000 engine dynamometer, let alone the cost of a building to put it in. For less money than it takes to rent dyno time for one day of engine testing, you can get a number of your own computer programs. The results are fairly accurate, but what has been even important to me was the ability to make comparisons quickly. Another appealing thing about these programs is that someone like you and I can operate them. You don't have to be a math genius or NASA engineer. All you have to be is someone who enjoys engines and the excitement of finding ways to make them go faster.

You can make your own sources of data by keeping track of how you built and tuned an engine and how the motor performed on any given day, at any given track, under whatever weather conditions existed. Keeping track of this kind of information is very helpful when you get far away from your baseline tuneup and things start to go wrong. A record of what you have done previously can get you back on the right track quickly. Over time these "racer's logs" will grow and provide valuable information. The point is that you can use information to get where you want to go and make your plans and goals a reality.

I also know that there is a sizeable group of racers that object to the growing use of computers, on-board sensors, throttle stops, delay boxes and other electronic devices. These racer's feel that as more and more of these electronic aids are used, there is less importance given to driver consistency. In the not too distant future, they feel, you won't need a driver in the car. Economics and electronics will dictate your chances in competition. As a result of these feelings "No electronics" classes have been intro-

RAT MOTOR Performance Tips & Tricks

THE SMALL STUFF

This is a reminder to take care of the small things which enable you to get the most out of your engine's power potential. There are thousands of things to check when you build a performance or race motor. Everything is important to get right, but here a some things which frequently get forgotten in the excitement of the pavement wars.

1. FASTENERS
Get in the habit of properly tightening every nut, bolt or screw you put on the car at the time you put it on and before you move on to some other fastener. All of us have hand started a bolt for example and then before finishing with it, were distracted by something else. The fastener you were working on gets forgotten. Obviously, there are some fasteners, like head bolts, that need to be started and torqued in sequence together. But finish that job before you move on to something else.

2. SAFETY
Make sure that the safety equipment you use is up to date and in good shape. The braking systems should work properly and be able to safely stop the race car at whatever speeds you operate it. Front and rear suspension should also be in good repair. Get under the car and look for cracked or bent hardware, worn bushings or missing nuts or bolts. Make sure the tires are in good shape.

3. CARB JETTING AND POWER VALVES
Bolting up a carburetor without checking on the sizes of jets or condition or size of the power valve can have an effect on power. If the jetting is off (even just one or two sizes richer or leaner) from the optimum for your particular engine setup, it can hurt best power. Likewise an incorrect power valve (in a Holley) can cause the motor to run rich or lean out. Take the time to experiment with the jet sizes and hook up a vacuum gauge to the motor and then get a power valve that opens at a level one point below your vacuum reading. While you are at it make sure the accelerator pump arm(s) is correctly adjusted. There should be .015" of free play between the tip of the arm and the bottom of the diaphragm on the accelerator pump, while the throttle is being held in the full open position.

4. IGNITION CURVES AND TOTAL IGNITION TIMING
Often a motor is built and a stock distributor dropped in with no attempt to get the ignition curve setup for the new engine parts combination. (See the sidebar on Ignition Curves.) How quickly the ignition curve occurs and at what rpm will have an effect on the best power the motor can make. Dialing in the advance curve can make a big difference. Also check to see that the ignition curve is operating. Don't just set the timing on the front balancer. Bring the motor up to 3000 rpm while watching the timing tab with your timing light. Figure out thru experimentation what amount of total ignition advance your motor likes best. Once the ignition advance curve is worked out, you can try different increments of total timing by changing the amount of initial advance you set on the front balancer. If you have a good timing light that allows you to view total timing, you can set the timing that way. While you are at it take a look at the plug wires. Make sure end boots are on tightly and that nothing is burning off the wire jackets.

5. VALVE LASH
Get in the habit of checking valve lash, even if you are using a hydraulic cam. Do it the evening before you go to the track. The lash will then be correct on race day and it gives you an opportunity to check for damage or wear on the pushrods, valve springs, etc. and look for any tiny bits of metal in the rocker arm valley that might be an indication of trouble. You can spot problems before things get worse.

6. FLUIDS
Check the coolant, oil, and fuel levels and respective filters. Check the trannie and rear end fluids. Look at the belts and hoses. Leaks should be fixed. Aside from the trouble that low fluid levels can cause, track officials take a dim view of any indication that something is leaking onto the track surface.

RAT MOTOR Performance Tips & Tricks

ONE CHANGE AT A TIME

An acquaintance built an engine for a friend of his. He put a high lift/long duration cam in the motor and since he heard you could advance the cam timing for more "bottom end" power, he advanced the cam 5 degrees. (He never looked at the cam card. If he had he would have discovered that the cam had already been advanced 5 degrees on the intake side when it was made.) Since this novice engine builder had also heard that it was good to increase the rocker arm ratio for more valve lift, he put on a set of higher ratio rocker arms. You know the rest of the story. The very first time the motor was revved, all of the valves hit the pistons (not enough piston to valve clearance, it hadn't been measured), two valve heads punched out two pistons and their respective cylinder barrels cracked and one other valve head broke off it's stem and exited it's cylinder via the water jacket in the combustion chamber roof. Within 24 hours, the owner of the car and now worthless engine, sold the car and bought another vehicle. Now this story hasn't been told to embarrass the engine builder. It's an illustration of how things can go wrong when you make changes to a parts combination without thinking about what will really happen. Bigger is not always better and whatever you do, check the clearances on the motor to see if parts you are using will work together.

When you make changes to an engine tuneup or change other parts in the engine or vehicle, try to make one change at a time. I know its difficult sometimes to take a patient approach, but if you make a number of changes all at the same time, you may not know which change helped you or hurt you. For instance, if you make two or more changes at the same time and then check the performance, you may find that performance increased, decreased or stayed the same, but you won't know what did it. One of the changes may have helped and the second didn't, cancelling out the improvement by the first change. If the changes hurt performance, things can then spiral down pretty quickly as you make even more changes in an attempt to recover lost power. If you make changes, one at a time, but lose power you can quickly return to where you began. Also do yourself a favor. If the car or boat isn't performing the way you think it should, your first step is to be honest with yourself and determine if you have done anything incorrectly or if you can change the current tuneup, chassis setup or other aspect of the vehicle or improve yourself as a driver, using the parts you have.

All too often people blame some other person or some part first and then think of another part to buy, bolt on and go faster. These folks go thru different camshafts, carbs, or entire cars and never seem to get anywhere except possibly closer to the poorhouse because of the amount of money they spend. There are certainly parts that can be purchased which will make any motor or vehicle improve its performance, but you should satisfy yourself that you have done everything that you possibly can with the parts you already have.

I've talked to folks who want to know why their particular combination isn't running as fast as they think it should. Asking some preliminary questions about ignition timing, advance curves, carb jetting, carb size, gear ratios, converter stall speed, valve lash or even tire pressures sometimes gets an answer of "I don't know" or "I didn't check it." Often a part is bolted on and absolutely no attempt is made to "dial in" the combination. Never mind that the carb may be jetted improperly, just get a "bigger" carb, "it'll go faster with a bigger carb." With some folks my first question is "what have you changed recently?"

Engine builders around the country have had the same sad experience with enthusiasts who take a well thought out combination and a perfectly assembled engine and turn it into a pile of trash, because the buyer wouldn't leave anything alone and made a number thoughtless "improvements".

duced at various race tracks. I think they are right. After all, the person using the electronics is doing so in order to gain an advantage or to remain competitive with other racers who are using them. If a racer chooses not to use them because of philosophic or economic reasons then there are really only two things that can happen. He can compete in a "No electronics" class or go ahead and compete against the electronics user. Since the use of electronics does not absolutely guarantee a trip to the winner's circle, "non-electronic" racers still have a competitive chance. This is an argument that may continue for some time, but it is not a new one. After all, questions concerning fairness in racing have been going on for a long time. The same argument occurred when the first guy showed up with a supercharger on top of his engine instead of just a carburetor or the first racer to use nitrous oxide pushed a button for more power.

As technologies change or new ideas appear on the scene, people will make decisions based on them. Racing associations and race track owners will make decisions, I hope, that will ensure that anyone who wants to can compete fairly and without a built-in economic disadvantage. At the same time room has to be made for new ideas and advances in technology. After all isn't that what all of this is about?

WHICH ENGINE SHOULD I BUILD?

For a high performance application, this is a basic question. Cost, parts availability and rarity, race rules, vehicle restoration requirements, maintenance costs, ease of repair, overall weight, longevity and reliability, amount of power required and a realistic appraisal of the engine needs and suitability for the required job should all be answered before you make your decision. It makes no sense to build an engine that operates best at 7500 rpms, if all you are looking for is a motor that will push your truck and trailer down the road at 2500 rpms. Likewise trying to build 8000 rpm race motor with parts designed for a low rpm grocery getter, is courting disaster. A four bolt block may be stronger, but a used two bolt case may make more sense in your application, since they also are quite strong and usually more available and at a lower cost. If weight reduction is a major concern, aluminum parts and even aluminum blocks are available.

A particular race class may dictate that you use a particular cubic inch motor and some individual horsepower ratings may be more desirable than others in order to take advantage of class weight breaks and to fall close to minimum class weight requirements. Searching for a very rare block casting for instance, may make sense if you are doing a restoration project, however many of the different big block castings have the same performance potential and are much easier to find.

Costs for any engine will grow depending on the level of performance you wish to obtain. Size will also play an important part. A 396 or 454 for a bracket drag racer can be built relatively inexpensively, however as cubic inches grow larger than 468 cubic inches or strokes increase from 4", the costs escalate. Horsepower increases above 600 cost. Similarly you have choices in induction systems. The costs for a conventional carburetor and intake manifold are still far cheaper than blower systems, EFI setups or alcohol injected inductions. That is not to say that these other systems can't make large amounts of power, they can and do. Large advances are being made in EFI systems that are very exciting, but there really hasn't been any reduction in system price. Nitrous still offers large increases in power at a relatively low cost and is, generally speaking, a cheaper power generating alternative to superchargers, EFI or turbos.

Two-bolt Mark IV blocks with cast crankshafts can handle horsepower levels to 600 for short durations such as drag racing. A forged steel crank and premium main cap studs can add to the bottom end strength of a two-bolt combination. At power levels above 600 hp block reliability can be enhanced by using a steel crank and a four bolt block. Cylinder barrel rigidity (the ability or the cylinder bores to stay round) and maximum bore size, can be increased by using a block with "siamesed" cylinder barrels like those found in all big block "Bowtie" castings and in the Gen.V and Gen.VI 502 blocks. Large bore and stroke combinations can be obtained using almost any block, however "Bowtie" blocks were originally designed to have large bores and have the ability to utilize long stroke cranks with little or no modifications. Trying to build a "stroker" motor out of a two-bolt production block is a waste of time and money and its reliability might be sus-

RAT MOTOR Performance Tips & Tricks

TRANSMISSIONS

Transmission choices depend once again on what your intentions are. There are some interesting choices for the street which will work with the Chevy big block. The now venerable Muncie four speeds are available, but are getting harder to find outside of specialty shops. The last ones were factory installed in 1974. Late model big block trucks could be ordered with 4 and 5 speed trannies.

The Borg-Warner T-10 and improved Super T-10's and Doug Nash 4 and 5 speeds are currently being made by Richmond Gear. Depending on the torque output of your motor and the type of vehicle use, you have a number of choices of gear ratios including overdrive on some. If you are racing the vehicle it is hard to beat the Liberty, Jericho or Lenco 4 and 5 speed transmissions for strength, ease of maintenance and the number of possible gear ratios available.

Automatic transmissions such as the 2-speed Powerglide, 3-speed Turbo 350 and 400's can be used on the street and if raced can be upgraded for additional strength. Racing with a Powerglide should be limited to vehicles under 2800 lbs. with driver, while the three speed autos will help launch a heavier car with an additional low gear ratio. The Chevrolet auto trannie cases had different engine mounting bolt hole patterns than those fitted to Buick, Oldsmobile or Pontiac (BOP) engines, however, there are a number of companies making easy to use adaptor plates that will allow you to use a "BOP" bolt pattern T-400, for instance, with a Chevy V-8. Don't forget to use a transmission cooler, it will help the trannie live. If you want to convert a Turbo-400 to a deep pan, GM/Delco offers the following parts over the counter: T-400 deep oil pan (PN- 8679994), extended filter intake pipe (PN-8629526), extended, shouldered bolt to hold the flat later style filters (PN-8633208), and the bolt spacer for use with the deep pan (PN-8629523). The deep pan will hold 2-3 more quarts of trannie fluid and help keep things cooler.

Later model 4-speed overdrive transmissions like the 700-R4, 4L60, 4L60E, 4L80 and 4L80E have been used by engine swappers who want the low first gear found in these transmissions for performance, while still retaining the overdrive final gear for decent mileage when on the highway. You have to be careful here as some of these overdrive automatic transmissions are not as strong as others. You need to make an honest appraisal of how much power your motor is making before you decide which trannie to use and any upgrades for reliability and strength that need to be made. A stock big block may work with an upgraded and strengthened 700-R4, but if you have large torque motor, a 4L80 series overdrive automatic may be a better starting place. Those transmissions that are computer controlled can be swapped into older vehicles thru the use of transmission controllers and wiring harnesses that are now available. Chevrolet, Painless Wiring, Accel, Art Carr and others have kits available to make these swaps and still retain the lock-up converters for street use, programmable transmission control modules and the necessary wiring harnesses and connectors. Art Carr also has new tail sections to mount the truck heavy duty 4L80E overdrive in cars and then use conventional cable driven speedometers. Chevy also has a tail section for 700-R4 trannies (PN-8673406) that will enable the 700-R4 to go into older musclecar chassis formerly using a T-400 cross member and original cross-member frame mounting holes.

RAT MOTOR Performance Tips & Tricks

REAR GEARS

After building a big block Chevy drag race motor, you may find that the engine makes very large amounts of torque and that you have trouble getting the vehicle to hook up. There may be suspension problems that need to be corrected or updated, but before you try that I have two suggestions. First try leaving the starting line at a lower rpm. This may help the car hook and a quicker ET and 60-foot time can result. The other is to try a lower (numerically) rear gear ratio. Many of us had our start racing small block motors which needed very tall rear gears to gain a mechanical advantage to help get the car moving and quickly get the motor up into a high rpm range where it worked best. With the large amounts of torque that are available with a big block, some bracket racers are overgeared and the motor overpowers the tires. A bracket racer I know, had a setup with a 29.5" diameter slick and was running a 5.13 gear, 4-link, a large 557 stroker motor and a Powerglide with a 5500 stall converter. The motor in a 3200 lb. car, ran well and at a track located at 5300 feet of altitude was running 10.10 ET's at 131 mph with 60-foot times at about 1.5 seconds. He was not happy because basically the car would violently smoke the tires thru first gear and obviously could do better. We talked and discovered that the 5.13 rear gearing had previously been used with a smaller production size motor (427) and had been left in the rearend when the new bigger motor was built. He decided to try a 4.56 gear, but was a little dubious about whether it would help. Next time out the picture changed. The car hooked up, lifted the front tires off the ground, ran a 1.32 60-footer with a 9.84 ET and 137 miles an hour. He was smiling when he came back on the return road. He next wanted to try a lower stall converter (4500) and possibly an even shorter 4.30 or 4.11 gear.

The point is to have the tires hook up and leave the line quickly. At high peak torque levels, a big block can power the car off the line without having to resort to a rear gear combination that works well with a small block V-8.

pect. If you want an engine with 500 or more cubic inches, start with a "Bowtie" or aftermarket case or a Gen.V 502, 502HO or Gen.VI 502HO block that was designed with those sizes in mind. You'll be happier.

Before you buy parts try to get some general idea of how much power you want or need and what it will take to get there in terms of parts combinations. Make some tentative decisions about the type of camshaft, heads and cubic inch displacement you will use. This is where one of the engine building computer programs comes in very handy, but you can still do some planning with your trusty pocket calculator. You can try a number of different combinations. On the computer, you can swap cams and compression ratios or a large number of other variables to see the changes that they produce before you start choosing or buying parts.

In the ENGINE MATH chapter of this book you will find some discussion of "Average Piston Speed". Take a look at the calculations. If you plan to operate the engine at an rpm which exceeds the safe rpm limits for your particular parts combination, parts breakage can follow. Once you know a safe rpm limit for your engine you can choose parts and build and operate the engine with those limits in mind. If your parts will only be good to 6500, running the motor to 7500 rpm may overstress the parts and you know what happens next.

We have all seen engine parts that have been extensively modified for high performance or race use. All of the modifications were done in an effort to create a part that would make more power. For years intake ports have been "hogged" out until the divider walls were paper thin, bores and strokes were increased beyond design intentions and a million other things tried in order to get more power without hurting durability. One of the reasons these extensive mods were done was the fact that a particular part or casting was being asked to do something or operate at a power level for which it was never intended. No other parts alternatives were available. As new parts have come on the market, many of the deficiencies found in earlier parts have been remedied. The newer design parts are made with performance or racing as a design parameter. Let me give you some more examples. With some older style heads it can be necessary to extensively modify the port volumes and contours in order to get the most out of them. Port volume was some-

The folks at Street & Performance with some of their inventory. These guys are a SEMA award winning company that love Chevy engines and are real innovators when it comes to complete performance electronic fuel injection systems. (Photo courtesy of Street & Performance)

times increased by welding aluminum on top of existing port floors and ceilings in order to create more room and material for grinding and additional porting beyond the original dimensions found on the heads. Later style race head designers, aware of what has been done previously, incorporate the latest features. Port volumes are increased, extra metal in the port wall surfaces was added and ports have been raised. Knowing that cylinder block and head decks may be altered by racers, more deck thickness is provided in the casting so that these deck reductions can be made more easily without making the deck surface too thin and prone to cracking. Rocker stud bosses on race heads have been made larger and stronger to withstand the high spring pressures of today's roller cam valvetrains. Cylinder walls have been made thicker on "Bowtie" blocks to maintain strength, even after large cylinder bore increases. Water jackets and oil passages that used to be a problem in heads and blocks have been moved so that when you enlarge a spring pocket on a head or grind a block to increase clearance for a stroker crank in a given area, hitting water or oil is no longer a problem. Some aftermarket blocks have very tall decks or have the cam tunnel moved upward, all in an effort to make mega-cubic inch motors easier to achieve. CNC machining has reduced or eliminated the amount of hand work required to get a part into a particular shape (particularly cylinder head porting) and can make the same modifications over and over in exactly the same way with exacting tolerances.

The big block Chevy is a formidable player, even in factory crate motor form such as this Gen. VI 502 version. You couldn't ask for a more thouroughly developed and totally reliable powerplant. (Photo courtesy of Chevrolet)

Now does all this mean that you have to go out and buy the latest parts in order to build a big block performance or race engine? No. I guess the point I'm trying to make is that you need to keep informed about what is going on in the performance industry. Sometimes you struggle with a problem and don't realize that someone else had similar difficulties and has come up with a solution. Successful manufacturers listen very hard to their customers and look for ways to make their existing products better. These people are always on the lookout for new products, new materials, manufacturing technologies and new ways of thinking about problems. Many manufacturers have the research and development facilities to explore new solutions. If you keep abreast of what is happening, you can make decisions on parts selections which will give you more freedom to experiment.

RAT MOTOR Performance Tips & Tricks

WHEN THINGS GO WRONG

Here's what can happen when a big block, cast piston, high rpms and a 16 year old enthusiast with a heavy foot get together.

It happens to everyone sometime. A part fails. A fastener isn't tight. Someone makes a mistake or forgets something. You build more power in the engine, but then start breaking transmissions or rearends. You strengthen those parts and then something else goes away. As the search for more power continues, you are essentially experimenting with your vehicle and yourself. If you lose a motor or something else, the real test comes in how you deal with it. The true loss occurs if you don't learn anything from it. Sure there is disappointment, anger and frustration, but you have got to get past those initial feelings and discover what happened and then do what you can to avoid those problems in the future. For those of you who are thinking that you have never broken anything or made a mistake, well you probably haven't been trying hard enough.

How To Build Max Performance Chevy Rat Motors

Engine Math

I've put this math section next because now is the time to get more specific about what you are going to build. Get your paper, pencil and a pocket calculator or computer and consider your options. If you have a computer, you can put these formulas on to a spreadsheet or database program that can do math. Then all you have to do in the future is plug in the variables. Keep in mind that most of these calculations will give you exact information. Other formulas like the one for HP, Torque, 1/4 ET and MPH will give you approximations. Now these approximations are subject to other factors not considered by the formula (such as vehicle aerodynamics). They do however, get close to the answer you need and are much better than guessing.

CUBIC INCH DISPLACEMENT

CID = bore x bore x stroke x 0.7854 x number of cylinders

$$CID = 4.280 \times 4.280 \times 4.00 \times 0.7854 \times 8$$

$$CID = 4.280 \times 4.280 \times 4.00 \times 6.2832$$

CID = 460.39268, which is what you get when you bore a 454 BBC .030" over and keep the stock 454 4.0" stroke.

ACTUAL STATIC COMPRESSION RATIO

You need to know the piston deck to block deck height, head gasket compressed thickness, combustion chamber volume, engine bore and stroke, piston dome volume (which may be positive if it has a dome or negative if it is a dished piston or has valve reliefs in a flat top). When you purchase pistons the specs include information on dome or dish volume, head gasket specs include the compressed gasket thickness and/or compressed gasket volume. These measurements are given in "cc's", not inches, so our equation will convert from cc's to inches by dividing the "cc" amount by 16.387.

You will have to measure the actual piston deck to block deck height. This is done with a dial indicator and a deck bridge. Just bring the piston up to top dead center and measure how far down the deck of the piston (not the dome) is from the deck of the block. Check various positions on the piston deck and remember that the piston can "rock over" at TDC. This measurement will usually run from .008" to around .025", but it may be different. If the block has been previously "decked" at a machine shop in order to insure flat surfaces or to reduce the deck height to "0", the deck height distance may be reduced from normal production tolerances. You can "cc" the combustion chamber volume in your head to measure that amount.

Okay, let's see what we found in an example 454 +.030" motor.

BORE	= 4.280"
STROKE	= 4.00"
MEASURED COMBUSTION CHAMBER VOLUME	=118cc
HEAD GASKET COMPRESSED GASKET THICKNESS (GM composition type gasket)	=.038"
MEASURED PISTON DECK TO BLOCK DECK HEIGHT	=.025"
PISTON DOME VOLUME (TRW #-2307 +.030" forged)	=+50cc

We need to convert the combustion chamber volume from cc's to cubic inches so—Chamber volume in cubic inches = 118cc's divided by 16.387 = 7.2008299" cubic inches.

We need to convert the head gasket compressed thickness to volume in cubic inches so—Head Gasket volume in cubic inches = 0.038" gasket thickness x 0.7854 x 4.280" bore x 4.280" bore = 0.5467163 cubic inches.

Now we need to convert the piston deck height into a measurement of volume in inches—Piston to Deck volume in cubic inches = 0.025" deck clearance x 0.7854 x 4.280" bore x 4.280" bore = 0.3596817 cubic inches.

Convert piston dome cc's to cubic inches—50cc's divided by 16.387 = 3.0511991 cubic inches.

Now let's finish everything up and remember to do those portions of the formula inside the brackets first and individually.

ACTUAL MEASURED STATIC COMPRESSION RATIO = 1 + (0.7854 x 4.280" bore x 4.280" bore x 4.00" stroke) divided by [(7.2008299" chamber volume + 0.5467163" head gasket volume + 0.3596817" deck height volume) - 3.0511991" piston dome volume]

CR = 1 + (57.549084) divided by [(8.1072279) - 3.0511991]
CR = 1 + (57.549084 cubic inches divided by 5.0560288 cubic inches)
CR = 1 + (11.382269)

CR = 12.382269 or 12.38 to 1

If you use a piston which has a dish or is a flat top with valve reliefs, then instead of subtracting the dome volume, you add the volume of the dish and/or valve relief volumes.

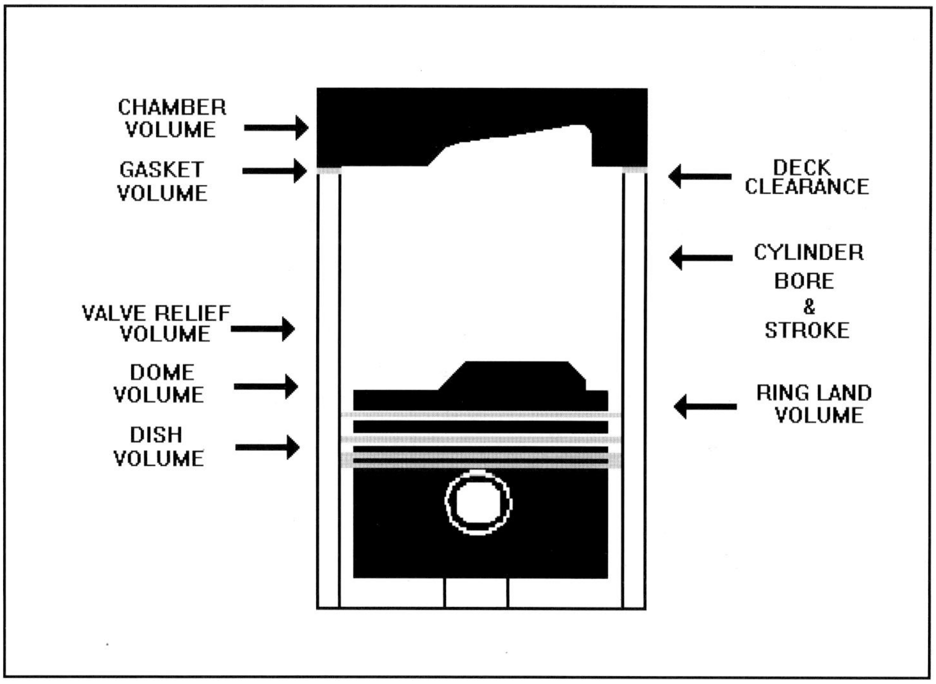

This illustration will help you understand the various clearances and volumes associated with computing actual static compression ratios. Simple compression ratio only involves squeezing one known volume into the space of a second known volume, but actual measured compression ratios require much greater precision.

MINIMUM REQUIRED CARB CFM SIZE

This is a formula that takes into account the cubic inches of a motor, the maximum rpm's that you are going to run the motor and the volumetric efficiency (VE) of the engine combination. On most street driven engines with stock, low performance parts combinations, volumetric efficiency of the motor will run from 75% to 85% . Motors that have some performance parts like performance intakes, increased compression, longer duration cams, tube headers and performance cylinder heads, will have higher VE's in the 85% to 95% range. Well thought out race engines can actually have VE's over 100%, but this is very difficult to do with a big block. Be realistic when you use this equation. Does your engine parts combination have a VE of 100% or is it really around 85% ? The only way to tell exactly is to use an engine dynamometer and measure the amount of air and fuel inhaled by the motor when it is running. Since most of us don't have a dyno and can't afford to rent time on one, this formula will help get your carb size in the ballpark.

REQUIRED CARB CFM = VE% x [(CID x MAX RPM's) divided by 3456]

CFM = 85% VE x [(460 cubic inches x 5500 rpms) divided by 3456]
CFM = 85% x [2530000 divided by 3456]
CFM = .85 x [732.06018]
REQUIRED CARB CFM = 622.25

A 650 CFM Holley or Quadrajet four barrel would work just fine on this street motor. Now this doesn't mean a larger carb wouldn't work. All this shows is what the minimum carb cfm requirements are for this application.

If VE and rpm's go up let's see what happens.

CFM = 95% VE x [(460 cubic inches x 6500 rpms) divided by 3456]

CFM = 95% x [2990000 divided by 3456]
CFM = .95 x 865.16203

CFM = 821.90392 or rounded off to 822

Since Holley doesn't make a 822 cfm carb, you could choose either an 800 cfm carb or one of their 830 cfm units.

If you were to use a split plenum intake manifold you would be able to use a carb with more CFM than indicated by this formula, because the cylinders on each side of the plenum only have 1/2 half of the total carburetor CFM to draw from. Likewise, if you had a race vehicle that never sees part throttle operation, you can increase the CFM size of the carb in order to reduce the high rpm, full throttle restriction caused by carb size. However, keep in mind that a smaller carb will generally keep velocities high and this will keep fuel well atomized.

HORSEPOWER and TORQUE, ELAPSED TIME, MPH

Here are two old standby, often quoted, equations for figuring horsepower and torque. There is a problem with using them however. Each equation requires that you know either horsepower or torque before you use them. You need a dyno run before you can accurately use the equation. I have a solution, but let's look at the formulas first.

HP = (RPM x TORQUE) divided by 5252

TORQUE = (5252 x HP) divided by RPM

See what I mean? I want to know the horsepower and torque of a motor but since I don't know the horsepower and torque, how can I plug in the required amounts to use the equations? If I had the actual dyno figures for horsepower and torque, I wouldn't need these two equations. By the way, the number 5252 is a constant in these equations. Because of this constant in the formula, torque and horsepower are always equal at 5252 rpms.

So what do we now? Well, try this if you can't afford dyno rental prices. Weigh your vehicle, with you in it, at the nearest truck scales. They usually only charge a few bucks for the privilege. Then take your vehicle to the nearest dragstrip. Make a number of runs, without changing anything, so that you get a pretty good idea of what the average ET and MPH are. Then try the following.
HP = WEIGHT x 0.4 x (1/4 mile MPH divided by 282)

An example would look like this:

HP = 2950 lbs. x 0.4 x (118 MPH divided by 282)

HP = 2950 x 0.4 x 0.4184397

HP = 493.75861 Round that off to 494

You can flip this equation around to find 1/4 MPH, now that you know HP and vehicle weight.

1/4 MPH = (HP divided by WEIGHT) x 282 divided by 0.4

1/4 MPH = (494 divided by 2950) x 282 divided by 0.4
1/4 MPH = 0.1674576 x 282 divided by 0.4
1/4 MPH = 47.223043 divided by 0.4

1/4 MPH = 118.0576 MPH You can round that off to 118 MPH.
Generally vehicle aerodynamics become a factor at speeds over 140. Now you can solve for 1/4 ET.

1/4 ET = (WEIGHT divided by HP) x 5.825 x 0.3333

1/4 ET = (2950 divided by 494) x 5.825 x 0.3333
1/4 ET = 5.9716599 x 5.825 x 0.3333
1/4 ET = 34.784918 x 0.3333

1/4 ET = 11.593813 You can round that off to 11.59 ET.

The use of these formulas does not take into account the variables that can be found at a dragstrip like track surface, weather, vehicle aerodynamics, driver's reaction and abilities, etc., but they can get you in the ballpark without having to take your motor to the nearest dyno facility. In the real world, this worked out pretty close to what we found on the track. A 2,950 lb. bracket car actually ran pretty consistently at 11.60 ET at 117 MPH at 5300 feet of track altitude. We put an alternator back on the car to help keep the ignition voltage up thru the entire quarter mile pass and changed the carb jetting and the ET dropped down to 11.40 at 118 MPH. By changing the horsepower number to 502 the 1/4 mile ET calculation showed an ET prediction of 11.409051 seconds and ultimately, the car ran 11.40.

You can solve that torque equation now that you know horsepower. This particular motor crossed the finish line at 6600 rpms.

TORQUE = (5252 x HP) divided by RPM

TORQUE = (5252 x 499) divided by 6600

TORQUE = 2620748 divided by 6600

TORQUE = 397.08303 ft./lbs. at 6600 RPMS

Now this torque figure is not the maximum amount of torque made by this motor. It is the amount of torque made by this engine at 6600 rpms. Maximum or peak torque will generally be found at a lower rpm figure.

VALVE AREA

You can figure out the square area of a valve if you know the diameter of the valve. It is similar to the bore and stroke formula we did previously to find cubic inch engine displacement, except in this one we don't use the stroke part of the previous formula. It can be used to find the surface area in square inches for anything that is round in shape, like valve surface, piston deck surface and clutch plate surface area.

VALVE AREA = Valve Diameter x Valve Diameter x .7854

VALVE AREA = 2.19" x 2.19" x .7854

VALVE AREA = 3.7668 sq. inches

You can use this figure to determine the ratio between the total surface area of the valve and the cubic inches of the motor. The more valve surface area you have the easier it will be to feed the cylinder barrels. You can take the total cubic inch displacement of your engine and divide it by 8 to get the total cubic inch displacement of an individual cylinder.
INDIVIDUAL CYLINDER SIZE = CID divided by 8

INDIVIDUAL CYLINDER SIZE = 454 CID divided by 8

INDIVIDUAL CYLINDER SIZE = 56.75 cubic inches

Now take that sum and divide it by the surface area of an intake valve.

56.75 cubic inches divided by 3.7668 square inches = 15.065

15.065 cubic inches of cylinder is fed

by each square inch of valve surface area or a valve to cylinder volume ratio of 15.065 to 1.

You can improve this ratio by using a larger diameter valve or by using a smaller displacement engine. If we kept the same valve diameter but used it on a 427 CI motor, the ratio would drop to 14.17 cubic inches fed by each square inch of valve surface, which is better. If we increased the engine size back to 454 cubic inches, but now used a valve that was 2.25" in diameter, what would happen? The cylinder would again contain 56.75 cubic inches and the valve surface area would go to 3.976 square inches. The ratio would then be 14.27 cubic inches fed by each square inch of valve area. Keep in mind that some of the later design Pontiac big block race heads will take an intake valve of 2.40" in diameter. On a 454 this 2.40" valve diameter gives a ratio of 12.54 cubic inches for every square inch of valve, but you really need large cylinder bores (like 4.500") and good heads to take advantage of an intake valve this big and not shroud the valve.

ROD RATIO

In order to find rod ratio divide the length of the rod center to center, by the stroke length of the crank.

ROD RATIO = Rod Length divided by Crank Stroke Length

ROD RATIO = 6.135" divided by 4.0"
ROD RATIO = 1.53375 or 1.53-1

PISTON SPEED

You can determine average piston speed with the following formula.

AVERAGE PISTON SPEED
= (Crank stroke x RPM) divided by 6

AVERAGE PISTON SPEED
= (4.0" x 6500) divided by 6

AVERAGE PISTON SPEED
= 26000 divided by 6

AVERAGE PISTON SPEED
=4,333.33 feet per minute

There are various opinions on what is a safe maximum average piston speed. It depends on what type of parts you are using in the reciprocating assembly. How strong are they? How light are they? Premium quality, lighter weight parts will be able to withstand higher stresses and loads. What type of use is the motor going to have? Street, drag race, endurance, boat? A safe maximum average piston speed with premium parts will be 4500 feet per minute. With that maximum in mind you can then determine the maximum rpm limits your combination can safely spin. Keep in mind this has nothing to do with the rpm level where your motor will make the most power. Your engine may make its best power at rpm levels lower than the maximum piston speeds shown by this equation. In addition, your particular parts combination may not use premium parts and therefore you will need to lower the maximum piston speed to 4,000 fpm or less. This is a formula that will help you determine what rpm levels are safe for your parts combination in terms of parts durability and reliability.

You can now take the MAXIMUM AVERAGE PISTON SPEED of 4500 fpm found above and put it in the following formula to determine the maximum rpms the parts can be safely run.

PISTON SPEED LIMIT
= (MAX AVERAGE PISTON SPEED x 6) divided by STROKE

PISTON SPEED LIMIT = (4,500 x 6) divided by 4.0"

PISTON SPEED LIMIT = 27,000 divided by 4.0"

PISTON SPEED LIMIT = 6,750 rpms

If we now change the crank stroke to the 3.76" stroke found in a 427 big block, what happens?

PISTON SPEED LIMIT = (4,500 x 6) divided by 3.76"

PISTON SPEED LIMIT = 27,000 divided by 3.76"

PISTON SPEED LIMIT = 7,180.85 or 7,181 rpms

The shorter stroke increases the maximum rpms the motor can be run before you hit the piston average speed limit. With a larger 4.250" stroke the 4500 fpm limit drops to 6353 rpms. Now it is true that some drag race motors, with extremely light and stronger premium parts, can exceed the 4500 fpm average speed limit and reach a maximum average piston speed of 6000 fpm. The parts experience maximum load for a few seconds and survive. But these engines are built by folks who know what they are doing and the parts get checked regularly at frequent intervals. Replacement is made before a part breaks. If a part fails, it gets fixed or replaced with a stronger part. The 4500 fpm limits I have given here are more conservative and will fit a broader range of circumstances. As improvements in metallurgy and technology occur the maximum safe average piston speed limits will rise. Remember this calculation concerns the maximum average speed of the piston. At certain points along its travel the PEAK PISTON SPEED will be much faster. Talk to your aftermarket supplier and determine which average piston speed and which peak piston speed can be used with their components. Remember that relative strength and weight of all the parts in the rotating assembly must be considered in your parts selection, pistons, piston pins, rods, rod bolts, stroke length and bottom end strength of the motor.

Finally, if you know the safe rpm limit of your reciprocating assembly, you can then make decisions about cam selection, heads, intake manifold, carb CFM, etc., based on this limit and maximize your engine power, durability and operating range. For instance, if you know that your limit will be, say 6700 rpms, you can build the motor to operate best below that figure. You can shift it at 6600 rpms and choose an engine parts combination that will give a horsepower peak at 6500. In a drag racer you can then also chose rear tire diameters and rear gear ratios that will put you across the finish line at 6600 rpm in high gear.

REAR GEAR RATIOS, TIRE DIAMETERS, MPH, RPMS

If you are going to go drag racing, one of the things that you have to work out is which rear gear ratio and what rear tire diameter you need. The idea is to go across the 1/4 mile finish line in high gear and slightly above the rpms where your motor is making the best horsepower. For example if the motor makes its highest horsepower at 6500 rpms, you want to cross the finish line at 6600 rpms.

Let's say the mile per hour that you are shooting for is 117 mph and you have a tire diameter of 29.0". What rear gear ratio do you need to use?

REAR GEAR RATIO =
(RPM at finish line x Tire Diameter)
(MPH x 336)

REAR GEAR RATIO =
(6600 x 29.0") divided by (117 mph x 336)

REAR GEAR RATIO =
191400 divided by 39312

REAR GEAR RATIO = 4.86:1

If you switch this formula around you can solve for each of the elements. If you know the rear gear ratio, tire diameter and finish line rpms, you can figure out what the mile per hour will be.

MPH =
(RPM at finish line x Tire Diameter)
(Gear Ratio x 336)

MPH =
(6600 x 29.0") divided by (4.86 x 336)

MPH = 191400 divided by 1632.96

MPH = 117.21 or 117 MPH

If you know the mile per hour, gear ratio and tire diameter, you can predict what the engine rpms will be at the finish line.

RPM = (MPH x Gear Ratio x 336) divided by Tire Diameter

RPM = (117.21 x 4.86 x 336) divided by 29.0"

RPM = 191399.24 divided by 29.0"

RPM = 6599.97 which rounds off to 6600 RPM

The required tire diameter can be found if you know the mile per hour you are seeking, rpms at the finish line you want and the rear gear ratio you have or plan to use.

TIRE DIAMETER =
(MPH x Gear Ratio x 336)
RPM at finish line

TIRE DIAMETER =
(117.21 x 4.86 x 336) divided by 6600

TIRE DIAMETER =
191399.24 divided by 6600

TIRE DIAMETER =
28.999" and that rounds off to 29.0"

A drag race slick will have a nominal diameter. With the low air pressures run in these race tires, the diameter will probably be less than nominal at the starting line, but at the finish line the tire diameter is generally more than the nominal tire diameter because of the centrifical forces acting on the tire which make it "grow" taller and therefore larger in diameter.

You can use the above formulas for a street operated vehicle to determine what tire sizes, gear ratios, rpms or mph you want to go. For instance let's say you want to know what tire size you need to go 65 mph on the highway. You want to keep the rpms reasonable, say no more than 3000 rpms at 65 mph and you have a 3.73 rear gear ratio.

TIRE DIAMETER =
(65 mph x 3.73 gear ratio x 336) divided by 3000

TIRE DIAMETER =
814632 divided by 3000

TIRE DIAMETER =
27.15" or rounded off 27"

All of the above formulas assume that your transmission gear ratio in high gear is 1.0-1. If you use an overdrive transmission with a high gear ratio of say 0.85-1, just multiply that by the rear gear ratio. For example if you had a 3.73-1 rear gear, multiply that by the overdrive final transmission gear ratio (0.85) which gives an effective reduction of the final gear ratio to 3.17-1 when you are in overdrive. Using the examples above, if you switched the transmission from a Muncie 4-speed to an 0.85-1 overdrive trannie, at 65 mph the motor would drop to 2550 rpm in overdrive with 27" diameter tires.

FUEL VOLUME

How much fuel volume do you need? One of the things that must be considered is what type of fuel are you going to use? Pump gas, high octane race gasoline, alcohol or nitrous? High rpms, using nitrous or alcohol will make large demands on the fuel delivery system.

For a gasoline fueled, non-turbo, non-supercharged, non-nitrous engine, try these formulas:

REQUIRED AIR CFM =
(ENGINE CID x MAXIMUM RPMS)
3456
x VOLUMETRIC EFFICIENCY

REQUIRED AIR CFM =
(460 x 7000) x .95%
3456

REQUIRED AIR CFM =
3220000 x .95%
3456

REQUIRED AIR CFM =
931.71296 x .95%

REQUIRED AIR CFM =
885.12731 CFM

Now that you know the amount of air that is required for this motor, you can pick the correct size carburetor. Using this figure you can then also find the capacity required for maximum power gasoline fuel volume with this size motor operating at this maximum rpm. This calculation assumes a gasoline maximum power air to fuel ratio of 12.5-1 at sea level (14.7 psi air pressure) and 59 F degrees outside air temperature. Gasoline weighs about 6 lbs. per gallon.

REQUIRED FUEL FLOW IN GALLONS PER HOUR =
AIR CFM x 4.38 x 0.08
6LBS. PER GALLON

REQUIRED FUEL FLOW IN GALLONS PER HOUR =

$$\frac{885 \times 4.38 \times .08}{6}$$

REQUIRED FUEL FLOW IN GALLONS PER HOUR =

$$\frac{310.104}{6}$$

REQUIRED FUEL FLOW IN GALLONS PER HOUR =
51.684 GALLONS PER HOUR

With this gasoline fuel volume requirement now known you can use the required fuel pump, fuel line sizes and fuel tank your engine combination and engine operating speed needs. Keep in mind that in this area you should build in some extra capacity in case something happens to restrict fuel flow like a clogged fuel filter or a reduction in battery voltage that would slow an electric fuel pump. Most performance and race fuel pumps will provide more than the amount needed in the above example. For instance the Holley "blue" electric pump (Holley PN-12-802) has a capacity of 110 gallons per hour. Pumps of even larger capacity are also available.

With an alcohol (methanol) motor you will need more than twice (125% more) the volume amount shown above because the best power air to methanol fuel ratio falls to about 6.0-1, instead of the 12.5-1 ratio for best power with gasoline. Multiply the above fuel flow by a factor of 2.25 and you will get 116.289 gallons per hour using the above example. Again you should use a fuel delivery system that can provide more total flow than is required in order to build in an extra margin of volume capacity and remember to provide a fuel bypass return line for the unneeded fuel.

Blower, turbo and nitrous motors will also need additional fuel delivery capacity depending on the amount of boost or nitrous used. Follow the manufacturer's recommendations and set up your fuel system to deliver the required amount.

BLOWER DRIVE RATIOS

On Roots style positive displacement blower drive pulleys using teeth and a cogged drive belt you can count the number of teeth on the crank pulley and the number on the blower pulley to determine the drive ratio and whether the blower is underdriven or overdriven.

BLOWER DRIVE RATIO =

$$\frac{\text{Number of teeth on crank pulley}}{\text{Number of teeth on blower pulley}}$$

A decimal amount less than 1.0 will give an underdriven ratio, while an amount more than 1.0 will determine a blower rpm that is overdriven in relation to engine rpms.

BLOWER DRIVE RATIO =

$$\frac{\text{32 teeth on crank pulley}}{\text{34 teeth on blower pulley}}$$

BLOWER DRIVE RATIO =

$$\frac{32}{34}$$

BLOWER DRIVE RATIO = 0.941 to 1

In this case the blower speed is being run at 94.1% of the engine speed and is therefore underdriven. You can alter the drive ratio by changing the pulleys to different tooth counts and change the amount of boost pressure either lower or higher. A rule of thumb is that for every 3% change in blower drive ratio, you get a 1 or 2lb. change in boost pressure. Keep in mind that a blower does not have to be overdriven to make power.

A number of centrifugal supercharger systems use a ribbed belt, instead of a cogged belt, that slips when boost pressures exceed a certain amount and prevent too much pressure. Pop-off valves can also be adjusted on blower systems to open when predetermined boost pressures become excessive.

EXHAUST TUBE HEADER DIMENSIONS

Header sizes for your big block can be generally determined with the following formulas. You can tune them to work best at maximum torque or maximum horsepower rpms. You need to know at what rpm maximum torque or peak horsepower are made. A computer dyno program or actual dyno run can help you find the rpm levels for peak torque and peak hp. Headers for a race vehicle depend on a number of factors. You can experiment with tube lengths and collector lengths easily, however once you chose a tube diameter or collector diameter, you are stuck with that diameter unless you change to another set of headers. There is an interrelationship between the primary tube diameter and the primary tube length. If you change one the other needs to be changed. This relationship will allow you to "tune" the headers by using different lengths of primary tubes and collector tubes while keeping the same tube and collector diameter. You need to know the engine rpm, engine cubic inch displacement and either primary tube length or primary tube diameter.

PRIMARY TUBE DIAMETER =

$$\sqrt{\frac{\text{CID} \times 1900}{\text{Primary Tube Length X RPM}}}$$

PRIMARY TUBE DIAMETER =

$$\sqrt{\frac{460 \times 1900}{32'' \times 6800}}$$

PRIMARY TUBE DIAMETER =

$$\sqrt{\frac{874000}{217600}}$$

PRIMARY TUBE DIAMETER =

$$\sqrt{4.0165441}$$

PRIMARY TUBE DIAMETER =
2.0041317" or 2", the closest common tube size.

To find tube length, use the following.

PRIMARY TUBE LENGTH =

$$\frac{\text{CUBIC INCH DISPLACEMENT} \times 1900}{(\text{Diameter} \times \text{Diameter}) \times \text{RPM}}$$

PRIMARY TUBE LENGTH =

$$\frac{460 \times 1900}{(2'' \times 2'') \times 6800}$$

PRIMARY TUBE LENGTH =

$$\frac{874000}{4 \times 6800}$$

PRIMARY TUBE LENGTH =

$$\frac{874000}{27200}$$

PRIMARY TUBE LENGTH =
32.132352" or rounded off to 32"

How To Build Max Performance Chevy Rat Motors

Cylinder Blocks

BLOCK IDENTIFICATION

Currently there are three families of Chevrolet and GMC/Truck big blocks. The first is the Mark IV which was used in production vehicles from 1965 thru 1990. The Mark IV's were used in passenger cars and trucks until 1975, after that the Mark IV was only factory installed in trucks. Factory installations of the Mark IV's ended in 1990. The Mark V or Generation V (hereafter abbreviated as "Gen.V") began production in 1991 and runs thru the present. The Gen.V big block production motors have only been factory installed in trucks or available as "over the counter" assemblies. In 1996, Chevrolet introduced the Gen.VI or Mark VI engine block. This block retains the one-piece rear main oil seal, but corrects some of the shortcomings of the Gen.V block. The water jacket core holes have been reshaped and deck material added so that any Mark IV or Gen.V production head will also fit. The mechanical fuel pump boss and clutch pivot stud boss, not found on the Gen.V cases, have been refitted to the Mark VI 454 and 502 HO crate motor blocks.

Block casting numbers, applications, casting dates, VIN and engine assembly codes are located in various places on the block itself. These numbers can help you identify exactly which block you are looking at. The accompanying chart of block casting numbers will help you. The photos will show you where to look on the block for the casting identifiers and differences.

Chevrolet and GMC/Trucks used different cubic inch sizes of big blocks. From 1965 thru 1990, the Mark IV motors for production vehi-

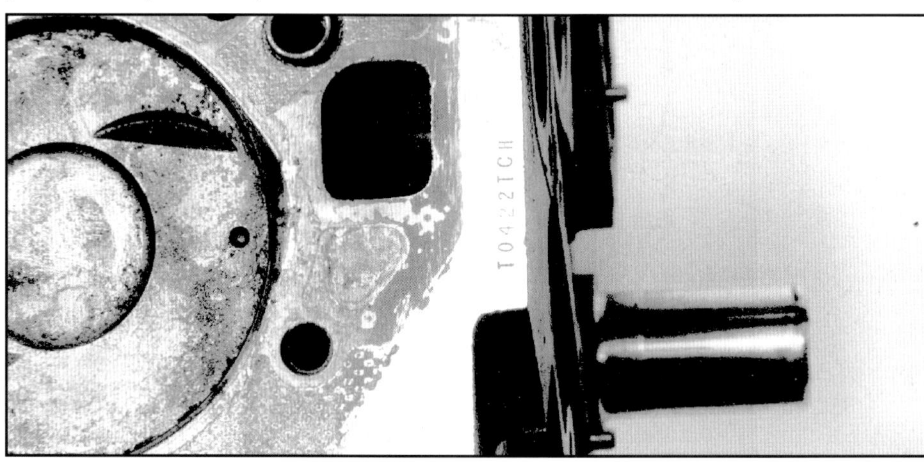

Stamped block assembly code is "T0422TCH" on this Gen.V 454 and tells you that it was made in the Tonawanda factory (T) in the fourth month (04 which is April) on the 22nd day. "TCH" is the "suffix code" which tells you what type of vehicle and year model, engine type and transmission that was originally assembled with this production engine. This stamping may be found on the passenger side block deck or on a block ledge above the timing chain cover mounting surface or on the rear flange of the block adjacent to the oil filter mounting boss.

Block casting date (D27 71) is found on a rear ledge on the block. This indicates when the block itself was made, not the entire engine assembly. Some very early Mark IV blocks have the block date code on the passenger side of the block, in the middle near the side oil pan rails.

"3969854" is a block casting number found on a ledge at the rear of the block. These block casting numbers can help identify the type and size of engine block you have.

You may find block casting identifiers like "HI PER", "HIPERF", "PASS", "TRUCK", "MAR" or "MARINE" cast into the block. Some help you identify the engine type, others like "HI PER" and "HIPERF" are found on all big blocks whether they had two or four bolt main caps or were used in low or high horsepower applications.

BIG BLOCK Quick Facts & References

BLOCK CASTING NUMBERS

CASTING NUMBER	CID	YEAR	NOTES
340220	427T	68-85	4-bolt
345014	454	70-86	
346236	454	75-76	2-bolt
359070	454	70-90	
361959	454	73-90	2 or 4 bolt
364776	427T	68-85	4-bolt
364779	366T	68-90	4-bolt
399204	509	70-71	4-bolt, Alum., CanAm, 4.5" bores, Cylinder liners
399289	454	73-79	
399293	366T	77-78	4-bolt
473478	427T	77-90	4-bolt
3782870	427T	68-76	4-bolt
3824553	366T	66-67	4-bolt
3855961	396	65-66	2 or 4 bolt
3855961	427	66	2-bolt
3855962	396	65-66	4-bolt
3855977	366T	66-73	4-bolt
3869942	427	66-67	2 or 4 bolt
3902406	396	67	2 or 4 bolt
3904351	427	67	2 or 4 bolt
3904354	366T	66-76	4-bolt
3916319	366T	68	4-bolt
3916321	427	68	2 or 4 bolt
3916323	396	68	2 or 4 bolt
3925521	427T	68-85	4-bolt
3935439	427	68-69	2 or 4 bolt
3935440	396	68-69	2 or 4 bolt
3937724	366T	68-85	4-bolt
3937726	427T	68-84	4-bolt
3946052	427	69	4-bolt, Alum. ZL-1
3946053	427	97	4-bolt, Alum. ZL-1, 2nd version
3955270	427	69	2 or 4 bolt
3955272	396	69	2 or 4 bolt
3955274	366T	68-85	4-bolt
3955276	427T	68-73	4-bolt
3963512	427	69	2 or 4 bolt
3963512	454	70-76	2 or 4 bolt
3969852	366T	68-84	4-bolt
3969854	396	69	2 or 4 bolt
3969854	402	70-72	2 or 4 bolt
3969858	427T	68-84	4-bolt
3999289	454	72-79	2 or 4 bolt, some "CE" replacement blocks had four bolt main caps
3999290	396	68-69	2 or 4 bolt, Truck
3999290	402	70-72	2 or 4 bolt, Truck
3999290	402	72	2 or 4 bolt, Passenger
3999293	366T	68-90	4-bolt
3999294	427T	68-84	4-bolt
10051107	454		4-bolt, Bowtie, Mk.IV, Short deck, CNC prep, 4.25" or 4.495" bores
10068286	454	90-91	4-bolt, Mk.IV, Short deck
10114182	454	91-up	Gen.V, 4-bolt
10114183	366T	91-up	Gen.V, 4-bolt
10114184	427T	91-up	Gen.V, 4-bolt
10134366	454T		4-bolt, Bowtie, Gen.V, Tall deck
10185050	454		4-bolt, Bowtie, Gen.V, Short deck
10237297	454	96-up	4-bolt, Vortec 7400", "L-29"
10237300	502	96-up	4-bolt, Gen.VI, 4.466" bore
12550313	454	91-up	4-bolt, Gen.V crate motor, Gen. VI 4.25" bare block, fuel pump boss
14015443	454	87-90	4-bolt, Mk.IV Truck, Motorhome
14015445	454	78-90	2 or 4 bolt
14044807	454T		4-bolt, Bowtie, Mk.IV, CNC prep, tall deck, 4.25" bore or 4.495" bores
14096859	502		4-bolt, Gen.V, HO
24502504	454		4-bolt, Bowtie, Gen.V, race prep, 9.8 short deck
24502506	454T		4-bolt, Bowtie, Gen.V, race prep, 10.2 tall deck
24502572			4-bolt, 4.5" bore Olds DRCE 2, 9.5" deck height

T = Tall deck, 10.2" deck height.

Short-deck blocks have a 9.8" deck height.

This 1965 396 block casting #-961 was a two-bolt low deck variety that was used in a number of vehicle applications as you can see from these casting identifiers found at the rear of this core.

Some early big blocks have the block casting date on the passenger side of the block. This block was cast on "I 27 5", November 27,1965.

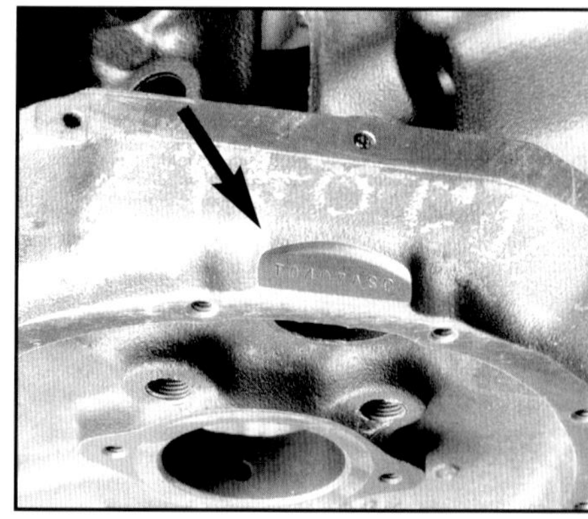

This Mark IV 454 block has the assembly code (T0407ASC) stamped into a ledge above the timing chain cover surface.

On this Mark IV 454 block, part of the VIN is stamped on to the front of the block deck on the passenger side and the assembly code is stamped into the ledge above the timing chain cover surface.

BIG BLOCK Quick Facts & References

MARK IV / GEN V DIFFERENCES

	MARK IV	GEN. V & GEN. VI HO	GEN. V CNC RACE PREP BOWTIE
Rear Oil Seal	Two–piece	One–piece	Two–piece
Main Oil Feed	Left side oil pan rail	Camshaft tunnel	Camshaft tunnel
Mechanical fuel pump boss	Yes	GEN. V - No GEN. VI - Yes	No
Clutch Z–Bar pivot stud boss	Yes	GEN. V - No GEN. VI - Yes	Yes
Adjustable Valvetrain	Yes	No	Yes
Rocker Covers	Stamped steel	Cast aluminum	
Oil Pans	Must use pan for two piece oil seal	Must use pan for one piece oil seal	Must use pan for two piece seal

The above are some of the major differences between the Mark IV, Gen.V and Gen.VI motors. Other differences are noted in the text.

MAIN CAP BOLT LENGTHS

The main cap bolts differ between the Mark IV, Gen.V and Gen. VI four bolt main cap engines. There are no two bolt Gen.V or Gen VI engines.

	MARK IV	GEN.V & GEN. VI
Inner Bolts	1/2"–13 x 3.34" (PN–3909834)	1/2"–13 x 3.52" (PN–10106461)
Outer Bolts	1/2"–13 x 2.60" (PN–3859927)	1/2"–13 x 2.78" (PN–10106460)

cles included 396, 402, 427 and 454 cubic inch motors. Medium and large heavy duty trucks were available with "tall deck" Mark IV versions of the 366T and 427T engines. Beginning in 1991, Gen.V 454's for production light duty trucks were available, 366T and 427T Gen.V "tall deck" blocks have been used only in medium and heavy duty trucks. Remember that a number of other manufacturers used GM big block powertrains for recreational vehicles (RV's), industrial pumps and generators, buses and specialty truck applications. These can serve as additional sources for good used engine cores. In 1996 Chevy introduced the "Vortech 7400" 454 and Gen.VI 454HO and 502HO motors and made them available as "over the counter" crate motors.

Chevrolet, GMC/Trucks or GM/ Performance Parts will sell you a new bare block. Mark IV 4.250" bore four bolt production bare blocks and Mark IV "Bowtie" blocks in short and tall deck versions are available. Gen.V 366T and 427T tall decks and 454 and 502 short deck production blocks all have 4-bolt main caps, (there are no 2- bolt Gen.V or Gen.VI blocks). There are four versions of the Gen.V "Bowtie" cases, two short decks and two tall decks. Gen.VI blocks have four bolt main caps and one piece rear main oil seals.

The Mark IV motors, including Mark IV "Bowtie" cases, all used two- piece rear main oil seals, with crankshafts and rear main bearing caps to match. Gen.V and Gen.VI production blocks use a re-designed one-piece rear main oil seal and consequently have a crankshaft and rear main bearing cap

BASIC BIG BLOCK SPECIFICATIONS

CI	MODEL YEAR	MAIN CAP BOLTS	BORE	STROKE	CYLINDER VOLUME	ROD LENGTH / ROD RATIO
366T Cast iron tall deck truck block	66–91	4	3.935"	3.76"	45.726"	6.135"/1.63
366T Gen.V tall deck truck block, one piece rear main seal	91–95	4	3.935"	3.76"	45.726"	6.135"/1.63
366T Gen. VI	96-	4	3.935"	3.76"	45.726"	6.135"/1.63
396	65–69	2&4	4.094"	3.76"	49.496"	6.135"/1.63
402	69–72	2&4	4.125"	3.76"	50.248"	6.135"/1.63
427 Cast iron block	66–69	2&4	4.250"	3.76"	53.340"	6.135"/1.63
427 Aluminum block with cast iron liners, ZL–1, very rare!	69	4	4.250"	3.76"	53.340"	6.135"/1.63
427 New aluminum block with liners	98–	4	4.250	3.76	53.340	6.135"/1.63
427T Cast iron Mark IV tall deck truck block	66–90	4	4.250"	3.76"	53.340"	6.135"/1.63
427T Gen. V tall deck truck block, one piece rear main seal	91–95	4	4.250"	3.76"	53.340"	6.135"/1.63
427T Gen. VI	96-	4	4.250"	3.76"	53.340"	6.135"/1.63
430 Aluminum race block with cast iron liners, extremely rare!	69	4	4.440"	3.47"	53.726"	6.135"/1.76
454 Mark IV	70–90	2&4	4.250"	4.0"	56.745"	6.135"/1.53
454 Gen.V, one piece rear main seal	91–95	4	4.250"	4.0"	56.745"	6.135"/1.53
454 Gen. VI, one piece rear main seal	96-	4	4.250"	4.0"	56.745"	6.135"/1.53
465 Aluminum race block with cast iron liners, extremely rare!	70–71	4	4.440"	3.76"	58.216	6.135"/1.63
495 Aluminum race block with cast iron liners, extremely rare!	70–71	4	4.440"	4.0"	61.932"	6.135"/1.53
495 Aluminum race block without iron liners, extremely rare!	72	4	4.440"	4.0"	61.932"	6.135"/1.53
502 Gen.V HO, one piece rear main seal	92–95	4	4.466"	4.0"	62.659"	6.135"/1.53
502 Gen. VI HO, one piece rear main seal	96-	4	4.466"	4.0"	62.659"	6.135"/1.53
509 Aluminum race block without iron liners, extremely rare!	72	4	4.500"	4.0"	63.617"	6.135"/1.53
509	97	4	4.500"	4.0"	62.659"	6.135"/1.53
509T Mark IV cast iron block	97	4	4.500"	4.0"	62.659"	6.135"/1.53

to accommodate this seal change that are different than the crank and rear main cap found on the Mark IV engines. However, the two Gen.V "CNC Race Prepared" Bowtie blocks use an adaptor that enables the buyer to use the old style two- piece rear main oil seals and Mark IV cranks. The change to a Gen.V one-piece rear oil seal also caused redesign of the crankshafts, blocks, rear main caps, gaskets, oil pan and timing chain covers found on Gen.V motors. There are a number of other differences between the Mark IV, Gen.V and Gen.VI which concern the interchangeability of parts between them. Some parts will interchange and others will not. These interchanges will be discussed in each section of this book.

ENGINE SIZES AND TYPES

There is an old proverb concerning engine performance and it is still true today, "There is no substitute for cubic inches".

When you bore a case, you need to stay within the piston sizes available from either Chevrolet or aftermarket sources like TRW,

BIG BLOCK Quick Facts & References

BIG BLOCK CHEVROLET GENERAL SPECIFICATIONS

Configuration	Canted ohv V-8 Cast Iron or Aluminum blocks & heads
Cylinders (front to back)	Drivers side 1-3-5-7 Passenger side 2-4-6-8
Firing Order	1-8-4-3-6-5-7-2
Bore Spacing	4.84"
Crankshaft	5 main bearings, thrust on #5 (rear) cast or forged crank
Crankshaft Diameters	Main Journal (1-4) 2.7494" (5) 2.7488" Rod Journal 2.20"
Connecting Rods	Forged steel, 6.135" length centers Big end diameter - 2.345" Rod bolt diameters-3/8" or 7/16"
Pistons	Forged & cast aluminum, slipper skirt
Lubrication	Full pressure oiling
Bores	3.935" - 4.500"
Strokes	3.47" - 4.0"
Displacements	366 - 510 cubic inches
Compression ratios	8.25-1 to 12.5-1
Factory Induction	2-bbl., 4-bbl., 3x2 bbl. and Throttle Body and Port Fuel Injection
Horsepower	215hp to 465hp (factory ratings)
Weight	approx. 500 lbs. all aluminum or 750 lbs. all cast iron, complete

PART NUMBERS & CASTING NUMBERS

Factory part numbers, occasionally change for any given part. There might be a major or minor revision to the design and this causes a change in part number. The cost may change or the supplier of the part might be new and different. This also will generate a new part number. The point is that part numbers, over time, can change for the same part.

Many parts also get a casting number placed on the part itself. For most parts, once it receives a casting number the number stays with the part and it can be recognized as such. In a few very rare instances, Chevy ground off existing casting numbers and replaced them at the factory with stamped in casting numbers.

Factory casting numbers can run to eight or nine digits so it is common to abbreviate them and only use the last three digits when speaking or writing in reference to them. GM used different numbers for the final three digits in casting numbers so rarely is there a number duplication in the same type of part. Keep in mind that some part casting numbers were used in a number of different model and year applications. Some numbers are difficult to see if the engine is still in a vehicle. Look closely when you are noting casting numbers. Sometimes a "3" can look like an "8" or a "6". Occasionally a casting number digit is missing at the end of a number.

Speed-Pro, JE, Wiseco, Arias, or Ross. Most machine shops will bore your block .020", .030", .040", .060" or .070" over the stock size. However, some early blocks can be bored 0.155" over size, if there was no core shift when the block was made. "BOWTIE" cases can be bored very large for those of you who want a 500+ "mountain" motor between your frame rails.

The standard deck height for a short deck Chevy big block is 9.80" which is the distance from the centerline of the crankshaft axis to the top of the block deck. Tall deck 366 and 427 truck blocks and "Bowtie" tall decks, have a deck height which is .400" taller at 10.20". Some tall decks, available from Oldsmobile or aftermarket sources are even higher.

Gen.V production cases and Gen.V "Bowtie" blocks do not have any provision for mechanical fuel pumps, so an electric fuel pump must be used. Gen.V production cases do not have any bolt bosses to mount a clutch "Z-bar" stud, so if you plan to use a manual transmission and mechanical clutch linkage keep this in mind when you make block selection. Gen.V "Bowtie" cases do have provisions to mount the "Z-bar" ball stud to the block. Gen.VI blocks have the mechanical fuel pump and clutch linkage mounting bosses.

A cast iron bare block weighs from 225 to 275 lbs. while an aluminum Olds DRCE bare block

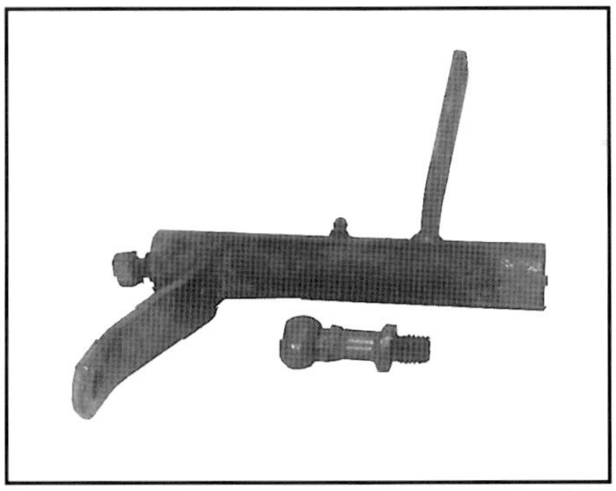

If you are using a manual transmission and you use a clutch "Z-bar" that will mount one end to the engine block, make sure the block you use has the bolt boss for the ball stud. Gen.V production blocks do not have these bolt bosses to mount the ball stud. Mark IV and Gen. VI and "Bowtie" block do have the bolt bosses.

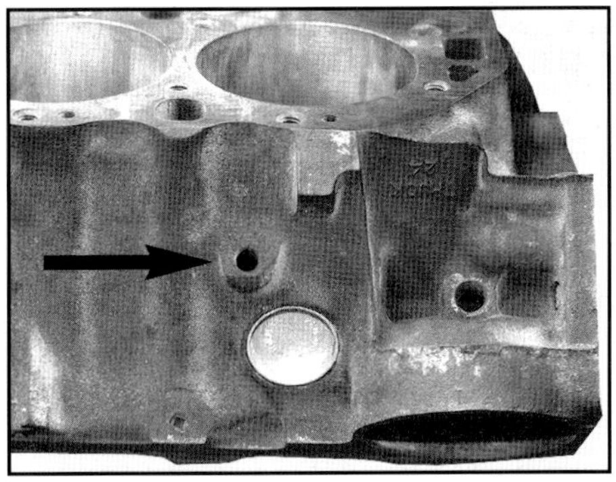

Arrow points to the bolt boss to mount a ball stud for a clutch linkage "Z-bar" or cross shaft. This tall deck "TRUCK" block has only one. Other blocks have as many as three different bolt boss locations on them. Gen.V production blocks don't have any.

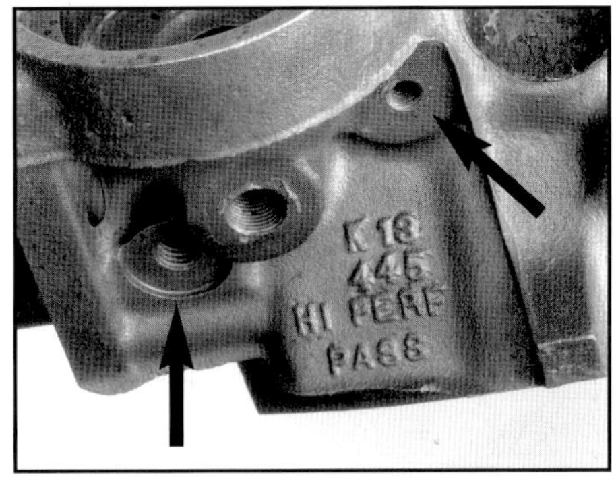

This Mark IV 454 4-bolt block has bolt bosses (arrow) for the block mounted clutch linkage ball and stud. If you are going to use mechanical clutch linkage make sure your block has the bolt boss in the position you need.

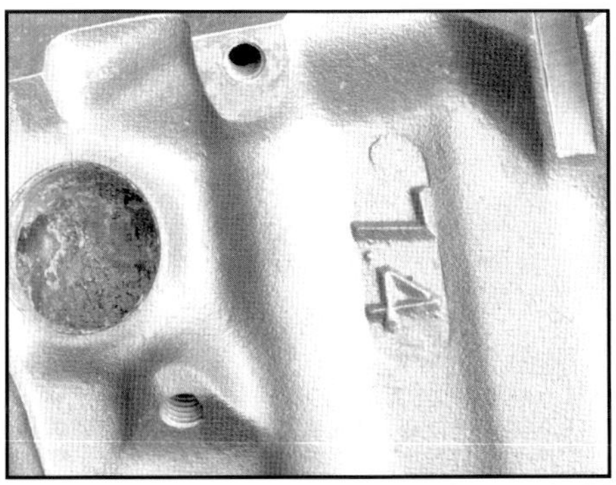

"7.4" cast on the side of this Mark IV 454 block is not a date casting. It is the size of this motor in liters.

Mark IV and Gen.VI blocks do have a boss for a mechanical fuel pump.

weighs approximately 157-160 lbs.

The new Chevrolet aluminum block weighs approximately 110 lbs.

Chevrolet originally did make a very limited amount of aluminum big blocks in the late Sixties and early Seventies. Most of these blocks used cast iron cylinder barrel liners and some did not have liners, but used specially coated pistons that could run directly on aluminum bores. The famous ZL-1 427 and certain other aluminum blocks for Can-Am road race and off-shore boat racing teams were available. These original ZL-1 blocks are extremely rare today and unless you are a wealthy collector of exotic cars and engines, not worth pursuing. In 1996, Chevrolet has reintroduced the ZL-1 aluminum big blocks for sale, and GM/Performance Parts does offer a Oldsmobile DRCE aluminum block with three different deck heights. Other aluminum blocks are available from after-market suppliers such as Donovan Engineering, World Products "Merlin" or Rodeck.

REMEMBER: IF YOU ARE BUILDING A MOTOR THAT WILL BE USED ON THE STREET BE CONSERVATIVE AND LIMIT THE BORE OVERSIZE TO .060".

THIS WILL GIVE YOU A RELIABLE ENGINE THAT WILL PROVIDE THOUSANDS OF TROUBLE FREE MILES. OVER BORING A PRODUCTION STYLE BLOCK LARGER THAN .060" IS POSSIBLE, HOWEVER, YOU RISK MAKING THE CYLINDER WALLS TOO THIN. "BOWTIE" BLOCKS ARE SPECIFICALLY DESIGNED TO TAKE LARGE OVERBORES AND STILL HAVE SUFFICIENT CYLINDER WALL THICKNESS.

366T-Mark IV and 366T Gen.V blocks are used in large, medium and heavy duty trucks and have tall decks and exceptionally thick

A dial bore gauge will help determine cylinder bore size and bore taper on a block.

cylinder walls along with four-bolt main caps. They have a standard bore of 3.935" and can be bored .125" over. The Gen.V 366T has a one-piece rear main seal. They have limited performance applications because even with a .125" overbore the possible cubic inches are relatively small. They have been made from 1966 thru the present.

396-Mark IV blocks have been used in cars and light duty trucks and have a standard bore diameter of 4.094". They can be bored .155" over the standard bore size

Gen.V 454 four-bolt blocks have redesigned oil passages which run along the camshaft tunnel. Notice the "MV" casting which is an abbreviation for "Mark V". These blocks use a one-piece rear main seal and matching components.

Gen.V production blocks do not have a boss to mount a mechanical fuel pump. You will have to use an electric or belt driven fuel pump with these blocks.

Lifter valley in a Gen.V block has plugs in the redesigned oil passages that run alongside the cam bore.

The bottom end of Gen.V and Gen. VI production blocks feature 4-bolt main caps and use a one-piece rear main seal and different crank, oil pan and timing cover which are different from those used on production Mark IV blocks.

to 4.250", IF THERE IS NO CORE SHIFT. If you do bore a 396 block .155" over and you use the 366T, 396, 402, 427T or 427 crank, which all have the same 3.76" crank stroke, you will have a 427 cubic inch motor. You may need to rebalance the rotating assembly if you mix and match parts. They had two or four bolt caps and use a two-piece oil seal. 325, 350, 360, 375 and 425 horsepower versions were made. The 396 was factory installed from 1965 to 1969.

402-Mark IV cases were installed in cars or light duty trucks, have a standard bore of 4.125" and can be bored to 4.250", if there is no core shift. Two or four bolt blocks were made. They used a two-piece rear main oil seal. Many of the 402 engines were called "396's" or "400 4-barrel" motors and this caused some confusion, but all were 396 engines bored .030" over by the factory. 210, 240, 300, 330, 350, and 375 horsepower engines were made from 1969 to 1972.

427-Mark IV cast iron production blocks, used in passenger cars, with short decks, have a standard bore of 4.250". They can be safely bored .060" over. Two or four bolt main cap versions were made. 427 Mark IV blocks made after 1969 have additional clearances in the bottom of the cylinder barrels which will accept 4" stroke cranks with no alteration. 335, 385, 390, 400, 425, 430 and 435 horsepower engines were installed, at the factory, from 1966 thru 1969.

427 "ZL-1"-Mark IV aluminum blocks were available in a very limited number of production ZL-1 Camaros and two Corvettes in 1969. Over the counter aluminum blocks were offered to racers for a few years in the early 1970's with iron cylinder liners and without liners. Then they were discontinued. In late 1996, twenty years after the last ZL-1 was made, Chevy is reintroducing the aluminum block. It has updated splayed steel 4-bolt main caps and uses a Mark IV style two-piece rear main seal and cap. The deck height is 9.80". Overbores are limited to .030" with cast iron liners. The block weighs approximately 110 lbs.

427T-Mark IV and Gen.V 427T heavy duty truck engine blocks have tall decks, very thick cylinder walls, have a standard bore of 4.250" and can be bored .125" oversize. They have four bolt main caps. Gen.V 427T's have one-piece rear main seals. These blocks are very interesting because with the taller decks and thick cylinder walls, large bore and stroke combinations can be obtained. They have been used from 1966 thru the present.

454 and 454HO-Mark IV (PN-10069285 four-bolt) and production Gen.V 454 blocks (PN-12508097 fitted short block) have a standard bore of 4.250 and can be safely bored .060" oversize. Mark IV 454's had two or four bolt main caps and use two-piece oil seals. All production Gen.V and Gen.VI 454 and 454HO motors have four bolt main caps and have one-piece rear main oil seals. The Gen.V cases have no provision for a mechanical fuel pump or clutch "Z-bar" stud bosses. Horsepower ranged from 215 to 460 versions made by Chevrolet and GMC/Truck and have been in production use since 1970. The Gen.VI 454HO crate motor can be bored .060" and has four bolt main caps. A one-piece oil seal is used. The Gen.V 454HO and the Gen.VI 454HO are only available over the counter.

502 and 502 H.O.-Gen.V cases (PN-14096817 bare block) and Gen.VI blocks have "siamesed" standard bores of 4.470" and can be safely bored .060" oversize. They have four bolt main caps and one-piece rear main seals. These blocks have "blind" tapped cylinder head bolt holes. The Gen.V block does not have a boss for a mechanical fuel pump, nor is there a bolt boss to mount a clutch "Z-Bar" ball and stud. This Gen.V block motor was introduced in 1993 and has not been factory installed in any production vehicles. It is available "over the counter" in three different long block versions and as a "bare" block. The Gen.VI block, introduced in 1996, does have provisions to mount a mechanical fuel pump and a clutch linkage pivot stud. The Gen.VI 502HO is also available over the counter.

BOWTIE BLOCKS

Currently six different "Bowtie" bare engine blocks are available from Chevrolet. Two of these are Mark IV style blocks in either short or tall deck versions and use two-piece oil seals. The other four are Gen.V style blocks and come in short or tall deck versions. Remember that Gen.V versions have no provision for the use of a mechanical fuel pump and that some blocks use a one-piece rear main seal and others use a two-piece rear main oil seal, matching crankshaft, timing cover and oil pan.

MARK IV BOWTIE BLOCKS

(PN-10051106) is a bare Mark IV short deck Bowtie block. It uses a two piece rear main oil seal and matching crankshaft. It has siamesed cylinder walls, 4-bolt straight main caps, "blind" cylinder head bolt holes and reinforced deck surfaces. The main oil passages run along the drivers side pan rail. This block has provisions for a mechanical fuel pump and a mounting bolt boss for clutch "Z-bar" linkage. The case is also drilled for oil cooler line inlet and outlet. The rear main cap bolt holes are reinforced, as are the bottoms of the cylinder barrels. The cylinder barrels are thicker than on a Mark IV production block.

The bores are semi-finished at 4.250" and need a final honing.

BIG BLOCK Quick Facts & References

"GENERATION VI BIG BLOCKS"

In 1996, the "Generation VI" or "Mark VI" big block Chevrolet engine was introduced. Either a Gen.VI 454HO or 502HO are being made available as " over the counter" crate engines. The Gen.VI engine block corrects a number of deficiencies found in the Gen.V production motors. On the Gen.VI, the water jacket core holes have been changed so that Mark IV or Gen.V style heads can be used with the Gen.VI blocks and will not leak. The mechanical fuel pump boss has been put back on the block and the bolt bosses for the clutch "Z-bar" pivot stud are also found on the Gen.VI. These four bolt blocks have windage trays and have oil cooler line ports mounted just above the oil filter mounting boss. A standard deck height of 9.800" is used and the 454HO cylinder barrels are non-siamesed, while the 502HO continues to use siamesed cylinder barrels.

The one-piece rear main oil seal found on the Gen.V engines has been retained along with the related Gen.V style crank and oil pan. However, the timing cover for the Gen.VI HO motors is made from a composite material and only has six bolts holding it on instead of the ten found on Mark IV and Gen.V motors. This timing cover has a notch instead of a timing tab and the new Gen.VI HO crate motors come with a degreed vibration damper that has timing marks every two degrees.

The Gen.VI 502T Truck Conversion (PN12360893) has small oval port heads and a mild hydraulic roller camshaft and is 50 state legal as a replacement engine for '88 to '94 big block trucks. The 502 HO crate engines will have rectangular port, open chamber, cast iron heads, non-adjustable drive train, and the split plenum, aluminum, high rise manifold long used by Chevy. A forged steel crank that fits a one piece rear main oil seal and 7/16" heavy duty "dot" rods are also standard equipment.

BIG BLOCK Quick Facts & References

"BOWTIE BLOCKS"

Chevy "Bowtie" blocks have many additional features when compared to stock production blocks. If you are building a high performance or racing engine, here are some of the things you will find:

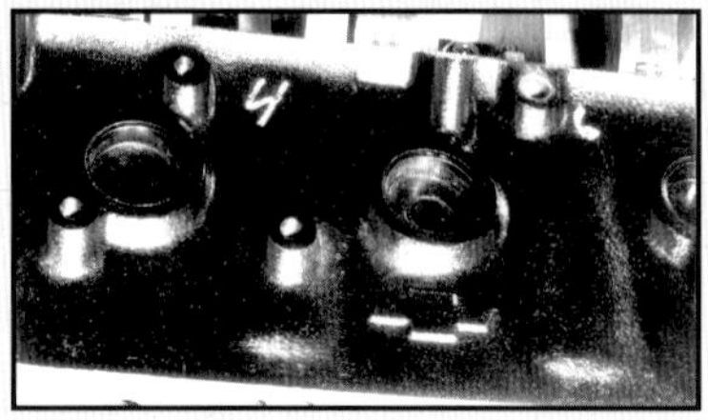

- **Short or tall decks**
- **Siamesed cylinder bores**
- **Four bolt main caps, straight or splayed**
- **Thicker cylinder barrels**
- **Reinforced cylinder barrels, main bearing bolt bosses, deck surfaces**
- **Thicker deck surfaces**
- **Fully machined main bearing bulkheads**
- **Provisions for oil coolers**
- **Redesigned water jackets to allow grinding for long stroke cranks**
- **Bores possible from 4.125" - 4.600"**
- **Strokes possible from 3.76" - 4.600"**
- **"Blind" cylinder head bolt holes**
- **Additional cylinder head bolt bosses in the lifter valley on Gen.V CNC "Race Prepared" Bowtie blocks**
- **Cast iron block weighs between 269-272 lbs.**
- **"Blueprinted" CNC machining on Gen.V "Race-Prepared" Bowtie blocks**
- **Water jacket holes in deck surface will seal with Mark IV heads**
- **Better uniform cylinder cooling**
- **Redesigned oil passages in the Gen.V "Bowtie" blocks**
- **Provisions for dry sump oiling on Gen.V CNC "Race Prepared" Bowtie blocks**
- **Larger valve lifter bosses in Gen.V "Bowtie" blocks**

The cases can be safely bored to a maximum of 4.560". The deck height on this block is 9.8". A "Bowtie" logo is cast into the side of the engine block case. The bare cast iron block weighs approximately 269 lbs. Maximum designed crank stroke in this block is 4.250".

(PN-14044808) is a bare Mark IV tall deck Bowtie block. The deck height is 10.2". It also uses a two-piece rear main seal and matching crankshaft. It has all of the other features found in the short deck Mark IV Bowtie block and weighs approximately 272 lbs. The block can be bored from 4.250" to 4.560". Maximum designed crank stroke in this block is 4.250".

Mark IV cylinder heads can be bolted to these two Mark IV blocks and the water jacket core holes in the decks will match. Do not use production style Gen.V cylinder heads or Gen.V head gaskets with Mark IV blocks. There will be a water leak into the lifter valley.

Mark IV oil pans and timing chain covers for use with two-piece rear main oil seals are required with these two Mark IV Bowtie blocks.

GEN.V BOWTIE BLOCKS

(PN-10185049) is a bare Gen.V short deck Bowtie block. It has a deck height of 9.8". The block has four-bolt ductile iron main caps, but requires the use of a one-piece rear main oil seal and matching Gen.V crankshaft, timing chain cover and oil pan. The block has no provision for a mechanical fuel pump and while some early Gen.V Bowtie blocks had no provision for a clutch pivot ball stud, all Gen.V Bowtie blocks now do have the bolt boss provision to mount a clutch pivot ball stud.

This case has siamesed cylinder walls which are extra thick (the minimum wall thickness at 4.250" is .500"). The barrel walls themselves are oval shaped to insure consistent thickness. When bored to its maximum safe limit which is 4.600", the cylinder walls will have a minimum thickness of .325". The blocks come with 4.250" diameter, semi-finished cylinder bores.

Because there are thicker cylinder walls in this block some distributor shaft gears may touch the block near the #-8 cylinder barrel when they are installed. If this happens grind the block where the gear touches or chamfer the bottom of the gear. Also some aftermarket distributors use an "O" ring to seal the distributor shaft. These "O" rings may not seal the oil passages properly. You may have to raise or lower the distributor accordingly.

"Blind" cylinder head bolt holes are used and the block deck surface is reinforced. The water passage core holes in the block deck surface match those found on Mark IV cylinder heads. There is a vent hole in the front and rear portion of Gen.V block decks that is used in the block casting process. You must use the proper head gasket or tap and plug these vent holes which will seal them and prevent unwanted bypassing of coolant circulation. If you use a Mark IV head on these blocks, use a Mark IV head gasket and plug the deck vent hole. If you use a Gen.V cylinder head, use a Gen. V style head gasket. The Gen.V gasket will block the vent core hole.

The main oil passage was moved from the Mark IV position

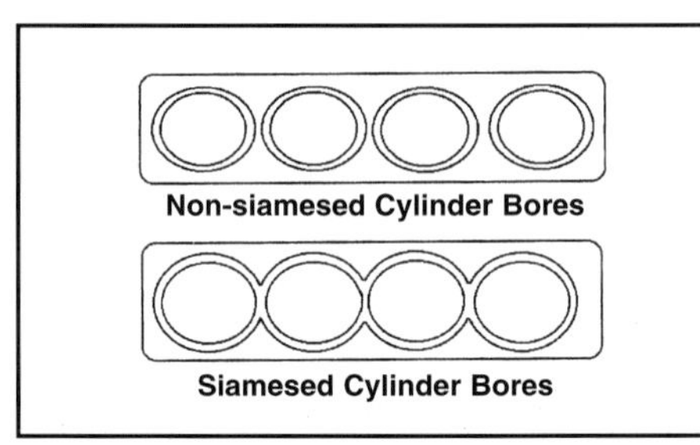

Gen V & GenVI 502s, Mark IV "Bowtie" and Gen V "Bowtie" blocks have "siamesed" cylinders to maximize cylinder bore sizes and bore strength. They also have "blind" head bolt holes.

Gen.V Bowtie CNC "Race Prepared" tall deck block, casting #-24502506, has a number of special features including additional head bolt bosses in the lifter valley. Notice the small, round water holes found on all Bowtie block decks.

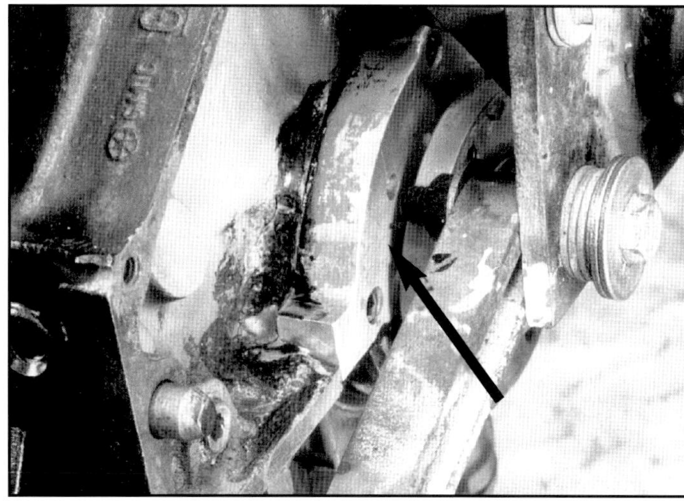

Gen.V Bowtie CNC "Race Prepared" short and tall deck blocks come with a special rear main seal block adapter to mount a two piece seal and then use a Mark IV style crank.

Casting "506" Bowtie CNC "Race Prepared" Gen.V tall deck block comes with steel, splayed, four bolt caps and studs. This block easily swallowed a 4.5" stroke crank.

"504" and "506" CNC "Race Prepared" blocks, unlike other Gen.V blocks, uses an crank seal adapter that allows the use of a two- piece crank rear seal and Mark IV style crank and oil pan. Notice the Gen.V style oil filter mounting boss, but the oil cooler ports are not located in the pan rail on this block. Oil cooler passage is located above the filter boss, similar to Mark IV four bolt blocks. CNC "Race Prepared" Bowtie blocks also come with splayed steel 4-bolt main caps. Finally, do you recognize the oil pump on this 557 cubic inch torque monster? That's right, a stock GM big block pump with an aftermarket screen.

and now runs along the camshaft tunnel and gives priority oiling to the main bearings. This allows for better oiling and also allows more grinding along the pan rails for longer stroke crankshaft clearance and makes possible the use of splayed main bearing cap bolts. This case however is delivered with four-bolt straight caps.

The lifter bosses are larger and allow an engine builder to increase and/or offset the lifter diameter by remachining these bosses if desired.

Mark IV or Gen.V style cylinder heads can be used with this block, provided that you use a head gasket which will match the style of cylinder head and block the correct water core holes found in the front of the block deck surface.

This block requires the use of a Gen.V style timing chain cover and oil pan designed for a Gen.V one-piece rear main oil seal and matching crank.

There are provisions for oil cooler line inlets and outlets in the oil pan rails on the driver's side and there are some provisions for dry sump oil inlet lines in the rear of the block.

The front timing cover cavity is more shallow on Gen.V Bowtie blocks. This was caused by moving the front of the case slightly forward to improve the water cooling around the #-1 cylinder. The overall length of the case was kept the same front to back, but this change altered the depth of the timing chain area. When installing any timing chain or especially a timing belt, make sure there is at least 1/16" clearance between the chain and the block face and between the timing assembly and whatever timing cover you use.

The designed maximum stroke for this block is 4.600" and it weighs approximately 269 lbs.

(PN-10134367) is a bare Gen.V tall deck Bowtie block. It has a deck height of 10.2" and uses a one-piece oil seal. It has all of the features of Gen.V Bowtie block

BIG BLOCK Quick Facts & References

CRATE ENGINES

GM has many different crate engine options including new and rebuilt stock big block motors from Goodwrench and AC/Delco dealers. Chevy can sell you two Gen.V HO (High Output) long block crate motors either in 454 cubic inches (PN-10185058) or 502 (PN- 10185085). These two Gen.V HO engines come with aluminum high rise intake manifold, square port Gen.V heads with non-adjustable hydraulic flat tappet valvetrains, steel cranks, 7/16" "dot" rods, six-quart truck oil pans, windage trays and include the water pump, valve covers, flexplate and front balancer. The hydraulic flat tappet cams for these two HO crate motors are different. The 502 HO cam (PN-14096209) has a valve lift of .500"/.500" with duration at .050" numbers of 220/220. The 454 HO camshaft (PN-10185060) has a lift of .510"/.510" with a slightly longer duration at .050" of 224/224 degrees. The 502 HO uses a siamesed cylinder barrels with bores of 4.466" and a 4" stroke. The 454 HO uses a different block with non-siamesed bores of 4.250" and a 4" stroke. The 502 HO short block assembly is made under (PN-10185059).

Photo courtesy of Chevrolet

Chevy also has a complete Gen.VI 502 (PN-12360893) that uses the same short block as the Gen.VI 502 HO, but comes with small round port heads (PN-12552888) with a non-adjustable Gen.V valvetrain and 2.07"/1.72" valves. The camshaft is a hydraulic roller tappet with .480"/.483" valve lift and a duration at .050" of 204/209 degrees. With the 118cc open chamber round port heads compression is 8.75-1. No intake manifold is provided, but they do include a 6- quart truck style pan and windage tray. This motor is a stump puller and makes 338 horses at 4500 rpm and 512 ft./lbs. of torque at 2800 rpm. Two versions are available, (PN-12360893) is an emissions legal replacement motor for 88-94 454's in trucks. You can pull the ignition, exhaust, intake and fuel injection systems off of your '88-'94 454 and bolt them on to this Gen.V 502 and be street legal. (PN-12360898) with square port heads and hotter roller cam is an off-road version intended for early emissions exempt vehicles.

Gen.VI 454HO and 502HO crate engines were made available in 1996. The four bolt engines use one-piece rear main seals and internal parts similar to the Gen.V HO crate engines. However the block castings have bosses for a mechanical fuel pump and clutch linkage pivot studs, not found on production Gen.V blocks. In addition these Gen.VI engines have a hydraulic roller cam and the block decks have been changed so that Mark IV and Gen.V and Gen.VI cylinder heads can be used without the possibility of a water leak into the lifter valley.

While the L-88 427 and LS-6 and LS-7 454 complete Mark IV engine assemblies are no longer being made, Chevy still stocks the individual parts that enable you to build one.

GM/Goodwrench and AC/Delco dealers can deliver new and reconditioned stock performance Mark IV and Gen.V motors. They carry excellent guarantees and can become the basis for a high performance combinations.

These days if you don't want to be build your own engine, you can buy a complete ready to run motor. Many mail order sources exist which can provide anything from a bone stock big block Chevy to an all out race motor built to your specifications or the supplier's. (You can even rent a Pro Stock drag race engine if you have the bucks.) Many of these motors are warrantied and are broken in and dyno tested before shipment to the purchaser. Some of the national suppliers can be found in the Buyers Guide list of this book or you can get a copy of S/A Design's HIGH PERFORMANCE CRATE MOTOR BUYER'S GUIDE by John Baechtel.

(PN-10185049) except with the taller deck. There is also an additional pressurized oil passage along the top front of the block for use as required. This block weighs 272 lbs.

(PN-24502500) is a bare Gen.V, CNC "Race Prepared" short deck Bowtie case. This case shares the same features as other Gen.V Bowtie blocks but also has some additional items. The block is a separate casting from other Gen.V " Bowtie" blocks and is "blueprinted" to factory specifications with very tight tolerances. They have steel "splayed" four bolt steel main caps with studs and 12-point nuts. These Gen.V CNC "Race Prepared" blocks have a rear seal adaptor that allows the use of two-piece rear main oil seals, Mark IV crankshafts, oil pans and timing chain covers.

Eight additional bolt bosses are cast into the lifter valley to provide additional cylinder head bolt locations. Many high performance cylinder heads have provisions for these additional head bolts and this provides greater gasket sealing with six head bolts around each cylinder bore.

The deck surface thickness is left at maximum and only enough material is removed at the factory to insure flatness.

This block can be bored to 4.600" and can use a maximum crank stroke of 4.600". This block weighs approximately 278 lbs.

You can use Mark IV or Gen.V style cylinder heads with this block and then use a head gasket that matches the head.

The oil cooler line ports are removed from the oil pan rail and are found above the oil filter boss. The main oil passage has been moved to the cam tunnel and is enlarged to 9/16" diameter.

(PN-24502502, Casting #-24502506) is a bare Gen.V, CNC "Race Prepared" tall deck Bowtie block. The deck height is raised to 10.2". It uses a two-piece rear main oil seal, and matching crankshaft, timing chain cover and oil pan. It weighs approximately 288 lbs. and has the same features as the "CNC Race Prepared" short deck Gen.V Bowtie block.

OLDSMOBILE Drag Race Competition Engine (DRCE1) cast iron big blocks, have deck heights of 9.925", (which is .125" taller than a "short deck" Chevy block) and siamesed bores of 4.440" and can be bored to 4.600". First introduced in 1983, DRCE blocks can use the Mark IV Chevrolet big block rotating assemblies. They can accommodate a maximum stroke of 4.600". These blocks use an Oldsmobile style rear main oil seal and have 4-bolt splayed nodular iron main caps. Two versions are made in cast iron. One block (PN-22521056) has a head bolt pattern for Oldsmobile DRCE heads (PN-22530959) and the other DRCE1 block (PN-24502486) has a head bolt pattern for Pontiac Super Duty big block heads (PN-10045427 or 10049875). Both blocks have a bellhousing bolt pattern for Buick, Oldsmobile or Pontiac (BOP) transmissions. Both of these cast iron blocks each weigh approximately 262 lbs.

In 1996 an improved version of the cast iron DRCE block (DRCE2 PN- 24502572) was introduced for Pro-Stock drag racing. This CNC-machined block has a deck height of 9.450" and the extra thick deck can be cut to 9.30". Bore spacing is 4.900" and cylinders come with 4.500" bores. The block can bored to 4.650" and even with this bore size the cylinder walls will be a minimum of 0.300" thick along the major thrust surface. The block comes with 4-bolt splayed caps and the camshaft tunnel is raised .600" over the stock height. There are eight additional cylinder bolt bosses in the lifter valley for additional head sealing. The lifter bores and head bolt holes are not drilled so that an engine builder can place them where he wants to. The distributor hole and boss have been moved rearward for additional valvetrain, induction system and distributor cap clearance. These blocks weigh approximately 11 lbs. less that other cast iron DRCE1 blocks. Strokes up to 4.600" can be used.

THE NEW "ZL-1" ALUMINUM BLOCK

"Good things happen to those who stand and wait." Casting #3946053, 110 pounds, PN-12370850, the ZL-1 block is being reintroduced over the counter from Chevrolet. The legendary aluminum big block has been updated for its second life. Bores will be 4.250", with reinforcement ribs in the lifter valley. Provisions are still there for dry sump oiling systems and the block will get updated splayed steel four bolt main caps. The bottom end of the block casting has also been beefed up with more material and strength. A two-piece, Mark IV style rear main oil seal is also used. There is a boss for a mechanical fuel pump and bosses for a clutch "Z-bar."

The head bolt holes are "blind" in the new block and there are additional head bolt hole bosses in the lifter valley so that each cylinder bore can have six head bolts surrounding it.

The original ZL-1 block of the early Seventies weighed 98 pounds! Whoever at Chevrolet or GM who thought of this and made it happen, should be named Chairman of GM's Board and Chief Executive Officer with a stock option plan that makes Ross Perot green with envy. Get your check book out. These alloy blocks will be pricey!

OLDSMOBILE DRCE1 aluminum big blocks, (PN-22536161) come in three deck heights (9.925", 10.50", 10.70"). They have standard siamesed bores of either 4.250" or 4.500" with ductile iron sleeves and your choice of two different honed surfaces. These blocks use an Oldsmobile style rear main oil seal. These are "Y-style" blocks that utilize side-bolted steel main bearing caps. These aluminum blocks weigh approximately 157 lbs.

MERLIN "SUPERBLOCKS", are available from World Products,Inc. Four different deck height blocks are available and they can use Chevy rotating assemblies and Chevy cylinder head bolt patterns. All of the blocks share a number of features including billet steel splayed 4-bolt main caps, ARP main cap studs, additional head bolt bosses in the lifter valley, reinforced lifter bores, extra thick bulkheads, relocated main oil galleries run along the cam tunnel, very thick cylinder walls and 200-230 cylinder wall Brinnell hardness. Also the block decks are squared to the crank centerline and the main bearing caps are line honed before shipment. The lifter bores are also indexed before shipment. Each block has a starter mount location on either side of the rear of the block and also has solid pan rails so that they can be ground for further crank stroke clearance if you need it. Cylinder bores are left .010" undersize when shipped so you can finish hone them whichever way you want. The oil filter boss has been moved 1/2" outward to provide additional room for large kickout oil pans. These blocks also have provisions to mount a mechanical fuel pump, side engine mount bosses and a

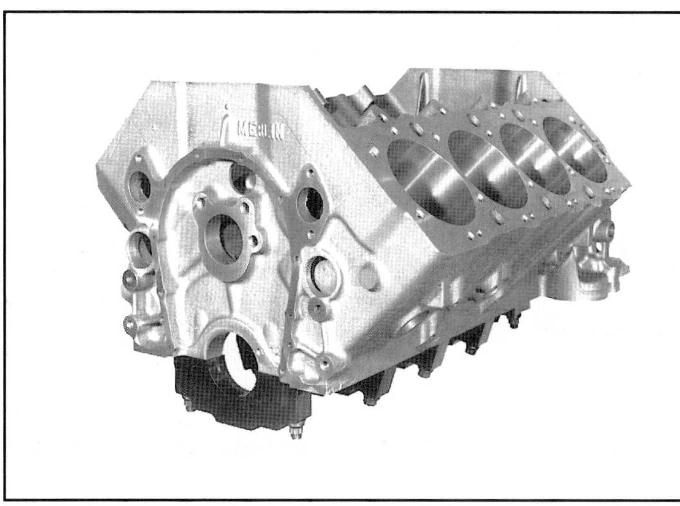

Merlin blocks from World Products are a popular choice among racers wishing to build engines in the 500 to near 800 cubic inch displacement range.These blocks are designed strictly for high performance racing applications.

Photo courtesy of World Products

clutch "Z-bar" ball and stud.

Merlin (PN-8009) is a custom deck (9.300") block.

Merlin (PN-8010) is a "short deck", cast iron block with a 9.80" deck height. The bore size is 4.430" and the cam tunnel is in the stock Chevy location and height. The block can be bored to 4.625". The pan rails are drilled for either a Chevy Mark IV oil pan or an Olds DRCE1 style pan.

Merlin (PN-8011) is a tall deck, cast iron block with a deck height of 10.2". It comes with a 4.430" bore and can be bored to 4.625". The cam tunnel is in the stock Chevy location. This block can also fit either a Mark IV Chevy oil pan or one for an Olds DRCE.

Merlin (PN-8012) is an extra tall deck, cast iron block with a deck height of 11.625". The bore is 4.490" and can be bored to 4.625". The cam tunnel has been moved upward .400" to provide additional cam lobe and crank stroke clearance. This block uses an Olds DRCE1 type oil pan. This block weighs 297 lbs.

All of these Merlin blocks can be used to build large stroker crank motors, including some 770 cubic inch combinations with the 11.625" deck height block (PN-8012). You will need to do a trial fit of the crank, rods and pistons that you are going to use. Put the crank in with one rod and piston at a time into each cylinder. If the block requires additional clearancing for your combination, you can mark the block on the pan rails and the bottoms of the cylinder barrels to see where you should grind the block. World Products recommends at least .030" clearance at all points of rotation. All other parts to finish the block assembly are available from Chevy or aftermarket suppliers. World Products supplies a list of those stock Chevy or specialty part numbers you will need and where to get them and instructions on block prep.

An aluminum version of the Merlin big block is also available.

Make sure there is enough block clearance for the spinning crank counter weights and rods, especially if you are building a stroker motor or are using aluminum rods.

MAKING A 454 OUT OF A 396

All production 454, 502 and some 427 blocks made after 1969, have extra clearances cast into the bottom of the cylinder barrels so that production steel connecting rods and 4" stroke crank counterweights can clear the bottom of the cylinder barrels. You can put a 4" stroke Mark IV crank into a 396, 402 or pre-1969 427 block, but you may have to grind extra clearance into the bottom of the cylinder barrels and possibly in the inner edges of the oil pan rails so that the 454 4" crank and rods will spin without hitting the case. I have found at least one 1966 396 block, casting #-3855961, which accepted a 454 4" cast crank, stock 3/8" rods, a .600" lift camshaft and an .155" overbore to 4.250" without having to grind any additional clearance into the block to get things to fit. I was all set to grind the additional clearances, but during the trial fit of components, it became clear that it wasn't necessary. There may be other 396/402 block castings where a 454 crank will clear without grinding, but you must do a trial fitting with your parts before any final decision should be made.

Only grind enough block material off to obtain a .050" clearance for the rotating assembly and remember in a production Mark IV block there is a water jacket near the bottom of the cylinder barrels and the main oil supply passage runs along the driver's side oil pan rail. Don't grind too much!

BLOCK DECK WATER CORE HOLES

The core holes in the decks of all Mark IV blocks and all "Bowtie" blocks are round. The core holes in the deck of a Gen.V production block are irregularly shaped and look completely different. You can use this deck surface feature to help identify block families. However, of more immediate concern is the fact that these different block core hole styles created some interchangeability problems that you must be made aware of. If you try to use a Mark IV style head on a production Gen.V 366T, 427T, 454 or 502 block, there will be a water leak from the deck surface to the lifter valley. Likewise if you use a Gen.V production head on a Mark IV block there will be water leaks. The core holes don't line up and the head gaskets for a Mark IV and Gen.V engine are different. Therefore, you must use a Gen.V head on a Gen.V production block along with a Gen.V style head gasket. Mark IV heads can be used on Mark IV production

Arrow points to Gen.V Bowtie front deck hole on each side that will be blocked by the Gen.V style head gasket when using a Gen.V style head. If you use a Mark IV style head and head gasket the hole at the front of the Gen.V Bowtie block on each side must be plugged to insure proper coolant circulation.

Irregular shaped water passage holes in the production Gen.V block deck surface are one reason that production Mark IV heads should not be used with these Gen.V blocks. The coolant holes don't match up and there can be leaks.

blocks and any "Bowtie" block with a Mark IV head gasket and on the Gen.VI 454 or Gen.VI 502 HO blocks. Chevy and a number of aftermarket sources are now making big block Chevy heads which will fit Mark IV, Gen.V & Gen. VI production blocks. These heads and interchanges are discussed further in the CYLINDER HEAD section.

There is another problem with core holes and this concerns all new Gen.V production blocks and Gen.V "Bowtie" blocks. When these blocks were designed an additional core hole was put into the deck of the blocks near the front head dowel pins. These holes allow gases to escape when the block is cast. If you use a Gen.V head gasket and Gen.V head, this hole in the deck is blocked by gasket material. Blocking these holes provides for the proper circulation of coolant thru the rest of the block. If however you want to use a Mark IV style head on a Gen.V "Bowtie" block, you must use a Mark IV style head gasket and then plug this hole in the block deck surface. Then you must punch holes in the Mark IV head gasket to match the rearmost Gen.V block core hole. If you don't get this right, water will enter the block from the front water pump holes and then exit thru the head gasket and head and then go right out the front of the intake manifold. Needless to say, overheating can be a fast occurring problem.

Gen.VI, also called Mark VI, blocks correct the core hole problem found on production Gen.V blocks. The water jacket holes on Gen.VI 454HO and 502HO blocks were redesigned to work with any style of big block Chevy cylinder heads in terms of matching and sealing water core holes.

OTHER BLOCK DIFFERENCES AND FEATURES

Certain interchanges are possible between the parts used in Mark IV, Gen.V and Gen.VI assemblies: camshafts, lifters, pistons, rings, connecting rods, timing chains and gears, water pumps, clutch and pressure plates, rod and cam bearings, motor mounts, starters, exhaust manifolds, intake manifolds, the front four main bearing caps (four bolt), distributors and accessory brackets.

All of the big blocks have their side engine mount bosses in the same location, except for the Oldsmobile DRCE blocks which have no mounting bosses on the side of the case. The DRCE block is intended for professional level racing and is expected to be used with front and rear engine mounting plates. Bell housing bolt holes in the block and the rear dowel pin locations are in the same Chevy position on all of the blocks except the DRCE1 block which has a pattern for "BOP" style transmissions. DRCE2 blocks have a bellhousing pattern to fit Chevy or "BOP" patterns.

This Mark IV 454 block was made in 1978. Arrow points to annular oil passage groove which will sit under the rear cam bearing when it is installed. '65 and '66 Mark IV blocks do not have this groove and a different grooved rear cam bearing must be used in '65 and '66 blocks.

This is the new oil filter mounting boss found on Gen.V and Gen.VI production blocks. Arrows point to oil cooler ports found on this block.

Freeze plug size changed from those used in the Mark IV blocks to smaller diameter ones used in the Gen.V and VI blocks. Incidently, the freeze plugs in a Gen.V or Gen.VI motor are the same as those used in the smallblock Chevy V-8 and are interchangeable.

The tall deck blocks require the use of longer pushrods and wider intake manifolds which are available from aftermarket manufacturers. Distributors that came with tall deck motors from the factory, have longer shafts. However, aftermarket intakes for tall deck blocks are available with lowered distributor mounting holes so that a standard deck, stock style distributor can be used.

Provisions for oil cooler lines are found in most four-bolt Mark IV blocks, Gen.VI HO and Gen.V "CNC Race Prepared Bowtie" blocks, just above the oil filter boss on the driver's side. Other Gen.V blocks, which are all four-bolt, have the cooler line ports in the bottom of the oil pan rail on the driver's side. Adapters are available from Chevrolet and aftermarket sources to add oil cooler lines to a block that does not have them. These adapters sandwich between the oil filter mounting surface and the oil filter and have inlet and outlet ports in them for oil cooler lines.

Cam bearings also need attention. In Mark IV blocks that were cast in 1965 and 1966, you need to install a grooved rear cam bearing and a cam with a groove cut in the rear cam journal. These grooves in the bearing and cam create an essential oil passage that was part of the original design incorporated into these blocks during these two years. If you use the wrong rear cam bearing, oil will not get up into the lifter valley or to the rocker arms. In 1967 and later blocks, an annular oil passage groove was added to the block itself lying underneath the rear cam bearing. The rear cam bearing and rear cam journal no longer used a groove cut into their surfaces.

The next change came when the Gen.V blocks came out in 1991. The main oil passage that ran along the oil pan rail on the driver's side of Mark IV blocks was relocated to run along the camshaft tunnel in Gen.V and VI blocks. Oiling priority was given to the main bearings with side passages going to the camshaft journals and cam bearings. With Gen.V and VI blocks you must make sure that the oil feed holes in the cam bearings match up with the feed passages in the block cam tunnel. This requires that the cam bearings be installed in Gen.V and VI blocks with the oil feed holes lined up in the "2 o'clock" position when viewed from the front of the block in the upright position.

Remember that depending on whether the block utilizes a one or two piece rear main oil seal, you must use the appropriate matching oil pan, oil pan gasket, timing cover, crankshaft and flywheel or flexplate. The Gen.VI block uses a new composite timing cover that is retained with six bolts instead of the ten found on earlier blocks. This plastic timing cover has a notch instead of a timing tab and should be used with a "degreed" vibration damper (PN-10216339).

TALL DECK BLOCKS

If you have access to a tall deck block, you should be aware that a tall deck block allows you to do a number of things. Tall deck production truck blocks were first used because the extra .400" of deck height allowed GM to use a four ring truck piston with a very thick piston deck. This type of piston was better able to withstand the abuse handed out to medium and heavy trucks with the tremendous loads put on them. Even with this four ring piston and increased deck height these motors still used the same stock length (6.135") connecting rod used in all other big block motors, although the truck rod forging had thicker webs on the big end while using a 3/8" rod bolt.

All of the truck tall blocks had other features such as very thick cylinder walls, four bolt main caps and most had provisions for external oil cooler lines. These features make them excellent cases from which to build a high performance engine. The extra deck height allows a builder to increase the rod length and/or stroke length and the extra thick cylinder walls can be bored out .125" oversize, if there was no core shift when the block was cast. Piston ring grooves can

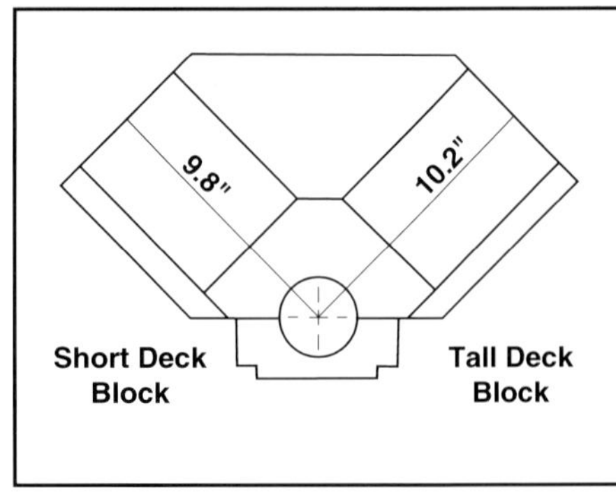

Big blocks have two different block deck heights. Heavy duty tall deck truck and "Bowtie" tall decks measure 10.2", while short-deck blocks are 9.8".

Tall deck "TRUCK" and "MARINE" blocks are have four bolt main caps and are a good starting place for a stroker or long rod motor.

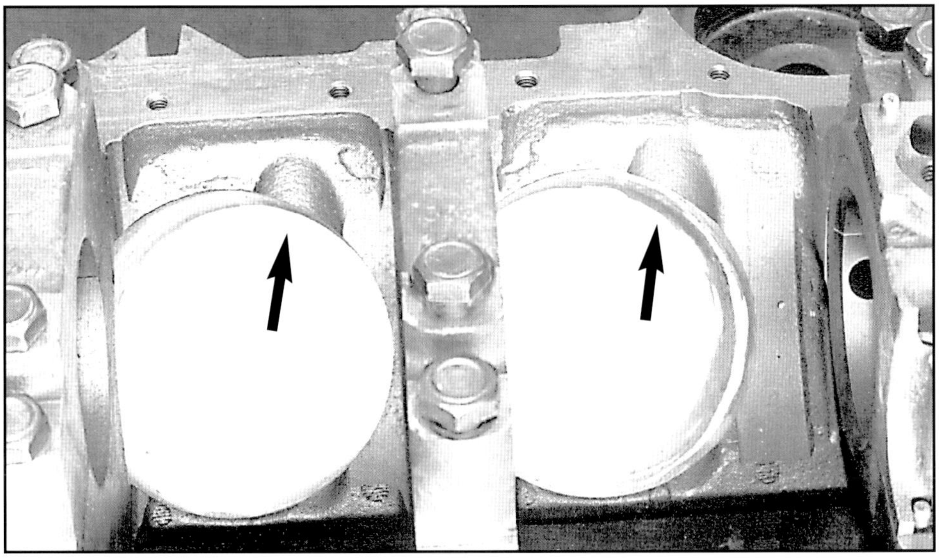

427 and 454 blocks from 1969 onward had extra clearance cast into the bottom of the cylinder barrels to provide clearance for crank counterweights and rods when using a 4" stroke crank. Some cases have clearance also cast into the oil pan rails for the same reason.

be moved up closer to the piston deck and the pin holes in the piston can be moved up. The 427T blocks start out with a 4.250" standard bore and with a .125" maximum overbore will be 4.375". Aftermarket pistons for production size blocks, are available off the shelf in .020", .030", .040", .060", .070", .090", .100" and .125" oversizes. Other piston sizes are available for Bowtie and other aftermarket larger bore blocks.

The crank stroke that you choose will be limited by how much increased stroke you can physically fit into the case and how much grinding on the bottom of the cylinder barrels and oil pan rails you can do without hitting water jackets or oil passages. Stock length four inch 454 cranks are usually no problem. 4.125" (1/8 inch) stroker cranks may require a little block massaging. Be careful of stroke increases beyond 4.125" in the tall deck truck cases. If you wish to use a stroke larger than 4.125", I suggest you use one of the "Bowtie" or aftermarket tall decks. The reason is that "Bowtie" blocks are a different casting, the water jackets have been relocated away from the lower cylinder barrel areas. The oil passages near the pan rails have been moved away or relocated in the cam tunnel and in some aftermarket tall decks, the cam tunnel itself has been moved upward in relation to the crank centerline. All of these changes were done to make it easier to increase stroke length to 4.250" and up to 4.600" in some blocks and still have the clearance necessary for the rotating assembly.

You can also increase the rod length because there is additional total deck height. Remember that when you are considering what to use, the parts lengths need to add up to the available block deck height. 1/2 the crankshaft stroke length, rod center to center length, piston compression height (also called pin height, the distance between the centerline of the piston pin to the deck of the piston) and the remaining deck height above the piston to the deck surface of the block, all have to be considered. Here's two examples, using a "Bowtie" tall deck block with a deck height of 10.2" and a cylinder bore of 4.350" (+.100" oversize):

1/2 Crank 4" (454) Stroke = 2.000"
Rod length (.400" long) = 6.535"
Piston compression height (standard 454 HP piston) = 1.645"
Block deck height above piston deck at top dead center = .020"
TOTAL 10.200"

The above combination measures 476 cubic inches and uses a rod which is .400" longer than stock to make up the extra deck height.

This combination has a rod ratio of 1.63-1 which by "coincidence" is the same rod ratio used in a stock 396 or 427.

Here's another where we increase the stroke and lengthen the rod over stock:
1/2 Crank 4.250" Stroke = 2.125"
Rod length (.250" long) = 6.385"
Piston compression height (standard 454 HP piston) = 1.645"
Block deck height above piston deck at top dead center = .045"
TOTAL 10.200"

This 505 cubic inch example uses a stroker crank that has a stroke .250" longer than a stock 454 crank and an aftermarket rod which is .250" longer than stock, while using a stock, off the shelf, 454 +.100 piston and a "Bowtie" tall deck block bored to 4.350". This combination has a rod ratio of 1.50-1.

Once you have determined the amount of deck height between the top of the piston deck and the block deck at top dead center and you know which head gasket thickness you will use, the block deck can be shaved down the appropriate amount to give you a minimum total of .035"-.040" clearance between the top of the piston deck and the surface of the head deck. The compressed thickness of the head gasket must be included in this calculation.

Other piston compression heights are available. When the piston is made, if you squeeze the piston ring grooves closer together and move them upwards closer to the piston deck, the piston pin can be moved upwards in the piston. This allows the use of an even longer rod or stroke because the pin compression height dimension in the above examples will be less. Generally the piston can then also be lighter. Many different combinations are available. Talk to your aftermarket supplier about what will work for you.

There are some other things to

Some Mark IV production cases have four-bolt main caps. All "Bowtie", and production Gen.V and Gen.VI big blocks have four bolt main caps.

Some Mark IV blocks have two-bolt main caps. These blocks are more than adequate for most sane street applications. A main cap stud kit will upgrade the bottom end for high performance street and some bracket drag race use.

keep in mind when you build a "stroker" motor. We have already discussed the possible clearance problems on the block. If you use aluminum rods, remember that the beams and big ends of aluminum rods are bigger and wider than those found on steel or titanium rods. They are bigger in order to maintain the strength equal to or greater than a steel rod. This also means that in order for an aluminum rod to fit inside the case, there has to be more room or clearance for the rod to rotate. Depending on the combination you are trying to build, it may be easier to fit a steel or titanium rod. Also aluminum rods stretch further than a comparable steel rod so provisions must be made to allow for this stretch. Some aluminum rod makers allow for stretch by making it slightly shorter in actual length than its nominal length. For instance, a nominal aluminum rod center to center length may be 6.135", but when actually measured it may be found to be 6.125", allowing for .010" stretch.

Long strokes over 4.00" may also require additional clearance in the oil pan rails of the pan. Aftermarket oil pan makers can supply pans for big block Chevys with additional rod clearance to make room for your combination. Finally, aluminum rods are intended to be used in short duration motors. They are not intended for street use, because over time the aluminum gets "work hardened" and will fail. If you are not willing to remove the rods to check for excessive stretching or work hardening, on a frequent basis, then use a steel rod.

Cylinder heads will fit these tall deck truck blocks, but you should note that intake manifolds for tall decks are wider side to side than low deck block intakes. Also the distributors that came on production truck tall deck engine had a longer shaft. Pushrods are .400" longer than those on short deck motors. If you are going to put a tall deck block into a passenger car type vehicle, make sure that there is room in the engine compartment for the slightly wider side to side dimension and the increase in deck height. The exhaust system you choose will also have to be checked because the exhaust ports on the motor will sit 0.400" higher than those found on a short deck motor.

BLOCK PREP

First the block has to be thoroughly cleaned. All the old freeze plugs, oil gallery plugs, deck dowel pins, cam bearings, bolts, and drain plugs are removed. Many shops now use a oven cleaning system that cooks the block and reduces the grease, sealant, residual gasket material, paint and oil to ash. The block is then put into a shot cabinet where the ash, rust and other corrosion is further removed. Finally the block is put in a high pressure washer to remove everything including any junk in the oil passages, leaving a clean, gray metal casting.

Blocks used in high performance engines need special attention. If you acquire a used block, determine whether there is any damage that will render the block useless. Obvious things like pieces blasted out of the side of the block, cracks in the pan rails, bottoms of the cylinder barrels, cracks in the cam tunnel that go into the water jackets or cracks in the outside walls of cast iron blocks may not be worth repairing. Some cracks in cylinder walls due to overheating, freezing or rod and piston damage can sometimes be repaired by installing a sleeve in the damaged cylinder bore. I usually draw the line at sleeving one cylinder bore on a block for street use. If two or more barrels are damaged its easier and cheaper to continue looking for an undamaged block. Engines that have had damage done to the main bearing block surfaces and caps (usually by a spun main bearing) can sometimes be repaired by align boring and align honing the main bearing block surfaces. Once again, I feel that this repair is a waste of time and money, go find an undamaged block. Some cracks may not be easy to see or find until the block is completely cleaned and magnafluxed or pressure tested.

Blocks that have previously been bored to their respective maximum bore sizes should also be rejected. Remember that when you bore a cylinder the cylinder walls get thinner. Thinner walls are inherently weaker than thicker walls. This is a trade off. Increasing the bore can help make more power, but it weakens the cylinder walls. That's one of the reasons that "Bowtie" block cylinder walls are so thick to begin with. Even with larger bores there is still enough wall material to provide sufficient strength.

Some engine builders sonic test the cylinder walls and deck thickness, particularly those blocks which will be used for serious racing. A minimum wall thickness of at least .200"-.300" on the major thrust surfaces, after boring, should be looked for on a race motor. Folks have bored blocks that have less finished wall thickness, but it is risky. Thin cylinder walls can be supported by "filling" the block water spaces, but this is done mainly with short duration drag race motors. Core shift or voids (air pockets) in the metal casting can be detected and a decision made whether the block is acceptable. Race blocks are also tested for hardness in order to insure maximum strength, but you need to make a decision on whether this is necessary for your situation.

Some builders prefer to debur the block, removing any excess casting flash on the outside edges of the block, water core holes and on the interior surfaces around oil drain back holes or surfaces.

Check all of the threaded holes in the block. If bolts are broken off in them, they will have to be drilled out and if necessary the threads sometimes replaced with Helicoils. All the threaded holes need to be cleaned out with a bottoming tap. Do not cut the threads, just get the accumulated junk out of the holes and threads.

There are some instances where the block water passages are filled with a concrete-like "block filler". Filling the block adds to the cylinder barrel rigidity and will help the bores stay round for better ring seal. The amount of weight added to the block by filling it, is offset by the weight of the coolant no longer filling or circulating in the block. The block can be half filled if you plan to run the motor on the street, since some coolant will still circulate thru the block. Other blocks are completely filled and are intended to be run only for short intervals and then allowed to cool down, such as in some drag racing classes.

Two bolt Mark IV cases can be machined and converted to use four bolt main caps. The main issue in this decision is whether the cost of an original four bolt block is more or less than the cost to convert the two bolt block to four bolt main caps. Most of the time its cheaper just to get the 4-bolt block in the first place.

Four bolt iron main bearing caps (numbers 1 thru 4) (PN-14015334) are available for this conversion and can be used on Mark IV or Gen.V motors with straight, not splayed, main cap systems. These main caps are semi-finished and will need to be fitted to properly register in the block and then align bored and honed to finished size and correct alignment with each other. The rear main caps are different on Gen.V and Mark IV blocks because of the difference in rear main oil seals used. The main cap bolts for Mark IV four bolt blocks are different lengths than those used for four bolt main caps on Gen.V blocks. Inner bolts on Mark IV's are 1/2-13 x 3.34" (PN-3909834), Gen.V inner main cap bolts are 1/2-13 x 2.52" (PN-10106461). The outer bolts for Mark IV's are 1/2-13 x 2.60" (PN-3859927), while the Gen.V outer bolts are 1/2-13 x 2.78" (PN-10106460).

Two bolt Mark IV Chevy blocks will withstand 600 horsepower applications for drag racing and are certainly durable for most street performance applications. Longer duration racing, large cubic inch and higher horsepower and torque combinations should utilize four bolt blocks. Large doses of nitrous or blower combinations should start with four bolt blocks.

When the decision is made to use a particular block, the deck surfaces are checked for flatness or warpage and the decks ground appropriately. Any plugs needed in core holes in the deck surface can be added before the block decks are surface ground.

Make sure that when your block is bored that the boring machine that is used registers the block off of the crankshaft centerline. Not all block boring machines do. This ensures that the new bores are perpendicular to the crank.

Machined correctly, the decks will be at right angles to each other and at ninety degree angles and equidistant and parallel to the crankshaft centerline. If the block and main caps are align honed or align bored, torque the oil pump housing onto the block. (Remember if the block is align bored and/or align honed, the block deck height will be reduced

You can make your own piston stop and it can be also used to mount a dial indicator to check piston to block deck heights or when measuring stroke lengths.

and you may need to use a timing chain that is shorter.) The cylinder barrels will be bored and honed with a torque plate and main caps bolted to the block to insure the roundness of the barrels while they are under stress, simulating the stress conditions when the cylinder head is bolted to the block. Final cylinder bore honing will be done with the pistons you plan to use in hand. Each piston diameter is measured and the correct piston to cylinder wall clearance added in during the final honing. Cast and hypereutectic pistons require less clearance than a forged piston. Each piston manufacturer has his own recommendations for this clearance. A chamfer will be added to the top of the cylinder bores, after boring and final honing, to help new pistons and rings slide easily into the cylinders.

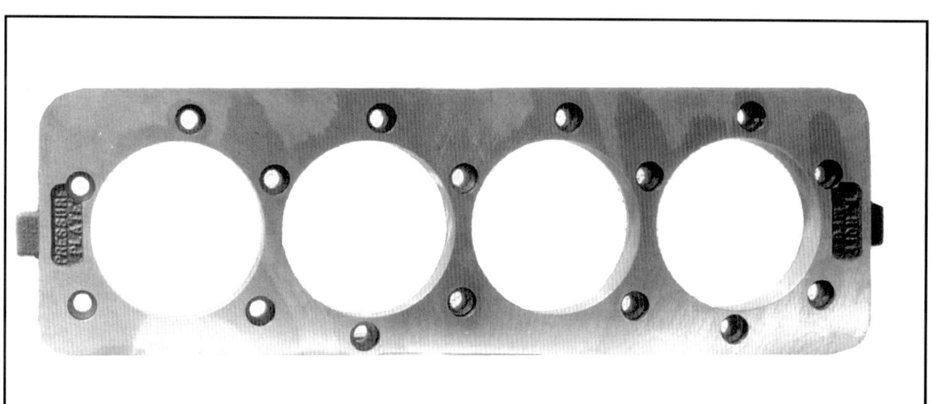

A torque plate should be bolted to the block when it is honed. Use a head gasket of the type you intend to use on the final engine assembly and torque down the main caps. The torque plate will simulate the stresses placed on the block when the cylinder head is torqued down. Torque plates will help keep the cylinders round during final honing and this will later help the ring seal.

At this point the block can be cleaned again and a trial fit of the rotating assembly can be made. One reason that the mock up is done is to check the deck height remaining between the top of the piston deck (not the dome) and the top of the block deck. This deck height measurement will determine whether any additional block deck surface should be removed. (Piston to block deck height for a race motor should be set as close to zero as possible to help resist detonation in the quench area and increase power levels.) Also check to see if there is any problem with the assembly parts. Actual stroke measurements can be made to see if there are differences between cylinders in actual stroke length. There should not be more than + or - .0025" difference between each of the individual cylinders on a performance motor. Valve to piston clearances can also be checked. If there is not enough clearance, the valve reliefs on the pistons can be deepened.

As spoken about previously, the new Gen.V "Bowtie" blocks have less depth between the timing cover bolting surface and the front of the case under the cover. This was done to improve cooling around the number one cylinder. It did not change the overall front to back dimension of the case. However, there may be a problem if you use a timing belt system. A clearance check should be made and appropriate amounts removed from the timing cover and/or the front of the cam thrust surface of the block to clear a belt driven timing assembly.

If you are going to be using needle bearing roller rocker arms and you wish to restrict the amount of oil going up to the rocker arm valleys, you can do it a number of ways. (Do not do this if you are using the stock style stamped rocker arms. Stamped stock rocker arms require more oil to properly lubricate the rocker arms and balls.) One way is to install screw-in oil restrictors in the oil passages at the rear of the block.

Another is to use pushrods with smaller internal oil passageways to limit the amount of oil that can travel thru them up to the rocker arms. You can also use a solid or roller lifter that reduces the amount of oil it passes up the pushrod. Restricting the amount of oil to the rocker arm valleys is a way to reduce the overall amount of oil that drains back onto the spinning crankshaft. This in turn can help reduce the amount of windage related power loses at high rpms. There will be more discussion concerning windage losses and gains in the LUBRICATION chapter.

If you are going to build a motor with extremely high cylinder pressure, generated by very high compression levels, large nitrous, turbo or blower systems, then you may also wish to have o-ring grooves cut into the block deck surrounding each cylinder bore to aid in head gasket sealing. This is only necessary with extreme combinations. A number of composition head gaskets featuring metal fire rings offer increased sealing even with very high cylinder pressures without cutting the block or heads for metal o-rings.

New freeze plugs, oil gallery plugs and the correct cam bearings are installed and properly positioned in their bores. The cam bearing installation can be checked by putting a known straight camshaft in the block and rotating it. Feel for any binding that would indicate trouble. Debris screens can also be epoxied into the lifter valley to help prevent damaged parts of the valve train also creating major damage to the rotating assembly. Some people also paint the inside surface of the lifter valley in order to aid oil drain back and to contain any possible grit that may work loose from the block surface.

When the block work is done, the machined surfaces of the block should be coated with some sort of rust preventative and after everything dries out, the block placed in a new plastic trash bag to keep everything clean until final assembly.

BIG BLOCK Quick Facts & References

BLOCK PREPARATION CHECKLIST

1) **Remove all freeze plugs, oil gallery plugs, drain cocks, cylinder head dowels, cam bearings and any miscellaneous bolts.**
2) **Pressure-spray clean the block to remove dirt and sludge. Some blocks are very grimy so don't hesitate to have it done twice if necessary.**
3) **Visually inspect the block for cracks in the cylinder bores, main webs and lifter bosses. Check closely for indications of prior abuse, including broken main caps and block mountlng bosses, stripped bolt holes, and cracks in the block surfaces.**
4) **Have the block pressure checked and Magnaflux inspected.**
5) **Deburr the block, making sure to break all sharp edges and radius all corners.**
6) **Carefully check all critical block dimensions to ascertain any need for special corrective machining. Write everything down in a note book.**
7) **Machine all areas that are out of spec. Align bore if required.**
8) **Bore and hone for the correct piston-to-wall clearance. Use torque plates for all high-performance use.**
9) **Chase all head bolt and main cap bolt holes. Use a bottoming tap on all blind holes.**
10) **Only install top-end oil restrictors if the engine will use roller-bear ing rocker arms; never restrict oil to conventional ball-and-trunnion rockers.**
11) **Using the trial assembly method, check the block for adequate crankshaft counterweight and connecting rod clearance. Grind if required.**
12) **Steam clean the block at the local car wash after all machining is completed and blow dry.**
13) **Lubricate all machined surfaces with light oil to prevent surface rusting.**
14) **Install new cam bearings and blow out any small chips that may be dislodged from the bearings.**
15) **Install new oil gallery plugs, rear camshaft plug, and all freeze plugs.**
16) **Do not store the block in a clean plastic trash bag until it has dried at least 3 or 4 hours after washing.**

SAFETY

Before you get started building your next Chevy powered rocket sled, take the time to make sure you are safe. Don't rely on a tech safety inspector at the track or anyone else to keep you safe. It's your responsibility. Use the right tools for the job and protect your eyes. Make sure that if you or someone else is under your vehicle, that it is properly supported. Use jackstands. Cinder blocks, milk crates or blocks of wood are not substitutes. When you use an engine stand make sure it will hold at least 1000 pounds. Some of the inexpensive engine stands being sold claim to be rated for 750 pounds, but sag under the weight of a 500 pound smallblock V-8. They won't safely hold a complete big block motor.

Keep a large, fully charged, fire extinguisher nearby. A fire that burns up your engine compartment is bad enough. A fire that starts in your car and then moves on to gut your shop, trailer and your home is worse.

Make sure that your race vehicle has all of the required safety equipment, seat belts, helmet, roll bar or cage, window net, fire extinguisher and that you have the proper clothing.

Gaskets & Fasteners

REAR MAIN OIL SEALS

Two different methods are used to seal the big block engine. All Mark IV production and "Bowtie" motors and Gen.V "CNC Race Prepared Bowtie" blocks use a two-piece oil seal. Gen.V "Bowtie" CNC "Race Prepared" blocks use a rear seal adaptor to mount a two- piece Mark IV style crank and seals. Gen.V production blocks, Gen.VI HO crate motors and two of the Gen.V "Bowtie" blocks use a one piece rear main seal (PN-6264902).

OIL PAN GASKETS

Two piece rear oil seal blocks require that four individual gaskets be used to seal the oil pan to the block pan rails, the rear main cap and the front timing cover. In a Gen.V or Gen.VI one piece rear seal case a different single gasket fits the oil pan to the motor. In order to use a single oil pan gasket, design changes were made to the rear main cap, the rear crankshaft flange and thrust surfaces, the timing cover to oil pan mating surface and the oil pan gasket sealing surface on the pan itself.

HEAD GASKETS

Three types of gasket materials are generally available, steel shim, composition and copper gaskets. Cast iron heads can use all three types of gasket materials. Aluminum cylinderr heads require the use of composition or copper gaskets. Various compressed gasket thicknesses are offered. Remember that you should have a minimum clearance of .035"-.040" between the top of the piston deck and the deck of the cylinder head when using steel rods. Aluminum connecting rods typically require a larger clearance.

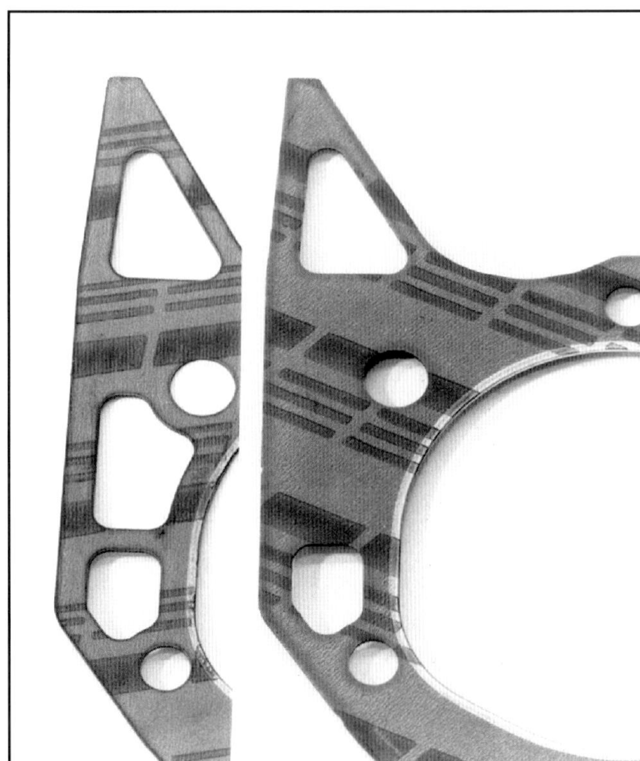

Fel-Pro head gasket (8180-PT, left) has a water core opening to match Mark IV production blocks and heads. The Fel-Pro gasket (8523-PT-1, right) blocks this core hole. This enables the (8523-PT-1) gasket to be used on Mark IV 4.250" bore blocks with matching Mark IV heads and also to be used on Gen.V production 4.250" bore blocks when using Gen.V production heads.

CYLINDER HEAD GASKET SPECS

PN	TYPE	COMPRESSED THICKNESS	FAMILY	BORE SIZE	
GM					
14015351	STEEL	.022"	Mark IV	4.250"	
10126768	COMP.	.039"	Mark IV	4.250"	
10159507	COMP.	.039	Mark IV	4.440"	
14097000	COMP.	.027"	Gen.V	4.250"	
14097001	COMP.	.040"	Gen.V	4.466"	
12363411	COMP.	.039"	Gen.V	4.375"-4.540"	
12363412	COMP.	.039"	Gen.V,VI	4.25"-4.37"	Marine
12363413	COMP.	.041"	Mark IV	4.375"-4.54"	
12363414	COMP.	.039"	Mark IV	4.25"-4.37"	
12555728	STEEL	.022"	Gen.V	4.250"	
FELPRO					
8180-PT-1	COMP.	.038"	Mark IV	4.250"	
8523-PT-1	COMP.	.038"	Mark IV	4.250"	extra holes
8523-PT-1	COMP.	.038"	Gen.V	4.250	
1017-1	COMP.	.038"	Mark IV	4.340"	C/I heads
1027	COMP.	.039"	ALL	4.370"	Alum. heads
1037	COMP.	.039"	ALL	4.370"	Marine
1047	COMP.	.039"	ALL	4.540"	Marine
1057	COMP.	.041"	"Bowtie"	4.630"	
1093	COMP.	.051"	"Bowtie"	4.620"	Round gasket bores

Aluminum heads require a composition style head gasket.

ket bore sizes are offered. Mark IV head gaskets have different coolant core hole passages than gaskets for Gen.V engines, blocking some passages and opening other coolant passages. Some block deck coolant core holes are round, while on Gen.V production blocks the holes are irregularly shaped on the block deck. Mark IV heads and head gaskets should be used on Mark IV blocks. Likewise Gen.V style cylinder heads and gaskets should be used on Gen.V blocks. There are exceptions. Mark IV heads can be used on Gen.V "Bowtie" and on Gen.VI HO blocks because the core holes in these block decks will seal the water core holes found on Mark IV heads. If you build this combination then use a Mark IV style head gasket and plug the core hole in the Gen.V "Bowtie" block deck located near the front head dowel pin location (see photo). Otherwise coolant entering the front of the block can bypass the rest of the block and exit out thru the head and intake manifold.

Steel shim gaskets and copper gaskets require the use of head gasket sealant. Composition gaskets do not require sealant.

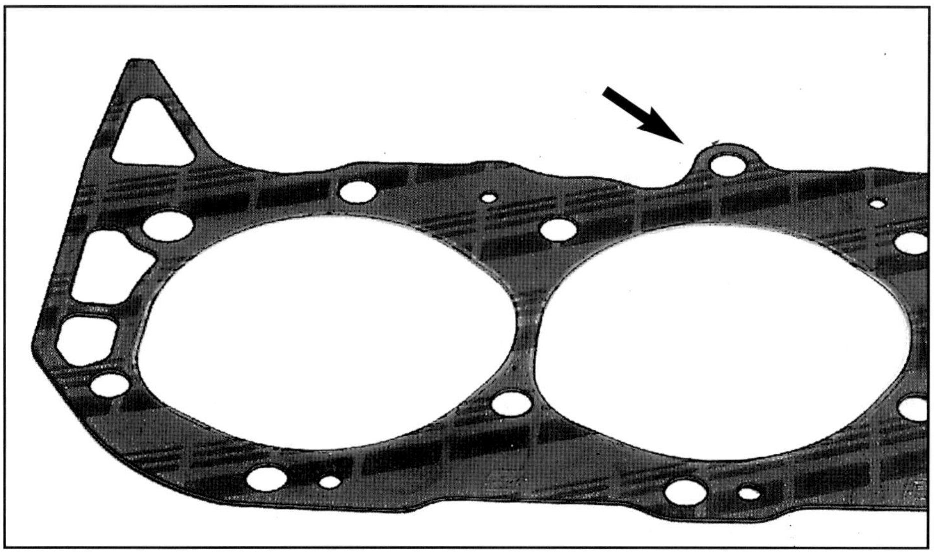

Fel-Pro composition head gasket (PN-8180-PT) for 4.250" bore Mark IV engines. It has two additional holes for head bolts that go into the lifter valley found on some high performance Bowtie and aftermarket blocks.

Head bolts should be completely cleaned and then torqued down in the proper sequence and to the correct torque values. Head bolts or studs, used on aluminum heads, need to have a hardened steel flat washer (PN-3899696) under each bolt head or nut, to prevent damage to the aluminum head surface. Head bolt threads going into water passages should be coated with sealant and head bolts going into blocks with "blind" holes should be coated with moly lube or oil. Some head gaskets do not need retorquing, but it is a good practice to retorque all of the head bolts after the motor has been heated up the first time and allowed to cool down. Remember, do not try to put Mark IV heads on a production Gen.V block and visa versa. You can use Mark IV heads on the new Gen.VI 454 and 502 HO blocks or on "Bowtie" blocks.

INTAKE MANIFOLD GASKETS

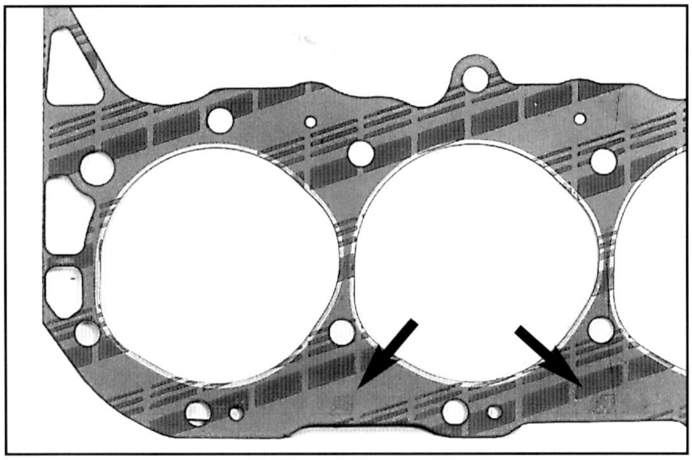

Fel-Pro (PN-8180-PT) is used on early Mark IV blocks that do not have the two extra water coolant holes in the deck.

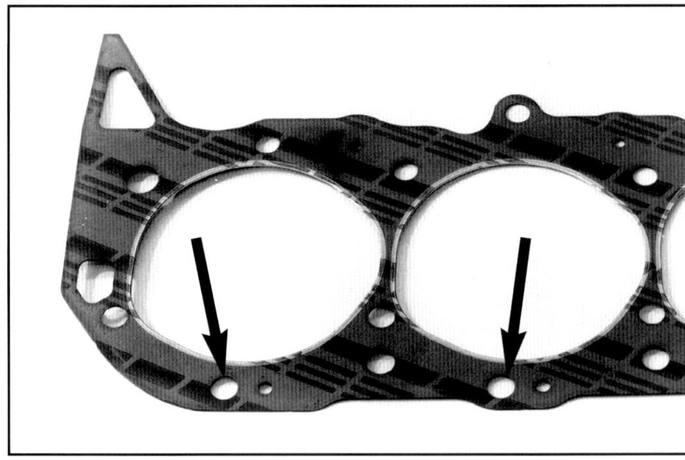

Fel-Pro head gasket (8523-PT-1) has additional water passage holes to fit Mark IV 4.250" bore blocks with these water passage holes drilled. This gasket will also work with 91-95 Gen.V production 454 blocks when using production Gen.V heads.

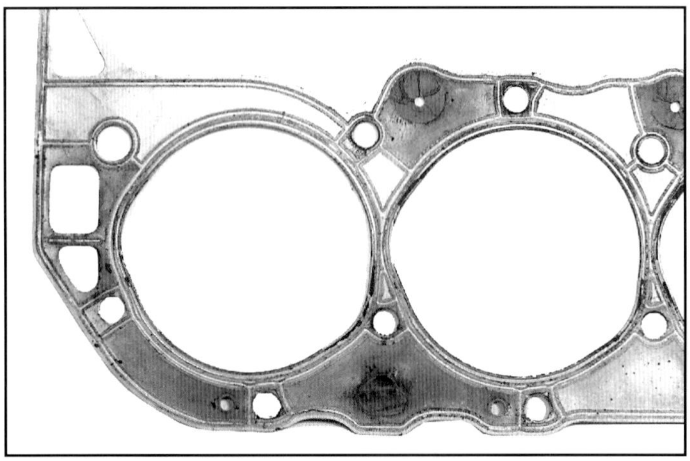

Production metal shim head gasket for Mark IV engines is .022" thick when compressed. These should not be used with aluminum heads.

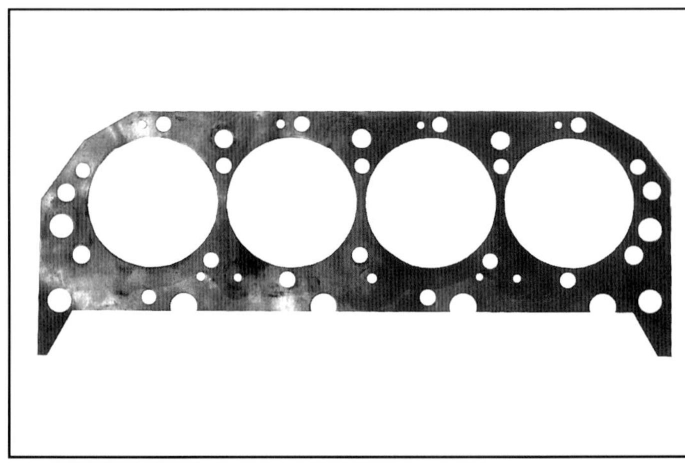

Copper head gaskets for big block Chevys are available in various compressed thicknesses from aftermarket suppliers and can be reused after cleaning and application of gasket sealer.

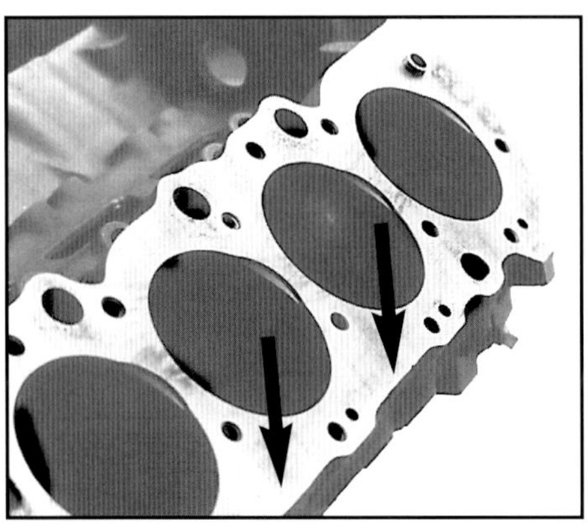

These photos show the different water core holes in the decks of various blocks. The left photo is an early Mark IV 396, the Mark IV 454 block (right) has additional deck core holes. The middle photo is a production Gen.V 454 and shows the shapes of the core holes on these blocks. These different core holes require head gaskets that match the holes on each different block. "Bowtie " blocks and production Gen.VI 454's and Gen.VI 502's have water core holes that allow the use of Mark IV heads if you want to.

These have remained the same and are interchangeable between Mark IV, Gen.V and Gen.VI motors. The intake port openings are either oval or rectangular to match the type of cylinder head and intake manifold you are using. Aftermarket gaskets are offered with various port dimensions to match those heads which have had extensive porting and enlargement or have "raised" intake port openings. Intake gaskets can also be obtained that do not have an opening for the heat riser passage. Closing this passage will keep the intake manifold a little cooler which in turn keeps the air/fuel mix cooler and denser. Blocking the heat riser passage will make cold starts a little more difficult. Some performance or race heads and intakes do not have a heat riser passage in the casting.

Some aftermarket intake gaskets have a bead of silicone sealant- like "Printoseal" material built onto them around the individual intake port openings to aid in sealing. These are nice since you can usually remove the intake manifold without tearing the gaskets or having excess sealant squeeze out into the port openings.

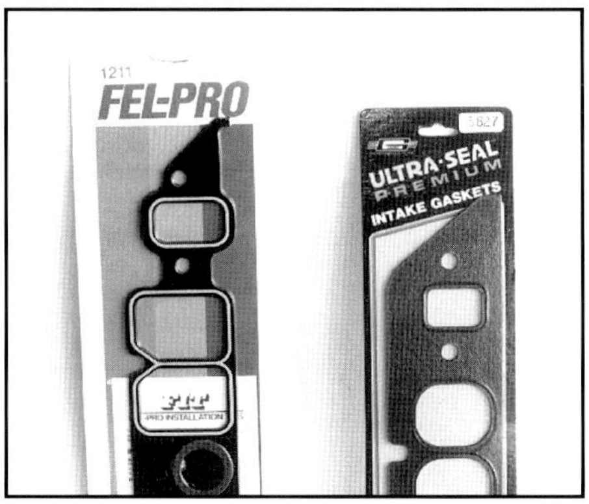

Both Fel-Pro and Mr. Gasket make a line of gaskets that use "Printoseal" technology. Seal is excellent and reduces the amount of sealant necessary.

CARB HEAT SHIELDS AND SPACERS

Chevy and aftermarket suppliers make heat shields which fit between the carb and the intake manifold. (Holley four barrel square flange GM PN-3999198 or Holley PN-108-70, Spreadbore flange four barrel GM PN-3969837 or Holley PN-108-51) These aluminum shields help prevent heat from getting to the carb fuel bowls and keep the fuel mixture denser. Some carb spacers will perform a similar heat insulating function if they are made from materials which do not conduct heat well, such as laminated wood or phenolic plastic.

Some intake manifold/carb combinations will cause a large 50cc Holley accelerator pump to hit the top of the intake manifold and prevent the carb from sitting flat on the intake. A small carb spacer or thicker carb gasket can solve this problem.

VALVE COVER GASKETS

The Mark IV engines used a stamped steel valve cover and gasket that were prone to leaking oil. The Gen.V and Gen.VI motors changed to a cast aluminum cover with a different gasket and machined the gasket surface on the Gen.V cylinder heads themselves so that a better leak free seal can be maintained. Tall valve covers are made so that if you have installed a high lift cam or a stud girdle, everything will clear. If you have installed a high lift cam and are trying to use the stock short valve covers, you will be able to get a little extra height by using a thicker valve cover gasket. The thicker gasket may give you enough height, if not, you will have to go to a "tall" valve cover. These thicker valve cover gaskets are available from aftermarket suppliers. Steel reinforced Mark IV valve cover gaskets are made by Chevrolet (PN-14085759) and they seem to last longer than other types.

SEALANTS

There is a need to be careful when you use sealants. Too much on a surface, especially an internal surface, can lead to the excess sealant being squeezed out when you tighten the fasteners down. This excess material can then break off and start circulating in the motor. Oil pickup screens and oil passages can be clogged up and prevent the circulation of oil. You can guess the result. Only use the amount required to hold the gasket down and seal the gasket surface. If you see any excess wipe it off.

If your motor is using an oxygen sensor be careful to use sealants that are "oxygen sensor safe".

BOLTS AND STUDS

The fasteners used by GM are adequate for most normally aspirated engine applications. If your combination is not going to exceed 12.5-1 static compression in a non-nitrous, carbureted engine, stock bolts can be used. If the static compression exceeds this figure or you are going to use healthy doses of nitrous-oxide, or a supercharger or a high boost turbo application, then consider the use of studs for cylinder heads and main caps. A two-bolt main cap engine can be strengthened on the bottom end by using studs in place of stock factory main cap bolts.

Cylinder head removal and placement is a little more difficult with head studs and may prevent you removing

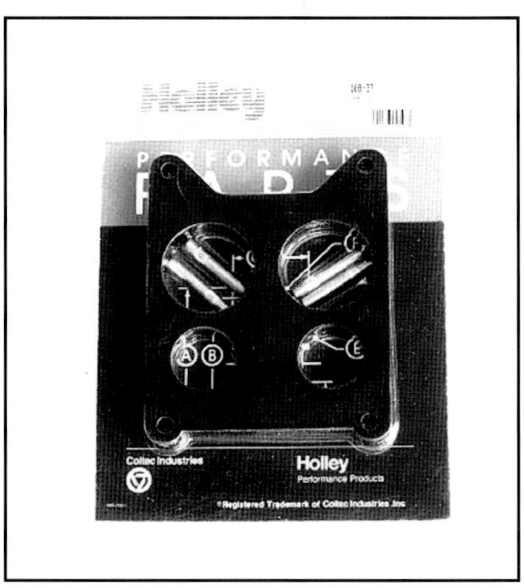

Various carb spacers are available from many companies. Four hole, single hole, wood, plastic and metal are made. Not all engines gain from the use of one. You'll have to experiment to see if your motor wants or needs one.

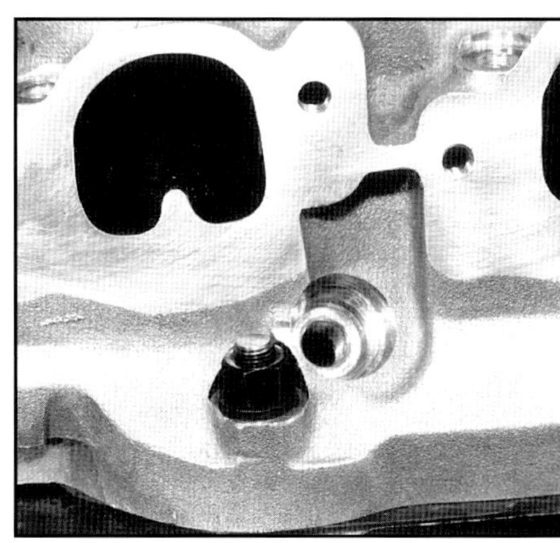

Exhaust port vane on an "861" Bowtie "W-port" head is seen. The port is raised .750" on this head to improve flow. Make sure that when you use head studs that the bottom row of studs do not stick up too high. They may interfere with the flange on your tube headers.

BIG BLOCK Quick Facts & References

HEAD BOLTS AND STUDS

PN	SIZE	APPLICATION
Long head bolt (24 required)		
10141204	7/16-14 X 4.06"	All Cast Iron blocks
Short head bolt (8 required)		
10141205	7/16-14 X 2.08"	Cast Iron or Aluminum blocks
Extra long head bolt (8 required)		
12337919	7/16-14 X 5.25"	Use with Bowtie head 861 or 862 which use extra long bolts next to each raised exhaust port
Long head stud (24 required)		
330861	7/16-14 X 5.14"	Use with CI or Alum. blocks
Short head stud (8 required)		
330862	7/16-14 X 3.13"	Use with CI or Alum. blocks
Hardened steel flat washers (32 required)		
3899696		Use on all aluminum heads
Stud hex nuts, 1038 steel (32 required)		
3942410		
12-point nuts, 4037 steel, magnafluxed (32 required)		
14044866		

the heads while the motor is in a vehicle. There may not be enough room in the engine compartment of some vehicles to slide the head completely off of the studs. Also, the use of head studs may cause a problem with the fit of tube headers. On some engine combinations the tips of the head studs on the bottom row may interfere with the flange of some headers. You may have to shorten the studs slightly to get the exhaust tube header flange to fit flush. A trial fit will tell you if you have to make some alterations.

Any fastener you use has to be spotlessly clean. If you are reusing a fastener make sure the threads are clean and undamaged. Look at the flats on the bolt head or nut. Are they rounded off? Are the threads damaged or corroded in any way? If they fail your inspection toss them away immediately. As you torque a fastener down, you can feel how it screws in. Some weak fasteners may feel differently as they tighten up. Those that feel "mushy" should be removed and replaced.

When you tighten a fastener down and torque it into place, you are actually stretching the bolt or stud. This stretch can be measured and is the preferred method of tightening rod bolts. A special measuring tool allows you to measure the rod bolt length uninstalled. As the bolt is tightened you can measure how much the bolt has stretched and lengthened. The proper amount of stretch is recommended by the manufacturer.

Stud kits and Allen bolt kits for Bowtie, "Symmetrical Port" Bowtie, and Pontiac "Spread Port" Super Duty heads are available from aftermarket sources. Most current high performance heads and blocks have provisions to use additional head bolts so that each cylinder has six head bolts surrounding it for additional gasket clamping seal. A production block has some cylinders that have six head bolts around them while other cylinders have only five.

MAIN CAP BOLTS AND STUDS

The main cap bolts for Mark IV four bolt blocks are different lengths than those used for four bolt main caps on Gen.V production blocks. Inner bolts on Mark IV's are 1/2-13 x 3.34" (PN-3909834), Gen.V inner main cap bolts are 1/2-13 x 2.52" (PN-10106461). The outer bolts for Mark IV's are 1/2-13 x 2.60" (PN-3859927), while the Gen.V outer bolts are 1/2-13 x 2.78" (PN-10106460).

Refer to the chapter on BLOCKS for information on the differences in main cap bolt lengths and the use of main cap studs.

THREAD LUBRICANTS

All fasteners need some type of lubricant on them to properly reach their clamping force. Obviously this is particularly important with cylinder head, main cap fasteners and rod bolts. Each fastener requires something on it.

MISCELLANEOUS BOLTS,

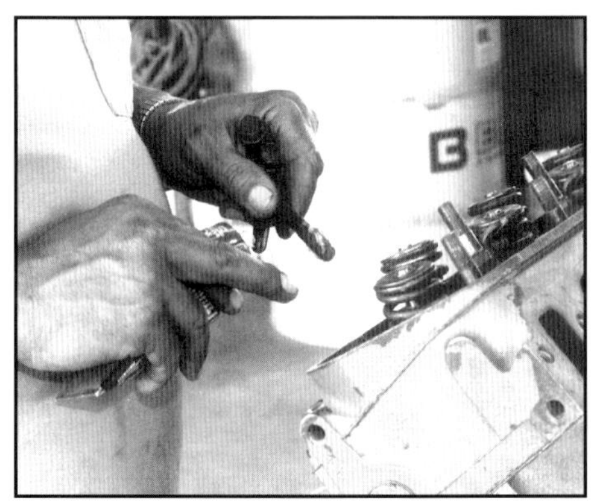

DOWEL PINS, ETC.

On the following page is a list of some of the small parts that often get overlooked and make you want to pull your hair out when you get ready to put an engine together and realize you don't have them.

BIG BLOCK Quick Facts & References

MISCELLANEOUS BOLTS & HARDWARE

Item	PN	Item	PN
Gen.V Balancer Bolt 1/2-20 X 1.50"	10126796	Oil Pump-Block Dowel Pin	10105879
MK IV Balancer Bolt 1/2-20 X 1.50"	9419218	Piston Stop, Spark Plug, Vented	12364087
Balancer Bolt Flat Washer	3864814	Rear Block-Transmission Dowel Pin	12338119
Cam to Gear Dowel Pin	10105879	Rear Main Cap O-Ring Seal, Gen.V	6264902
Cam Sprocket Bolt	9424877	Rocker Arm w/Ball, Gen.V	12523976 non-adjustable
Cam Sprocket Shim, .015" Thick	3975949	Rocker Arm Bolt, Gen.V	10114123 non-adjustable
Cam Thrust Button Spacer	12364086	Rocker Ball, Gen.V	10213466 non-adjustable
Camshaft Tunnel Plug, 2 7/32" OD	3859926	Rocker Arm Kit, Long Slot, Gen.V	12508879 non-adjustable
Crank Key, Gen.V (1 required)	10114166	Rocker Arm, Long Slot, MK IV	3959182 adjustable
Crank Key, MK IV (2 required)	106751	Rocker Arm Ball, MK IV	12338047 adjustable
Dipstick (Oil Level Indicator)	10218765	Rocker Arm Stud, MK IV	3921912 adjustable
Dipstick Tube (Oil Indicator Tube)	12551417	Rocker Arm w/Ball & Nut, MK IV	10112680 adjustable
Dipstick Tube with O-Ring	274244	Rocker Arm Nut, MK IV	3896648 adjustable
Flexplate Bolt	3727207	Rocker Arm Ball, Hardened, MK IV	5232762 adjustable
Flywheel Bolt	839756	Timing Cover-Block Dowel Pin	10105879
Flywheel Locator Dowel Pin	3701679	Gen.V Timing Cover Bolts	9439930
Freeze Plug, 1 3/4", MK IV	3743389	MK IV Timing Cover Bolts	9442008
Freeze Plug, Brass, 1 3/4", MK IV	3826963	Timing Pointer, 8" Balancer	3992073 MK IV
Freeze Plug, Brass, 1 5/8", Gen.V	3826504	Timing Pointer, Chrome, 8" Balancer	14044821 MK IV
Head-Block Dowel Pin	3894327	Valvetrain Organizer Tray	12364088
Lifter Bore Repair Kit Sleeves	14044875	Water Jacket Drain Plug, 1/4-18NPT	3889330
Oil Cooler Plug, Brass	444625	Windage Tray Main Cap Stud	10224104 Gen.V
Oil Cooler Bypass Valve	5575416	Windage Tray Main Cap Stud	3902885 MK IV
Oil Gallery Plug, 1/4-18 NPT	444777	Windage Tray Nut	9422297
Oil Pan Drain Plug, Magnetic	23011420	Windage Tray Washer	11503606

The biggest advantage to using factory chrome accessories is fit and finish. It always fits right and looks good. You should make use of these factory components wherever possible to maintain the "GM Look" and high reliability that these parts offer.

BIG BLOCK Quick Facts & References

CHROME STUFF

Item	PN
14" Open element "Classic" style air cleaner	12342069
Oil pan dipstick	12342008
Valve cover, short	12342093
Valve cover, tall	12342099
Timing cover, Mark IV	12342091
Timing cover pointer, Mark IV, 8" balancer	12342001
Timing cover hex bolts, chrome	12342040
Timing cover hex bolts, stainless steel	12342039
Hex bolt dress up kit, chrome	12342047
Hex bolt dress up kit, stainless steel	12342038
Valve cover "Bowtie" wing nuts	12341987
Water neck	12342024
Alternator brackets, use with tall water pump	12342061
Fuel pump block off plate	12341999
Mechanical fuel pump	12355618
Mini-Starter, High Torque	12363128

BIG BLOCK Quick Facts & References

TORQUE SPECIFICATIONS

BOLT	APPLICATION	TORQUE	LUBRICANT
1/4-20	Oil Pump Cover	80 in. lbs.	Loctite
	Oil Pan to Timing Chain Cover	80 in. lbs.	
	Timing Chain Cover	75 in.lbs.	
	Valve Cover	25in. lbs.	Antiseize
5/16-18	Oil Pan	165 in. lbs.	
	Cam Sprocket	20 ft. lbs.	Loctite
3/8-16	Intake Manifold	25 ft. lbs.	Oil or antiseize on aluminum heads
	Water Neck	20 ft. lbs.	Antiseize
	Exhaust Manifold	20 ft. lbs.	Antiseize
	Bell Housing	30 ft. lbs.	
	Water Pump	30 ft. lbs.	Antiseize
	Pressure Plate	35 ft. lbs.	
3/8-24	Stock Connecting Rod Bolt	50 ft. lbs. or 0.0055" stretch	Moly Lube
7/16-14	Cylinder Head		
	Long	75 ft. lbs.	Sealant
	Short	65 ft. lbs.	Sealant
	Alum. Head		
	Valley Stud	55 ft. lbs.	Sealant
	GM Head Stud	65 ft. lbs.	Sealant
	Rocker Arm Stud	50 ft. lbs.	Oil
	Oil Pump	65 ft. lbs.	Oil

Note: 502 and "Bowtie" blocks with "siamesed" cylinder barrels and "blind" cylinder head bolt holes do not require sealant on head bolts or studs, use oil instead.

BOLT	APPLICATION	TORQUE	LUBRICANT
7/16-20	Hi-perf Connecting Rod Knurled Bolt PN-14096148	73 ft. lbs. or 0.007" stretch	Moly Lube
	Hi-perf Connecting Rod Boron Bolt PN-3969864	67-73 ft. lbs. or 0.009" stretch	Moly Lube
1/2-13	Main Bearing Cap	110 ft. lbs.	Oil
1/2-20	Harmonic Balancer	85 ft. lbs.	Oil
14mm	Spark Plugs		
	13/16" Hex	25 ft. lbs.	
	5/8" Hex	15 ft. lbs.	

Note: Spark plugs used in aluminum cylinder heads should be coated with Antiseize.

If you use fasteners from an aftermarket source follow their recommendations for the amount of necessary torque or stretch.

Rod bolts should be carefully torqued to the recommended specs indicated in the accompanying chart. Be sure to use the proper lubricant. Torque the bolts equally in several steps to avoid stressing the components unevenly. These methods are appropriate for all but the most serious racing engines. Race engines should have the rod bolts torqued to the manufacturer's recommended bolt stretch as indicated on a rod bolt stretch gauge.

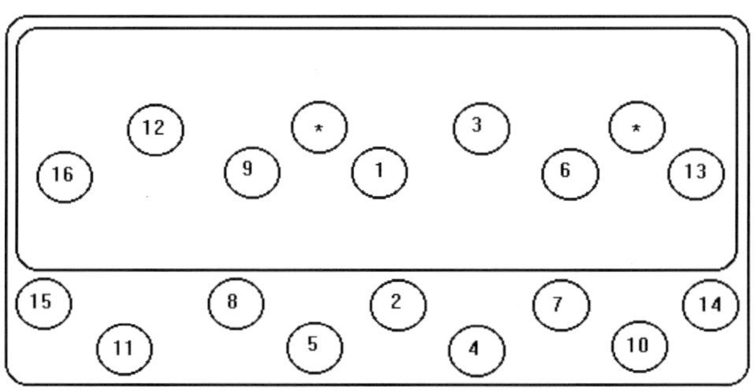

CYLINDER HEAD TORQUE SEQUENCE

(*) ADDITIONAL HEAD BOLT HOLES ON SOME PERFORMANCE HEADS AND BLOCKS

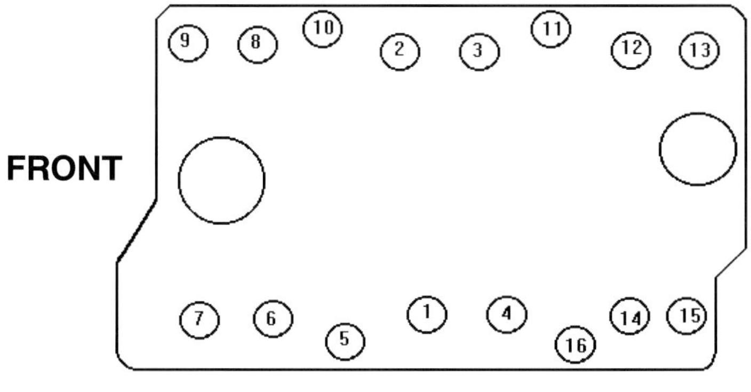

INTAKE MANIFOLD TORQUE SEQUENCE

Torque cylinder head bolts evenly in several steps to compress the gasket evenly. Follow the recommended torqueing pattern and pull all the bolts or studs to 50 foot pounds first. then sneak up on the final tprque spec, 10 foot pounds at a time. High quality gaskets such as the Fel Pro line don't require retorquing when carefully torqued in this manner.

Main cap bolts should also be torqued in several even steps to avoid pinching a bearing. Applying even torquing pressure assures maximum bearing contact with the housing bore and promotes optimum bearing crush.

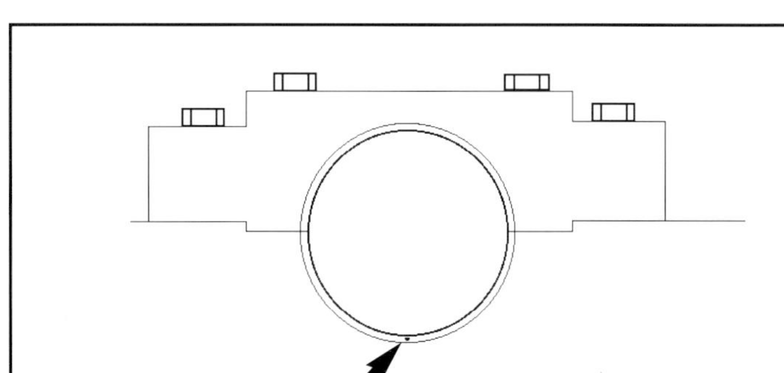

Cleanliness is essential. Even the smallest piece of dirt or other trash behind a bearing will alter the clearances. Also avoid using shop towels for wiping surfaces during assembly. You would be amazed how much lint they actually leave in the motor. Use paper towels instead.

How To Build Max Performance Chevy Rat Motors

Cranks, Rods, Pistons

Big block cranks used two different stroke lengths over the years, 3.76" and 4.0" and different ways to balance the rotating assemblies, internally and externally. Internal balance motors use neutral balance flywheels or flexplates and front harmonic balancers. External balance engines have additional weight added to the flywheels or flexplates and to the front vibration dampers. It is essential to use the correct combination of parts. All Gen.V and Gen.VI cranks, regardless of stroke, are externally balanced and use a one-piece rear main seal. External balance cranks can be made to internally balance by using heavy metal plugs inserted into the crank counterweights, but this is an expensive way to accomplish the conversion. Aftermarket suppliers also have balance plates which can be bolted in between the crank flange on Mark IV cranks and a 14" flywheel or flexplate that is neutrally balanced. This balance plate then acts as the external balance weight. You must still use an external weight front balancer when you use these aftermarket balance plates.

Mark IV 454 4" stroke forged cranks are available in two different steel alloys, the "1053" alloy forged crank (PN-3963523, casting numbers 3520 or 7416) or the stronger "5140" alloy (PN-3963524, casting number 3521). Both of these cranks have cross-drilled oil passages and are nitrided at the factory for increased surface hardness. This surface treatment only penetrates a few thousandths of an inch. If the crank journals are ground at some later time, you will have to have the crank journals nitrided again.

Be aware that cranks were balanced differently at the factory depending on which cubic inch size engine they were originally installed into. A 396 crank and a 427 crank, for instance, share the same stroke and will physically interchange in a block, but they are not balanced the same and in fact do not have the same size counterweights. The 427 crank has thicker counterweights. The Gen.V 454 and 502 4" forged steel cranks share the same casting number and 4" stroke, but they have different part numbers (PN- 14096983 is a "1053" steel alloy for the 454 and PN-10183723 is a "1053" alloy for the 502) because they are not balanced the same. Its okay to mix and match rotating assembly parts, but if you do, you must have everything in your combination rebalanced before final assembly. This includes crank, rods, pistons and pins, front damper, and flywheel or flexplate. This is especially true if you are using lightweight components such as aluminum rods, lightweight pistons or cranks.

In addition, Gen.V, VI cranks use different Gen.V, VI flexplates and flywheels depending on whether it is a cast crank or a forged steel crank. Consequently, Mark IV balancers, flywheels and flexplates are not interchangeable with Gen.V or Gen.VI parts.

In an effort to illustrate what happens if just the stroke on a motor is changed, here is a comparison using a motor with only the stroke altered.

The bore on the engine in the chart on the next page was kept at 4.500" and with the stroke altered that gave a total of 478, 509, 525, 541 and 573 cubic inches respectively. The heads

Photo on left shows a Gen.V and VI crank rear journal. This crank is used in one-piece rear seal blocks as shown in the middle photo. Right photo shows a Mark IV crank that needs a two-piece rear main seal and a Mark IV style block.

were fully ported with 2.30"/1.90" valves, 13.5-1 CR and a single plane manifold with a 1100 cfm four barrel carb. The cam was a Crane roller with .712"/.712" lift. Calculated peak horsepower was pretty much the same, however the amount of torque at any rpm range and at peak torque grew as the stroke was increased.

These examples can be created using either short, or in the cases of longer strokes, a tall deck case. In a short deck block the longest sort of practical stroke length is around 4.25". With a 6.385" (+.250") rod length and an aftermarket piston with a pin height of 1.290", you get the 9.8" total deck height of the short deck blocks. With a bore of 4.500" this will give you a 540.7 cubic inch motor with a rod ratio of 1.5-1. Make sure you do a trial assembly to check block clearance for the rotating assembly.

If you use a tall deck block it is easier to increase stroke and rod length (in order to keep acceptable rod ratios).

CAST VS. FORGED CRANKS

Chevy stock cranks are either nodular cast iron or forged steel. The parting line left from the molds is what to look for. A cast crank has a very thin parting line while a forged crank line is thick or wide. You can tap a crank lightly with something hard and with a cast crank, the resulting sound is like a thud. A forged crank rings like a bell and is quite distinctive. Of course you can also check the casting number.

Some people building a high performance motor feel that they have to have a forged crank, that cast cranks just don't cut the mustard. They would rather shell out the extra money and get a forged crank. It depends, up to a point, cast cranks are just fine. In a short duration bracket drag race engine, 600 genuine horsepower is around the upper limit for a cast crank. Almost any street application, short of insanity, can use a cast crank. At horsepower levels over 600 or if you are planning a nitrous or high boost blower or turbo applications, then move to a forged steel crank.

Good used cast cranks are more numerous and less expensive. You can take the money you save and spend it on something else. Just make sure that the crank is properly prepped, use a high quality vibration damper, (especially with a cast crank) and then good things happen.

These days there are a large number of forged cranks available from aftermarket suppliers that cover a large range of applications, stroke

POWER & TORQUE vs STROKE LENGTH COMPARISON

STROKE	3.76"		4.00"		4.125"		4.25"		4.50"	
	HP	TQ	HP	TQ	HP	TQ	HP	TQ	HP	TQ
RPM										
2000	175	459	188	494	195	512	202	531	217	570
2500	243	510	262	549	271	570	281	591	301	633
3000	306	536	330	577	342	599	355	620	379	663
3500	383	575	411	616	424	637	438	657	464	697
4000	474	622	503	660	518	680	532	699	559	734
4500	570	665	600	700	614	717	628	733	655	764*
5000	654	687	682	716*	696	731*	708	743*	729	766*
5500	726	693*	750	716*	760	726	770	735	783	748
6000	783	685	799	700	808	707	813	711	817	715
6500	822	664	831*	671	833*	673	832*	672	825*	666
7000	825	619	824	618	820	615	815	611	796	597
7500	832*	583	822	576	813	569	801	561	772	540
8000	810	531	788	517	774	508	758	497	719	472

Gen.V Bowtie CNC "Race Prepared" short and tall deck blocks use a special rear main seal block adapter to mount a two piece seal and then use a Mark IV style crank and oil pan.

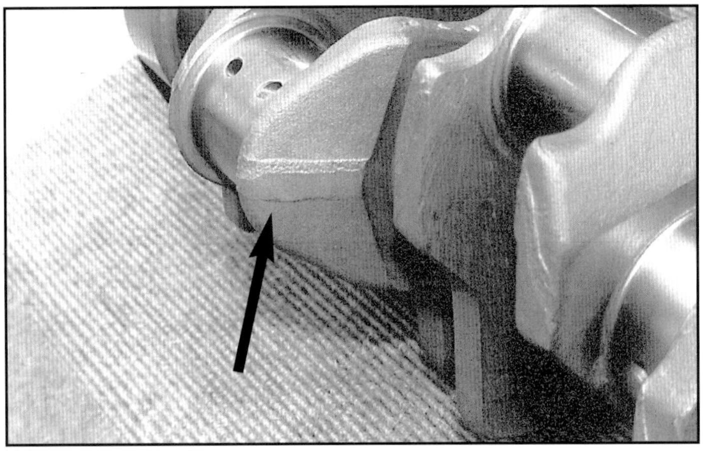

Forged crank above is identified by its characteristic wide, thick, parting line on the crank throws.

Cast crank has a thin parting line at the crank throws, making it easy to identify on sight.

lengths, steel alloys (most use a very strong 4340 alloy) and total weights. These aftermarket cranks can also be prepared in various ways. All of the counterweights can be "knife edged" to reduce weight and windage. Crank pins on the rod journals can be drilled to further reduce weight. Extra wide radius journal fillets add strength. Counterweights can be reduced in radius to reduce rotating mass. Special coatings to reduce oil cling, and chemical, vibration and heat treatments to further harden and stress relieve the metals, are available. Callies, Trick Flow, Lunati, Cola and BRC are companies that provide aftermarket, high quality forged cranks. Off the shelf aftermarket cranks with up to 4.750" strokes are available.

For those of us without fat wallets who can't afford an aftermarket crank, take heart. A stock Chevy steel crank will take a tremendous amount of abuse. For that matter, a cast big block crank can take a great deal of abuse also. It all depends on what your intentions are.

CRANK PREP

The following checks should be done on any used or new cranks. Take a good look at the entire crank for damage. Check the thrust surface and the journals for damage. Do any of the journals look blue from lack of oil and overheating? The standard journal diameter sizes are 2.75" on the mains and 2.20" on the rods. Measure the journals. Have the journals been cut down from their standard sizes? Are the journals worn or out of round? Check the crank key slots for damage or wear. Look at the threaded holes in the crank snout and the crank flange. Check the bore at the rear of the crank. Any damage to the pilot bearing hole?

The crank can be set up on V-shaped blocks so that it can be rotated and with a dial indicator you can check for straightness and runout. You can also slip the crank into an engine block with just the main bearings and see if it spins easily. If it is straight, clean it thoroughly and make sure you get into the oil supply holes with a brush. Gun bore cleaning brushes work fine in this application. After it is clean have the crank Magnafluxed for any cracks. Don't skip this step. You may not be able to see cracks with the naked eye. Magnafluxing will save you a lot of time and money. It is easy to replace a cracked crank with another, before you put it into a motor. Waiting until you build a motor, put it in a vehicle and running the motor for a time is another way to find a crack in the crank, but it is not easy and it isn't cheaper than Magnafluxing it before you use it. Some small cracks in journals may be removed when the journal is cut undersize, but this is one of those things where it "depends".

Crank stroke length can be checked to make sure that each rod journal is equal and the crank can be indexed to make sure that each throw is ninety degrees to each other in turn.

Some repairs can be made if necessary. Journals can be welded up and recut back to standard. The journal can also be chromed and built back up, then recut to standard. Some bent cranks can be straightened. Damaged threaded holes can be Helicoiled. Cracked cranks or cranks with damaged thrust surfaces should be sent to the nearest junk pile or metal recycler. Its all relative to how much you want to spend. Sometimes it is just cheaper to get another crank that doesn't need as much repair. On the other hand if you have a $1,500.00 Super Gas aftermarket crank with all the bells and whistles, then maybe its worth trying to repair it and cheaper than buying a new one.

Check the journal diameter on the crank and compare it to the inside diameter of the bearings to determine the bearing clearances.

That's a decision that you, your crank grinder and your significant other get to make.

Be aware that certain metal treatments such as nitriding help make the surface of the journals harder. These treatments do not go completely thru the metal but only penetrate a very thin layer beneath the journal surface. When you reduce the diameter of the journals by crank grinding the treated surface metals may be removed. The crank will have to be re-treated if you want those surface qualities. Some after market crank makers are using a nitiriding process that penetrates .015"-.017" deep so that if the crank is later cut .010", the surface penetration of the nitriding will still be .005-.007" thick and the surface will not need to be retreated.

Cross drilling the oil passages in the main journals will help any big block crank survive by insuring a steady pressurized oil supply to the bearings thru the entire rotation of the crank.

With all of that out of the way, the crank journals, if necessary, can be cut undersized and the journals polished. Make sure that the crank grinder leaves a large radius in the journal fillet. This is much stronger than a tight small radius in the journal fillet. The oil holes in the journals should be chamfered lightly to help distribute oil to the bearings.

Another common practice with stock external balance cranks, is to internally balance them by using heavy "mallory" metal plugs. When the rotating assembly is balanced the correct position for the weight is determined and holes are drilled into crank counterweights. The heavy metal plugs are then inserted into these holes to achieve an internal balance. Then a standard internal balance, front vibration damper and standard internal balance flywheel or flexplate can be used.

BIG BLOCK Quick Facts & References

BIG BLOCK CRANKSHAFTS

Casting No.	Engine CID	Notes
3.76" STROKE used on '65-'90 Mk.IV 366T, 396, 402, 427 or 427T Two-piece rear main oil seals		
6223	396, 402, 427	forged, 1053 steel
7115	396, 402, 427	forged, cross-drilled, nitrided, 5140 steel
3804816	396, 402, 427	cast
3856223	396	forged
3863114	402	
3863144	396, 402	
3874874	396, 402	
3879621	427	forged
3882841	396	forged, cross drilled, nitrided
3882842	427	forged, cross drilled, nitrided
3882847		
3882848		
3882849	427	forged
3887114	396	forged, nitrided, 5140 steel
3904815	396, 402	cast, nodular iron
3941180		
3942411		
3965746		semi finished, forged
3993804		raw, 5140 steel
3.76" STROKE used on '91-up Gen.V and Gen. VI 366T, 427T One-piece rear main oil seals		
10114186	427T	forged, 1053 steel
10114187	366T	cast
4.00" STROKE used on '70-90 Mark IV 454 or Mark IV 502 Two-piece rear main oil seals		
3520	454	forged, cross-drilled, 1053 steel
3521	454, 502	forged, cross-drilled, 5140 steel
3524	454	forged
7416	454, 502	forged, cross-drilled, 1053 steel
7463	454	forged
336782	454	cast
353039	454	cast
353638	454	
359730	454	
3962523	454	forged
3963523	454	
3963524	454	forged
3967416	454	forged
3967463	454	forged
3975945	454	
3963521	454	forged, cross drilled, 5140 steel
3965753		forged, semi finished, 5140 steel
3975495	454	
3993878	454	forged
10051169	454	forged raw casting
4.00" STROKE used on '91-up Gen.V and VI 454 or 502 One-piece rear main oil seal		
10114188	454	cast
14097044	454, 502	forged, 1053 steel

BEARINGS

It seems like everybody is making engine bearings these days. Each claims to have better qualities than others. For me, the most important thing is to get the clearances for the bearings correct. I prefer the tighter side of the clearances rather than running the bearings on the "wide" side. The bearings and their respective mating surfaces on the main and rod caps have to be spotlessly clean. One tiny piece of metal grit, dirt or lint from a shop rag on the back of the bearings or the caps can keep the bearing from sitting properly on these surfaces. Any measurement you then make will be incorrect and false.

Wear and the quality of the machining work that the crank, block and rods receive will also effect these clearances and your ability to maintain them when you actually assemble the motor. If the machine work was sloppy or incorrect the problems seem to multiply.

Oversize bearings are available in sizes .010", .020", .030" and .040". Bearing sets can also be ordered that are .001" oversize or undersize if you have a particular situation where you need them.

Keep in mind that if you have had your crank journals ground with large fillet radii, some stock bearings may not have enough clearance on the side edges to clear a large journal fillet radius. The bearing side edges may need to be chamfered slightly to obtain clearance. Aftermarket bearings are made with a large chamfer already on the bearings for this type of situation. Aluminum rods utilize a dowel pin to help locate and retain the bearing. Bearings can be supplied with these dowel holes already drilled in them or you can buy a tool fixture that will allow you to drill the dowel hole in the bearing shells yourself.

If you are racing, regular, periodic inspection of the bearings is something that you need to do. Observation of the wear pattern, if any, on the bearings can give you an

CLEARANCES

MAIN BEARINGS	.002"-.003"
CRANK END PLAY	.005"-.007"
ROD BEARINGS	.002"-.003"
ROD SIDE CLEARANCE	.015"-.025"
PISTON TO CYLINDER BORE	Follow manufacturer's recommendation, cast and hypereutectic pistons will have a small clearance and forged pistons will require more
FLOATING PISTON PIN CLEARANCE	.0005"-.0008" in piston, 0.0"-.005" end play, .0005"-.0007" in rod
MINIMUM PISTON DECK TO CYLINDER HEAD	.035"-.040" with steel rods
MINIMUM VALVE TO PISTON DECK	.045" on a stock motor, .080" on intake and .100" on exhaust for high rpm motors
VALVE LASH	Follow manufacturer's recommendation

indication of any problems. The idea is to catch and correct any problems before a major disaster occurs.

Rod bearings are available with various amounts of "eccentricity", that is when measured in the rod the bearing will be slightly wider in diameter at the parting lines of the cap than they are

Dial indicator mounted on the front of the block can establish the amount of crank thrust clearance when you assemble the short block.

when measured 90 degrees from the parting line axis. These can be helpful if you find that the wear pattern on previous bearings are too narrow or too wide.

HARMONIC BALANCERS

Various balancers, more correctly called vibration dampers, were used in big blocks. (One of my engine buyers refers to them as "harmonica balancers" and makes himself understood.) You must use the correct one for your engine combination or if you wish to have your "harmonica" in balance.

Gen.VI 454/502 HO motors are supplied with a "degreed" external balance vibration damper that works with the plastic, notched timing covers found on these engines.

The GM/Fluidampr is an "over the counter" piece that meets SFI 18- 1 standards and is test spun to 12,000 rpms and is accepted by major race sanctioning bodies as a safety device. This is a viscous fluid damper and dampens vibrations at all rpm frequencies, unlike the stock factory elastomer dampers which are tuned to a specific rpm range of vibrations. When the elastomer rubber dries out on a stock damper, the outer metal ring on the damper can move around the inner metal ring or move backwards or forwards in relation to the inner ring. This causes the timing mark on the damper to move from where it should be. The timing marks on a GM/Fluidampr won't move. Also because of the superior qualities of the Fluidamper, bearings last longer. High rpm motors suffer less from spark chatter because less crank vibration is transmitted to the cam. The outside ring on the Fluidampr is marked off in crank degrees for 50 individual degrees of ignition timing and then at 90 degree intervals around the rest of the damper which is helpful when setting rocker arm lash. If you are going to race a big block Chevy, you should replace the stock vibration damper with the Fluidampr. This is especially so if you use a cast crank or if you use a blower or nitrous.

Fluidampr makes three other dampers for internal balance Mark IV motors. PN-713282 is 6 1/4" in diameter and weighs 9.0 lbs. which is half of the 8" external balance GM/Fluidampr which weighs 17.3 lbs. Another Fluidampr is PN-712100 which is for internal balance motors, is 8" in diameter and weighs 15.1 lbs. PN-712440 has a 7 1/4" diameter, for internal balance big block engines.

On a motor that will be used in short duration races, like drag racing, a lightweight, small diameter balancer can be used to reduce reciprocating weight. If you are building a long duration performance motor, say for a sustained rpm marine application, then use the largest diameter, heavier damper. Stay away from the lightweight aluminum hubs offered by some. Testing has shown that they don't reduce crank vibrations very well.

BIG BLOCK Quick Facts & References

HARMONIC BALANCERS

PN	Engine/Family	Diameter	Balance
	366-427 Mark IV	7"	Internal
3879623	366 thru 427 Mark IV	8"	Internal
14097024	454 Mark IV	8"	External
14097023	454-502 Gen.V steel crank	8"	External
10101160	454 Gen.V cast crank	8"	External
10051171	454-502 Mark IV, Gen.V GM/Fluidamper	8"	External
10101159	427T Gen.V 91-94	8"	External
10227166	427T Gen.V 95-	8"	External
10216339	454-502 Gen. VI degreed, steel crank	8"	External

When you remove or install any front vibration damper, always use the correct tool. Never use a gear puller to remove a balancer and don't use a hammer to install one. Both of these improper methods can cause damage to the balancer.

Before you put a balancer on your new motor, take the time to mark the outside circumference of the balancer with lines indicating each 90 degrees of rotation if the balancer doesn't already have them. It helps when you want to adjust the rocker arms or bring a particular piston up to TDC. There are adhesive tapes that show degrees of rotation that can be used or you can measure the circumference, divide that figure by four and put the marks on. You can mark the balancer at the point of total ignition advance degrees. We often used 36 degrees of total advance. Your engine combination may run best with some other amount. 36 degrees on an 8" diameter big block damper was 2.51" from the mark at 0 degrees TDC. We could quickly confirm total advance and/or set the total ignition timing on the motor.

Stock cast iron vibration dampers can wear out. The rubber-like elastomer material between the two metal rings can dry out, crumble and begin to fall out in pieces. In any case the outer metal ring of the damper can start to move around in relation to the inner ring. A number of things can then happen, the outer ring can move backwards toward the timing cover and actually move enough to start cutting into the cover. The outer ring may rotate on the inner ring making it impossible to set the initial timing on the motor since the timing mark will move from its original correct position. The worst thing is when the outer ring moves forward and actually comes off. When it does you'll know it. As it takes out everything in its path, it will sound like a bomb going off under the hood. If you are lucky you may get to watch the damper ring go thru the hood on its way to the upper atmosphere. If you are unlucky a lawyer might catch the damper shrapnel with his face. Most racing associations don't like stock balancers because of what can happen. The NHRA for instance prohibits the use of a cast iron stock elastomer balancer if the ET on the vehicle is quicker than 10.99. That doesn't mean that you are safe if your drag race vehicle is slower than 10.99. The GM/Fluidamper meets SFI testing requirements and is an acceptable and safe alternative.

If you are having some mysterious parts breakage, like a valve spring or timing chain or main bearing wear that seems out of the ordinary, one of the reasons may be a balancer that is failing to properly dampen crank vibrations. There may be other things causing the breakage, but check out the damper, it might be the culprit. Remember that the stock cast iron balancers are "tuned" to work best at certain moderate street use rpms. You can use a different balancer that covers a wide range of rpm, like the GM/Fluidamper.

FLYWHEELS, FLEXPLATES AND STARTERS

This is another area where the combination of parts is a little complicated, but important to get right. Mark IV and Gen.V flywheels and flexplates are not interchangeable. It is further complicated by the fact that Gen.V and Gen.VI HO engines use different parts depending on whether the crankshaft is cast or forged. Starters are interchangeable between engine families as long as you use the right starter with the correct size flywheel or flexplate. Chevy V-8's used two different diameter sizes and tooth counts on flywheels and flexplates, 14" with 168 teeth on the ring gear and 12 3/4" with 153 teeth. They then made two types of starters to work with these two diameters. Each of these starters have their respective bolt patterns on them (see the photos) and most engine blocks fortunately are drilled for both starter bolt patterns. The starter bodies and solenoids are pretty much the same, but the noses of the starter are shaped differently for each one. The starter noses can be changed on the starter to switch them from one style to the other.

Flywheel pressure plate bolt patterns also differed with some being drilled for 11" diameter clutches and others for 10.4" and some have holes for both. Flexplates also have drilled holes in them to mount the torque converter. Some of the flexplates come with two different torque converter size bolt hole patterns, some have only one pattern.

Fluidampr vibration damper is available in different diameters for the big block and for either internal or external balance engines.

The flywheel and flexplate chart on page 56 is a list of what is currently available from Chevy.

If you have a high static compression motor, it is generally more advantageous to use the 14" flywheels and flexplates because they have a mechanical advantage and can turn the motor over easier. On the other hand if you are trying to reduce reciprocating weight, a smaller flexplate or flywheel, with a high torque, gear reduction starter may be better. The mini starters also weigh about 10 pounds less than a conventional size starter.

Most race sanctioning bodies have strict rules about the type of flywheels and flexplates you can use. SFI approved dampers, flywheels, flexplates, and bellhousing scatter shields are available on the aftermarket. Scatter shield bell housings, flexplate shields, trannie shields and blankets, drive shaft safety loops and safety approved front vibration dampers may all be required and the stock Chevy parts may not pass tech inspection. Check the rules and remember, if one of these parts come apart at 6000 or 7000 rpms, it can do horrific damage to your vehicle, to your body and maybe some poor innocent soul nearby. Don't even think about saving any money on safety equipment. Get the best stuff and check it periodically.

FLYWHEELS & FLEXPLATES

PN	SIZE/ TEETH	FAMILY	CLUTCH SIZE	NOTES
Flywheels				
14085035	12 3/4" 153	Mark IV	10.4"	Nodular iron, 15 lbs. internal balance Casting #-3856579
3963537	12 3/4" 153	Mark IV 454	10.4"	Nodular iron, 15 lbs. external balance Casting #-39355411
3991469	14" 168	Mark IV	11"	internal balance
3993827	14" 168	Mark IV 454	11"	external balance Casting #-3993457
14096987	14" 168	Gen.V, VI 454/502	11"	nodular iron external balance Casting #-14096805 use w/forged Gen.V, VI crank only
10101169	14" 168	Gen.V 454	11"	external balance use w/cast Gen.V crank only
Flexplates				
471529	12 3/4" 153	Mark IV		internal balance
471597	14" 168	Mark IV		internal balance
14001992	14" 168	Mark IV 454		external balance
10185034	14" 168	Gen.V, VI 454/502		external balance use w/forged Gen.V, VI crank only
10185035	14" 168	Gen.V 454		external balance use w/cast Gen.V crank only

Flexplate bolts are 7/16-20 X 27/32" (PN-3727207). Flywheel bolts are 7/16-20 X 1" (PN-839756). Don't use a flywheel bolt to hold on a flexplate or visa versa. Use the right one. Flywheel/crank flange locating dowel pins are (PN-3701679). Replacement flywheel ring gears with 153 teeth are (PN-3991407) and with 168 teeth are (PN-460583). Mark IV cranks have a 3.58" bolt pattern on the rear flange. Gen V and VI have a 3" bolt pattern.

STARTERS

PN-10496870 is for heavy duty use with 12 3/4" OD ring gears.

PN-1876552 is a heavy duty starter for 14" OD ring gears.

PN-12361146 is a high-torque, gear reduction, mini starter. It weighs one-half (10.5 lbs.) that of a conventional starter andl fits either 14" or 12 3/4" OD ring gears. It provides more clearance for headers and oil pans .

Externally balanced motors (Mark IV 454, production Gen.V and Gen.VI) use flexplates, flywheels and vibration dampers that have additional weight added to them to balance the rotating assembly.

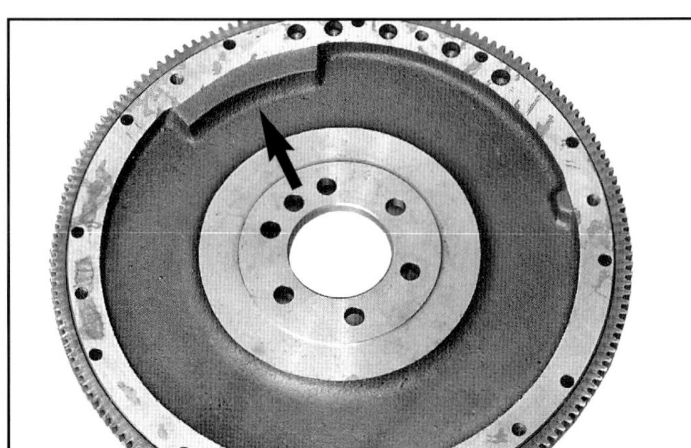

PN-12363128 is a chrome, high torque, mini-starter and will fit both 12 3/4" and 14" flywheels or flexplates.

PN-10455702 is another high-torque, gear reduction, mini starter originally used on '88 smallblock Corvettes. It weighs 10.14 lbs., but only fits 12 3/4" OD ring gears.

If you change flywheel or flexplate size, you don't have to change the entire starter. If it is a conventional size (not a mini starter) you can just change the starter nose to the appropriate one. (PN-1968122) is a starter nose for 12 3/4" OD ring gears and (PN-1984098) is for 14" OD ring gears.

Bolts for starters are unique because they have knurled sections and shoulders which are required to properly register the starter to the block surface. Always use the correct starter bolts. Three sizes are made: short 3/8-16 X 1 27/32", long 3/8-16 X 4 21/64" and a bolt used in 1966-68 427 heavy duty starter applications

This Mark IV 454 external balance flexplate carries the Safety Foundation, Inc. (SFI) sticker label showing date of manufacture and that it meets their specification 29.1 for flexplates. Racing tech inspectors look for these SFI labels when you bring your vehicle to the track if you are required to meet this safety spec.

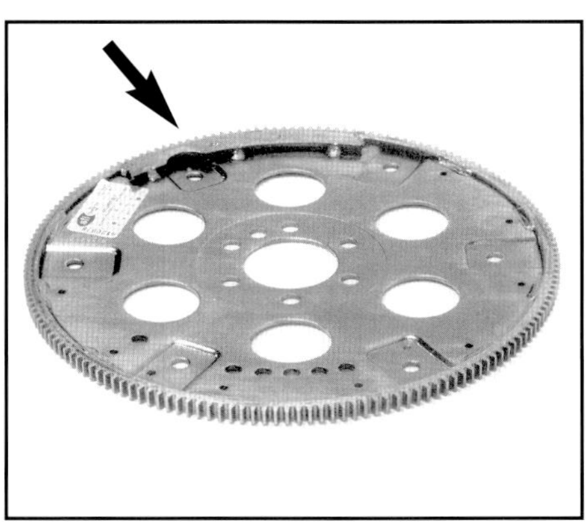

View of an externally balanced Mark IV 454 flexplate. Arrow points to the additional balance weight located on externally balanced motors.

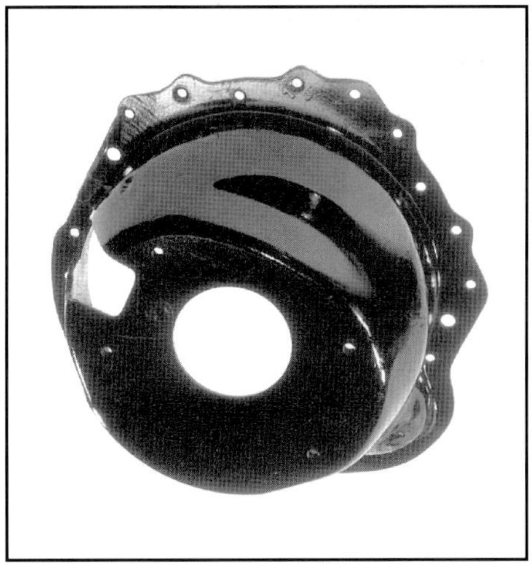

If you run a stick transmission you are going to need one of these to race at the track. A clutch "scatter shield" is required as a replacement for stock cast bellhousings. In the event of a clutch or flywheel failure, one of these and a matching engine plate, will contain it and save your legs.

which is 3/8-16 X 3 5/8". Mini starters come with their own bolts and starter shims as part of the assembly. Use a starter brace to help support the front of a conventional size starter motor.

If you are having trouble with starter heat soak (especially common after switching from stock cast iron exhaust manifolds to tube headers) you can use a starter heat shield to lower the radiant heat absorbed by the starter solenoid. Another aftermarket system uses a Ford style solenoid, mounted on the inside fender well and a small jumper wire to bypass the Chevy starter solenoid. It also solves the heat soak problem and gives you a more convenient location to hook up a remote push-button starter when you are working under the hood alone.

CONNECTING RODS

Connecting rods, rod bolts and the pistons and pins attached to them have to deal with the tremendous loads imposed on them by high rpms. Think about it, the reciprocating assembly in one cylinder, has a certain amount of weight and this weight is all moving in one direction until it reaches the limit of the crank stroke, then it has to reverse and go in the opposite direction. At that instant the piston and rod assembly are straining from inertia and momentum to keep going in the direction they were going, but now the crank (with the power generated by 7 other cylinders and their pistons) is trying to jerk the piston and rod in the other direction. The G-forces are tremendous. The piston tries to launch itself into the ozone layer. The piston pin is trying not to be broken in half. The rod wants to stretch like taffy or break and the rod bolts are thinking about doing the same thing. The rod bearing surface is trying to push a tiny .001" cushion of oil out of the way so that it can touch the crank journal face. You are asking the parts to stay together just this one time, on one stroke of the engine, now you multiply this and ask the assembly to do this 6500 or 7500 times a minute. You don't want any hassles, you just want to "firewall" it and go. Frankly, its amazing that the whole thing doesn't fly apart.

Maximum piston speed occurs before the piston is halfway down the cylinder and these maximum speeds increase with an increase of stroke length. G-forces, acting on the piston and pin, are also a factor. A lighter rotating assembly will accelerate faster and be better able to withstand the G-force loads placed on it at high piston speeds. Increasing the rod length will allow the piston to stay a little longer, (in terms of crankshaft degrees of rotation), in the area of top dead center. Increasing rod length will also decrease the amount of time spent at bottom dead center. Piston acceleration speeds also decrease when you use a longer rod. As mentioned before, rod angularity is reduced with a longer rod and there is a reduction in friction and bore wear along the thrust surfaces of the cylinder wall.

Maximum piston speeds with a 4" stroke crank and 6.135" length rods at 6500 rpms reach approximately 7,019 feet per minute. At 7500 rpms the peak speeds reach about 8,100 feet per minute. With a shorter 3.76" stroke crank and the stock rod length of 6.135" the maximum piston speeds are less. At 6500 rpms the maximum piston speed is 6598 fpm and at 7500 rpms the maximum piston speed is 7614 feet per minute which is just about 500 feet per minute less than that reached with the 4.0" stroke crank.

Obviously, anything that you can do to strengthen the reciprocating assembly will help it live at high rpms. The overall weight of the assembly can be reduced. This will reduce the total loads on the parts and will help the

STARTER TIPS

Two types of conventional starters for the big block. One on the left has starter bolt holes which are straight across for 12 3/4" flywheels or flexplates. Starter on right, with staggered bolt holes, is for 14" flywheels or flexplates. The starter noses themselves are slightly different and are interchangeable between starters.

Aftermarket lightweight, hi-torque starter will fit either size flywheels or flexplates and uses less amperage than a conventional starter.

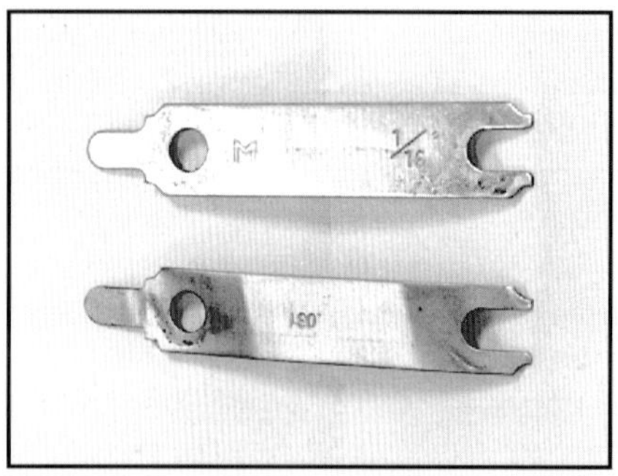

Two starter shims of different thicknesses help get the teeth on starter gears correctly meshing with flywheel or flexplate teeth. These shims are not always necessary, but may be required if excessive starter noise or drag is encountered.

The dual starter bolt patten found on most blocks allows the use of a 12 3/4" or 14" flywheel or flexplate with the matching starter.

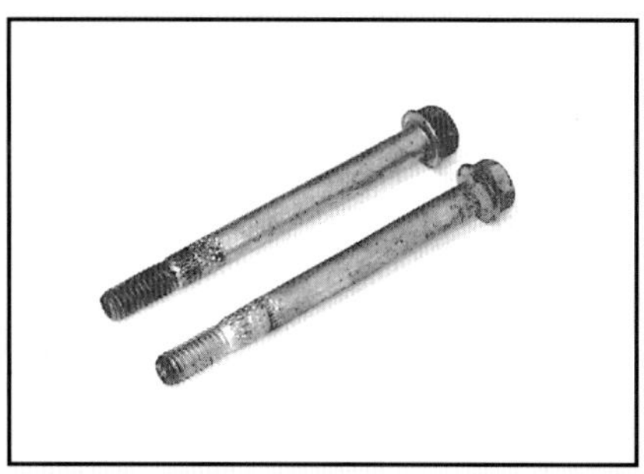

Starter bolts have correct shoulders and knurled sections. Proper starter alignment cannot be achieved without these knurled bolts. Your starter may still work, but it will eventually fail or cause damage to the ring gear on the flywheel or flexplate.

parts accelerate to these speeds quicker. If the total weight is reduced the maximum rpm levels and loads that the parts can endure can be increased. There are some compromises involved because at some point the particular part could be lightened further but then the weight reduction comes at the expense of strength and durability. How the engine is to be run is also a factor. A short duration drag race motor only has to endure the increased loads and stresses for a very short amount of time, whereas a boat or off road endurance race engine must put up with this type of punishment for much longer periods. Remember that a drag race motor, while it may be spinning at a rate of 7000 rpms, actually only runs all out like this for a few seconds. If the motor was able to accelerate to 7000 rpms right off the line (which doesn't instantly happen) and stay there for an entire 10 second run, it would actually turn over only 1167 times, but it would revolve 116.7 times a second. Each plug would fire every other revolution, 584 times during the run or 58.4 times per second. An endurance motor in a boat may be spinning at a rate of 5500 rpms, but may have to do this for a couple of hours at a time or 330,000 revolutions per hour.

Given the bore sizes, stroke lengths and the rod lengths used in the big block Chevy, its difficult not to have relatively heavy piston and rod assemblies when you compare it, for example, to a smaller bore, short stroke 327 Chevy smallblock.

Chevy makes five types of big block connecting rods. All of them can be used in a street driven performance vehicle. Some are stronger than others. All are forged steel and are the same length, 6.135" from center to center. The rods are interchangeable between Mark IV, Gen.V and Gen.VI engines and have a bearing bore diameter of 2.3247" to 2.3252". Most stock rods have a piston pin bore sized for pressed-on piston pins, while one rod was factory intended for use with floating piston pins.

Individual rod bolts can be purchased from Chevy: a 3/8" factory magnafluxed rod bolt (PN-3862720) was used on rod (PN-3856240) and can upgrade any rod with stock 3/8" bolts. 7/16" knurled shank bolt (PN-14096148) and nut (PN-3942410) can be upgraded to the heavier duty 7/16" boron steel bolt (PN-3969864) and 12-point 6304 steel alloy

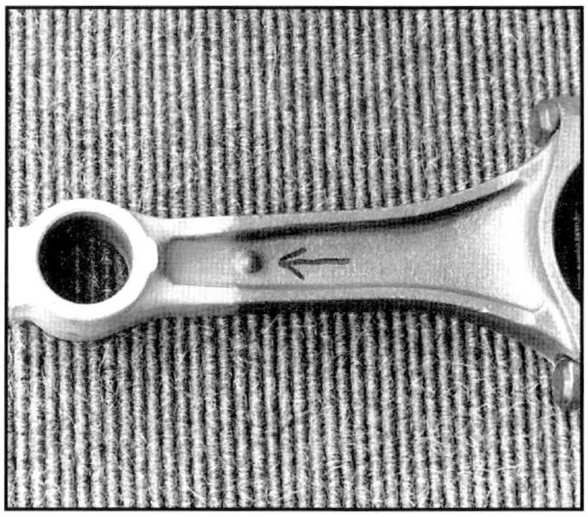

GM 7/16" heavy duty "dot" rod will upgrade the strength of your rotating assembly. Some early "dot" rods from the Sixties had a 3/8" rod bolt.

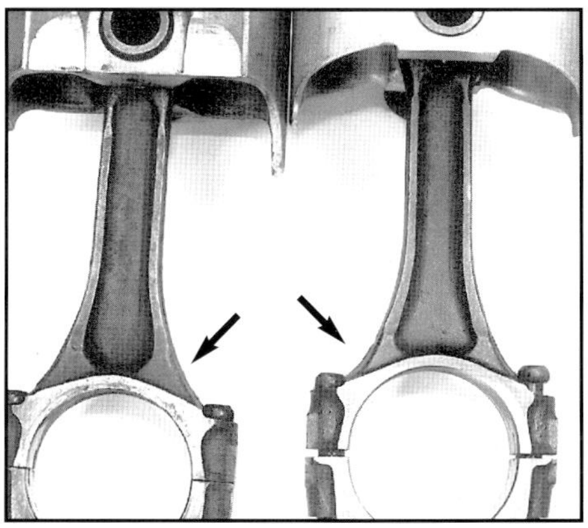

The truck rod (left) uses a 3/8" rod bolt and has thicker webbing on the big end when compared to the stock passenger 3/8" big block rod. Center to center length is the same on all production big block rods, 6.135".

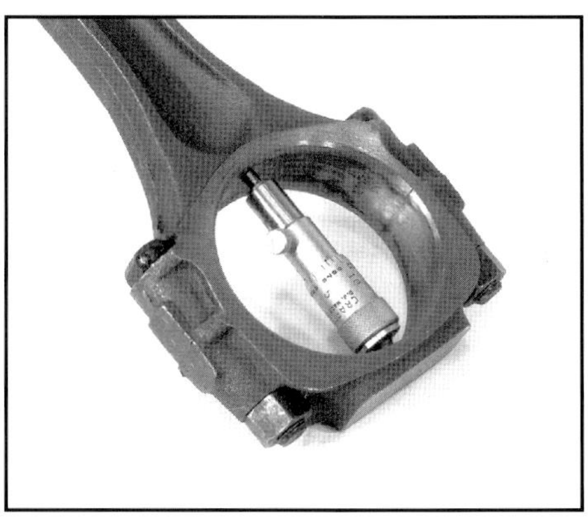

After torquing the rod bolts in a rod vise, you can measure the inside diameter in the big end. Then measure the thickness of the rod bearing inserts and determine the bearing clearance on the crank rod journals.

nut (PN-340289) or 12-point 4037 steel nut (PN-14044866).

There are a number of aftermarket companies specializing in fasteners, including rod bolts, such as Automotive Racing Products (ARP). Many different metal alloys, increased tensile strengths, an shank designs are available. Given

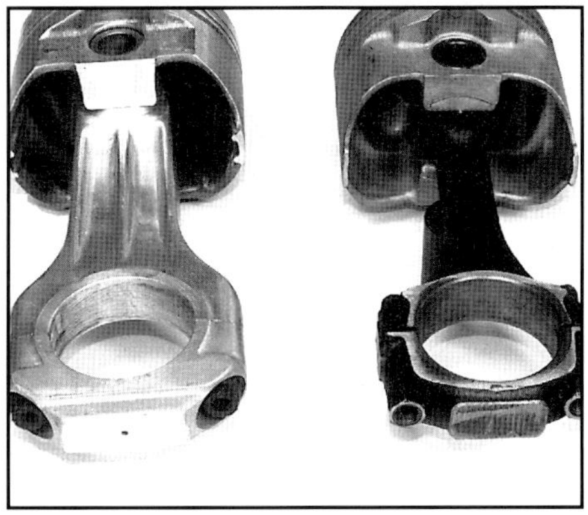

Careful attention to block clearance must be made when you use an aluminum rod in a race motor, since its dimensions are so much larger than those found with a steel rod.

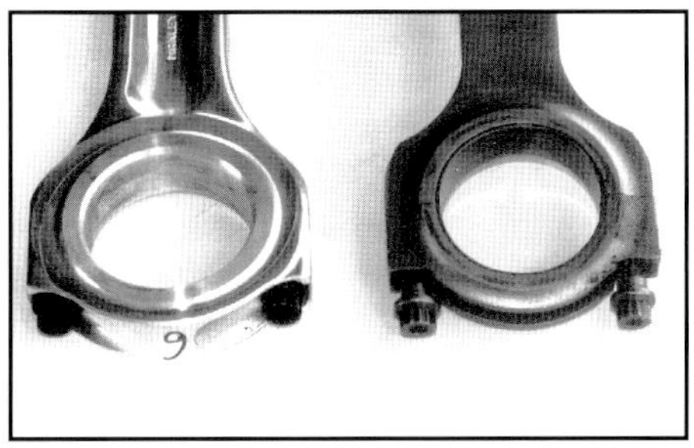

Manley H-beam steel rod (right) and a Manley aluminum rod.

A feeler gauge is used to check the rod side clearance during assembly. Too much clearance is an indication that the rod side surface is worn and the rods will have to be replaced. Too little clearance can be fixed by reducing the width of the rod appropriately to get the clearance you need.

that a rod bolt is probably the most stressed fastener in an engine, get the best, especially if you plan on using a 3/8" bolt rod in a bracket drag racing engine. Keep in mind that a number of aftermarket rod bolts have a radius under the bolt head that strengthens the shank of the bolt. If the bolts have this radius you must provide for it by chamfering the bolt holes on the connecting rod caps themselves.

Speaking of 3/8" bolt rods, I think they are okay to about 6500 rpms in a 396-454 size bracket drag race motor, if you install a premium aftermarket rod bolt

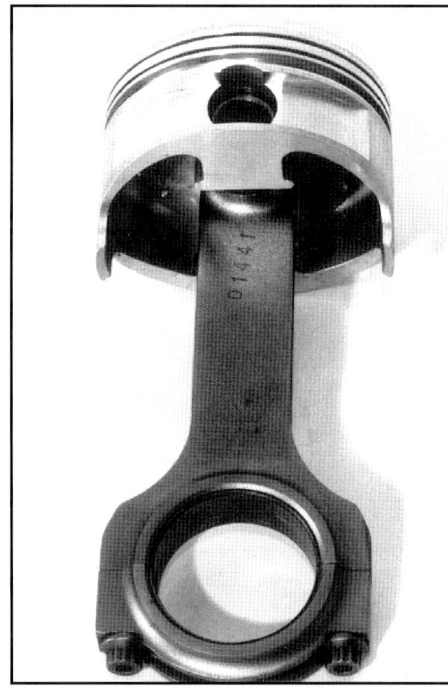

Ross lightweight race piston and 6.535" length Manley H-beam steel rod assembly makes for a strong rotating assembly that is easy to fit and clearance in a block.

and make sure you use a rev limiter. In a longer duration motor or to live at higher rpms, use a 7/16" bolt rod with the best bolts available. If weight is a racing consideration you can search for a lighter weight aftermarket steel rod or for drag racing only, move to aluminum rods, if you don't mind fitting them to your block, checking them often and then replacing them before they break.

Stock "dot" rods with quality 7/16" rod bolts can be used to 7000 rpms in a production size motor. Remember that the stock reciprocating assembly of a big block Chevy is heavy. Spinning the heavy weight of pistons and rods at high rpms is hard on the stock parts. If you build a high performance or race motor that will make more power above 7000 rpms, you should think about using aftermarket parts that are lighter and stronger than those offered by Chevrolet. Lighter parts will spin faster, but lightness must not be at the expense of strength. Aftermarket parts that meet these requirements are expensive. Most oval port head motors will make their maximum torque and horsepower at rpms less than 6500 rpms. If your engine's maximum power is made at rpms less than 6500 rpms, it makes no sense to spin the motor to higher rpms. It makes no more power, its hard on parts and hard on your wallet. Don't spin the motor any faster than you need to!

High quality aftermarket steel or aluminum rods are available from Manley, Carrillo, Eagle, Childs and Albert, Lunati, Oliver, Venolia and Bill Miller (BME). They offer many different off the shelf lengths or can custom make any length you want generally up to 6.800" center to center length or provide rods with blank holes for those who wish to do their own finish work. Talk to them about your combination and application. If you find yourself thinking about using aluminum rods in a street motor, forget it. Aluminum rods are strictly for short duration race motors and as such must be routinely removed from the engine and checked for stretch or work hardening. They are not intended for everyday street use.

Keep in mind that the "H-beam" or "I-beam" steel rods are not as large or wide as a similar aluminum rod. This makes them easier to fit into a block, especially if you are increasing the stroke or rod length on an engine. Aluminum rods, because of their size, will require more block clearance.

ROD RATIOS

There has been a lot of discussion about the best rod ratio for any given engine. A number of factors come into play such as cam timing, intake and exhaust runner length and volume, piston pin heights, block clearances, and deck heights. Rod ratio is determined by taking the length of the rod, center to center, and dividing it by the stroke length of the crank. One the next page is a list of the rod ratios for stock Chevy big blocks and some possible stroker crank/long rod combinations.

When you look at the list you notice that some of the combinations have the same rod ratios as others. One thing that increasing the rod length will do is to reduce the rod angularity in the cylinder bore. Reducing this angle lessens the amount of side force produced on the thrust surface of the cylinder walls. This in turn reduces the amount of friction and frees up a few horsepower. Also the piston itself tends to linger less at bottom dead center and stays longer at top dead center for a few more degrees of crank rotation than a combination using a shorter rod length.

Obviously before buying parts and building some of the combinations you will have to consider total block deck height, block deck clearance, piston pin height (also known as piston compression height), total clearance of the rotating assembly in relation to the block, crank stroke, cam timing and static compression ratio. When using a 4 1/8" or longer stroke crank, a longer rod can help restore and improve the rod ratio. Keep in mind that there is a limit to how far up the piston pin hole can be placed in relation to the piston deck.

It is possible to use an aftermarket rod that is .400" longer (6.535") than stock in a short deck motor with the appropriate aftermarket piston and a 4" stroke crank. It is accomplished by moving the piston pin hole up .400" towards the piston deck and rearranging the ring grooves to give a piston pin height of 1.245". JE Pistons is one company that can provide this type of light weight race piston off the shelf. This combination puts the oil ring groove partially over the piston pin bore. A support rail runs in the groove and holds the oil ring and oil ring side rails where they pass over the pin hole.

FLOATING PINS VS. PRESSED ON PINS

There really isn't that much difference between a rod that takes a pressed on piston pin and a floating pin. For me it is more a question of assembly and disassembly convenience. With a pressed on pin, the rod must be put in a "rod oven" and the small end heated. The press on pin has an interference fit in the rod, so when the pin end is heated, the piston and cooler pin can be fit on the rod. When the rod cools, the pin is held tightly by the rod and the piston swivels on the pin. When you have to remove a piston that utilizes a pressed on pin, a heavy hydraulic press is used to push the pin out of the rod, but as sometimes happens, the piston is damaged in the act of pressing the pin out of the rod and piston.

With a floating pin, assembly and disassembly is much easier. The pin rides on a small clearance in the rod end and can be slipped in and out of the rod by hand. The piston pin is held in the piston by placing circular retainers in grooves in the piston pin bosses at each end of the pins. These retainers keep the pin from moving side to side in the piston. To take the floating pin assembly apart, you remove the pin retainers and slide the pin out of the piston and rod. If you are going to be taking an engine apart frequently, either between rounds or between race seasons, this is one of those small things that makes life easier. If you don't expect to be taking a motor apart until it has 100,000 miles on it, a pressed on piston pin and rod assembly will do just fine.

Pin retainers, used with floating pins, basically come in two styles. Two C-clips (also called Tru-Arcs) on each end of the piston pin or a single Spirolox retainer on each end. Some engine builders feel that the Spirolox retainers are better, but they are difficult to remove once they are installed in their respective grooves. Which ever you use, don't reuse them once they have been removed during rebuilds. Always use new pin retainers. Most aftermarket performance piston makers make their pistons so that they can be used either with a pressed on pin or a floating pin. If you use a floating pin setup make sure that you place the pin retainers in and that there is a 0.0"-.0005" clearance between the end of the pin and the retainers. A larger clearance will allow the pins to pound out the retainers. Make sure that the piston you use has some provision for pin oiling in the piston pin boss. The pin should have a side clearance of .0004"-.0008" in the piston pin boss. Some pistons in race motors have floating pins that are retained by pin "buttons". Made of Teflon, these buttons fit into each end of the pin and actually ride close to the cylinder wall and prevent the pin from moving side to side in the rod and piston.

If you want to convert your stock rod from a press on pin to a floating

CONNECTING ROD RATIOS

The stock rod length used on all short or tall deck big block Chevy engines is 6.135" measured from center to center. Different strength rods and rod bolts have been factory produced, but they all shared the same length. Aftermarket suppliers make other rod lengths of various strengths and materials, steel, aluminum and titanium. Altering the rod ratio will have an effect on the breathing characteristics of any engine and the amount of time (in crankshaft degrees of rotation) the piston stays near top dead center (TDC) or bottom dead center (BDC) in the cylinder. A longer rod may also reduce rod angularity in the bore and thus there may be a reduction in friction and wear along the thrust surfaces of the cylinder bore when compared to other parts combinations.

Remember you may have to grind additional rotating clearance into the block in order to use some of these combinations. Always check for proper clearance before attempting to use a longer or larger connecting rod. Some combinations may only work with tall deck truck, Bowtie, Olds DRCE or aftermarket blocks. You will need to include one half the stroke length, the rod length, piston pin compression height and total block deck height from the centerline of the crank to the top of the block deck in your calculations to see if you have room. Rod ratio is determined by dividing the rod length, center to center, by the crank stroke.

STOCK BIG BLOCK CHEVY

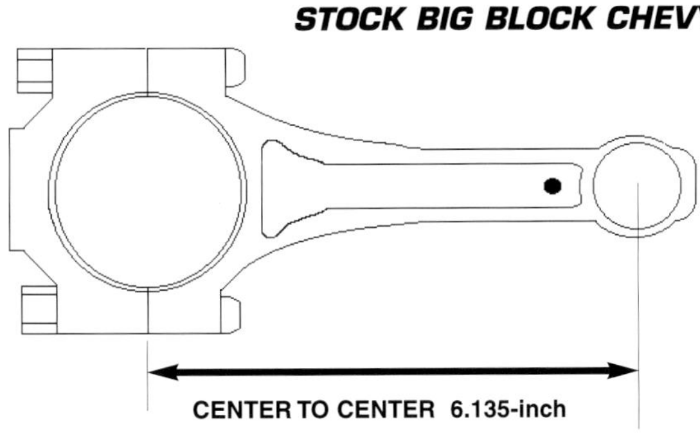

CRANK STROKE	ROD LENGTH	ROD RATIO
3.76"	6.135"	1.63-1
	6.385"(+.250")	1.70-1
	6.535"(+.400")	1.74-1
	6.700"(+.565")	1.78-1
	6.800"(+.665")	1.81-1
4.00"	6.135"	1.53-1
	6.385"(+.250")	1.59-1
	6.535"(+.400")	1.63-1
	6.700"(+.565")	1.67-1
	6.800"(+.665")	1.70-1
4.125"	6.135"	1.49-1
	6.385"(+.250")	1.52-1
	6.535"(+.400")	1.58-1
	6.700"(+.565")	1.62-1
	6.800"(+.665")	1.65-1
4.250"	6.135"	1.44-1
	6.385"(+.250")	1.50-1
	6.535"(+.400")	1.54-1
	6.700"(+.565")	1.58-1
	6.800"(+.665")	1.60-1
4.500"	6.135"	1.36-1
	6.385"(+.250")	1.42-1
	6.535"(+.400")	1.45-1
	6.700"(+.565")	1.49-1
	6.800"(+.665")	1.51-1
4.600"	6.135"	1.33-1
	6.385"(+.250")	1.39-1
	6.535"(+.400")	1.42-1
	6.700"(+.565")	1.45-1
	6.800"(+.665")	1.48-1

Other custom rod lengths are available from aftermarket connecting rod makers.

pin, have the pin end of the rod honed to give .00015"- .0008" clearance between the floating pin and the rod bore. To insure that oil lubes the floating pin in the rod, the small end of the rod should have a 1/8" hole drilled into the top of it for pin oiling. The 1/8" hole should be counterbored with a 1/4" bit to chamfer the opening. Debur the pin bore and then hone to the correct side clearance for the piston pin. Another alternative way to insure oil gets to the floating pin/rod surface is to place two very shallow oiling grooves into the inner bore surface of the small rod end bore. Each oiling groove should be 90 degrees to the top of the rod pin bore.

Some folks have the small end bushed when using a floating pin. Others feel that this reduces the overall strength of the stock rod end because it reduces the thickness of the metal on the rod surrounding the pin. They run a floating pin in the rod without using a bushing. Many aftermarket rods have more material on the small end and are fitted with a pin bushing. Since these rods were designed initially with a pin bushing in mind, there's no problem.

A floating piston with a pin bushing does have an advantage. If there is any wear, it occurs to the pin and the bushing. A worn bushing can be replaced without having to junk the rod.

ROD PREP

The first thing is to look at the individual rods to determine if any damage has occurred even if you are looking at brand new rods. Are the bearing bores of the large end scored, nicked or if used, have they turned blue or black from overheating due to a spun bearing? Does the small end show any damage? A used rod should be cleaned and Magnafluxed to check for any cracks. A rod should also be checked for straightness or twist. If the rod is acceptable, you can make a decision concerning whether the mold parting lines on the side beams of a stock rod need

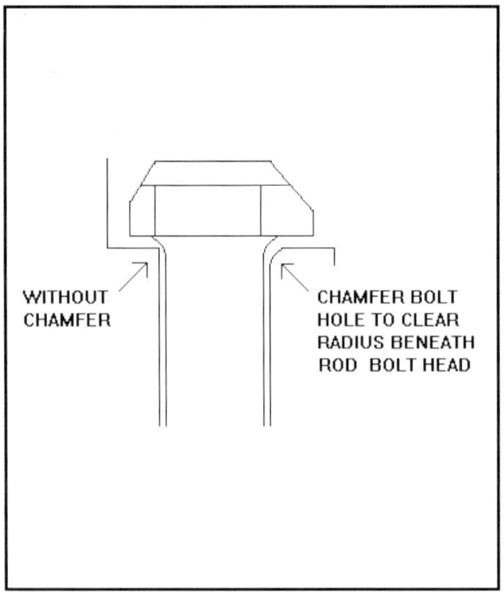

A number of aftermarket rod bolts have a radius from the stem of the bolt to the head of the rod bolt. This adds strength to that area of the bolt by reducing the chances of a stress related fracture. However, the hole in the rod for the bolt often is a sharp 90' surface where the radius on the bolt occurs. Chamfer the bolt hole lightly so that sharp edge is removed.

to be ground and polished. Keep in mind that all you are trying to do is smooth out the rough portions of the beam in order to help prevent stress cracks from beginning. You do not want to eliminate weight from the side beams, nor do you want to take so much metal off that you are actually compromising the strength of the rod rather than improving it. The rod beams can be shotpeened to further improve the surface and to help resist the beginning of cracks. The balance pads on stock rods can be reduced in size to reduce the overall weight of the rod, but you need to leave a little on the balance pads of the rod so that when the rotating assembly is balanced you have something to work with in order to match all eight rods.

Throw away used stock rod bolts especially if they are the stock 3/8" variety and install new aftermarket premium rod bolts. Don't skimp here on cost. The new bolts need to be installed properly with a suitable large press. They are not smacked in with a hammer.

The rod bearing bore diameter needs to be checked. The big end bearing bore should measure 2.3247"-2.3252" and is measured at 90 degrees to the parting line of the rod cap. Big ends that are out of spec can be resized. Many rods used in high rpm engines are deliberately resized and honed in a way that makes the big end diameter slightly larger when measured at the parting line as compared to the diameter when measured at 90 degrees to the parting line. As the rpms go up and the speed and g-forces increase on the rod and piston, the bore on the big end of the rod can be ovalized and is no longer round. If this ovalizing of the bore gets too large the "crush" on the rod bearing can weaken and the bearing can spin in the rod bore and then onto the crank journal surface. Bearing and crank journal failure are what happens next. In an effort to compensate for this high rpm speed ovalizing, the rod bore is finished in the above manner so that as the forces acting on the rod increase, the big end bore goes from a deliberate oval shape to a round shape.

PISTONS

Chevy has used three types of aluminum pistons, cast, hypereutectic and forged. Cast pistons are found in most stock production, low performance engines. They give long service in this type of application because of the tight piston to cylinder wall clearances allowed. Hypereutectic pistons are aluminum pistons made with additional silicon in the alloy and have more strength than a cast piston. Because they don't expand as much as a forged piston, hypereutectic pistons can be fit with tight piston to cylinder wall clearances and have the stability of a cast piston. Forged pistons are the strongest pistons and can take the most heat and pressure, but they expand more than other types of pistons. This requires that they have a larger piston to cylinder wall clearance when installed and they may clatter a bit when the engine is cold and the piston has not expanded completely.

Pistons decks are made with flat tops, domes or recessed "dish" shapes. Most have reliefs for valve clearance in their deck surface. These shapes and resulting volumes have a direct effect on the amount of static compression a cylinder has. Other engine dimensions are involved and you need to know what amount of static compression your parts combination will make. Engines running on the street with pump gasoline should not exceed 9.5-1 static compression ratios in order to avoid detonation. Bore, stroke, cylinder head chamber volume, head gasket compressed thickness, deck clearance and piston dome and valve relief volume must all be measured or known, in order to figure out static compression ratio. (See the ENGINE MATH chapter to see how to make these calculations.)

Factory installed pistons for each engine size had different pin heights as shown on page 67. Piston pin height also known as compression height, is measured from the center of the piston pin bore to the top of the piston deck, not the dome.

Domed pistons have two different dome styles to match either "open" or "closed" chamber big block heads. GM pistons with domes for "closed" chamber

BIG BLOCK Quick Facts & References

FLOATING vs PRESSED PIN RODS

PN	Floating or Pressed Pin	Rod Bolt Size	Note
3933174	Pressed	3/8"	Stock passenger car rod
3856240	Pressed	3/8"	'65-'69 early HP "dot" rod Magnafluxed, shotpeened
	Pressed	3/8"	Stock truck rod, wider beams thicker web on big end
10198922	Pressed Knurled Shank Bolt	7/16"	LS-6, LS-7, Gen.V, VI 454/502 HO magnafluxed, shotpeened "dot" rod, white color code 4340 steel
3969804	Floating Boron Steel Bolt	7/16"	L-88, ZL-1 427, 4340 steel magnafluxed, shotpeened "dot" rod, green color code heat treated

heads can be used with "closed" or "open" chamber heads. Pistons with domes for "open" chambers can only be used with "open" chamber heads. Some aftermarket domed pistons will fit both "open" or "closed" chamber heads. Check with them for their recommendations on domed pistons. Of course, flat-top or dished pistons can be used with either style of cylinder head.

In any case, make sure you figure out what your static compression will be when you make your parts selection!

If you are going to use a very high compression ratio, large doses of nitrous or a blower with high boost, cylinder temperatures and pressures will be high. A forged piston with a thicker deck or with the top ring groove lower in relation to the piston deck may help the piston survive. Check with your piston maker for designs made for these applications. High tech piston coatings can help the piston reject the increased temperatures, increase parts survival and keep the high temperatures and pressures in the combustion chamber where they can work on the top of the piston to produce power.

If you are assembling a high performance street or mild bracket race engine, you can use a hypereutectic piston with a pressed on pin. They are stronger than a cast piston, and can take more abuse. They are quiet, cost only a few dollars more than a cast piston and can be fitted with tight piston to cylinder wall clearances.

When building a very high performance street motor or a serious race engine, use a forged piston. This is especially true if you plan to use a turbocharger, supercharger, or nitrous-oxide with any static compression ratio or if you plan to use a carbureted engine with a static compression ratio over 10.5-1. These types of engines can develop high heat and cylinder pressures that might melt or break a cast piston. Forged pistons expand more than other types of pistons so a larger clearance must be used between the piston and the cylinder wall. Follow the manufacturers recommendations for piston to cylinder wall clearances.

When your block is being bored, bring the pistons you are going to use. The

BIG BLOCK Quick Facts & References

PISTON MATERIALS & APPLICATIONS

Cast pistons have been the standard automotive piston for decades. They are inexpensive and easy to produce. They have a thermally stable crystalline grain structure and often incorporate cast-in steel expansion struts that allows them to fit tightly in the bore for optimum stability and ring seal. Under normal use they will stand up well to tens of thousand of miles of use. However, they have limited speed, thermal, and detonation resistance. They should only be used in moderate performance engines where speed is limited and detonation is strictly avoided.

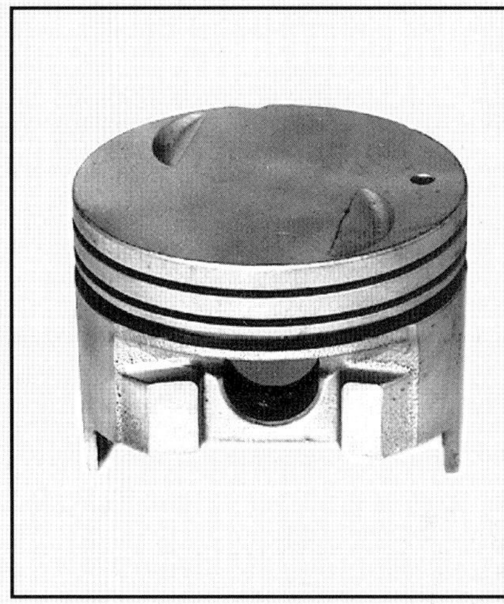

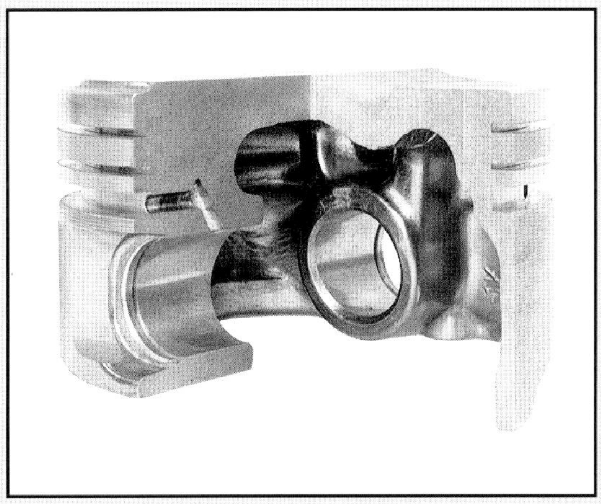

Hypereutectic® pistons are castings but have nearly 2-1/2 times the silicon of a standard cast piston for increased hardness and greater resistance to higher temperatures and cylinder pressures. They are dimensionally stable and require very little skirt clearance. In some instances they can be operated with less clearance than standard cast pistons. This feature keeps the pistons and ring package well stabilized in the bore and improves sealing and blowby control. While Hypereutectic pistons are well suited to street performance applications, they do not have the detonation and temperature resistance of forged pistons. They should not be used with more than very light nitrous-oxide injection loads, or with high-pressure turbo- or supercharging systems.

Forged pistons are manufactured with a forging die from a solid slug of heat-treated aluminum alloy. They possess the dense grain structure and metallurgical properties to stand up to severe use, including a degree of detonation resistance. However, forged pistons have less dimensional stability and require greater skirt clearance for reliable operation. Forged pistons remain the top choice when strength and durability are required for racing, turbocharged, supercharged, or nitrous-oxide injected applications.

Photo courtesy of Federal Mogul Corp.

TRW-L2519F is a fully machined forged 454 race piston for open chamber heads. Dome volume is 46.8cc and will give 12.34-1 compression with 118cc heads and .038" compressed gasket. Pin height is 1.645". (Photo courtesy of Federal-Mogul Corp.)

block will be bored out until the cylinder diameter is just shy of the desired size. The pistons diameters will then be individually measured and the final block honing will bring the cylinder diameter out to the correct diameter. This will include whatever piston to cylinder wall clearance that is required. Each piston is measured and fit to a particular cylinder barrel. Use a torque plate.

Cast and hypereutectic pistons generally need a .0007"-.002" cylinder wall clearance. Forged pistons, which have a greater expansion rate, may have a required clearance of .004"-.007". It is very important that you follow the manufacturer's recommendations for the proper clearance for a particular piston. The manufacturer will also state where on the piston skirt you need to measure in order to determine piston diameter which is usually 90 degrees to the piston pin centerline axis.

Increasing the compression on any motor will increase the engine's mechanical efficiency. Now you may be tempted to use a piston with the biggest dome that you can find and fit into the heads you plan to use. What will the static compression be? A street motor run on low octane, unleaded pump gasoline should have a static compression ratio lower than 9.5-1. In a race motor, a large piston dome has some disadvantages because the large dome inhibits good flame travel. The flame front must go over the top of the dome. Current thinking in high performance and race engine design is to increase static compression by using smaller cylinder head combustion chambers and flattops or the smallest piston dome volume possible to obtain high compression ratios and good flame front travel speed. "Bowtie" "symmetrical port" heads, which are the latest design race heads available from Chevy in 1996, have small 68 or 72cc combustion chambers. Pontiac Super Duty and many aftermarket heads are also available with small combustion chambers and thick head decks. These head decks can be shaved down to produce even smaller combustion chamber volumes. Very high compression ratios can be made with flattop pistons or small piston domes if you use a small combustion chamber volume. Flame travel is enhanced and more power is made when compared to large combustion chamber/big piston dome combinations.

Let me give you an example of a high static compression ratio combination in a drag race motor using a flatop piston and a small combustion chamber.

Mark IV "Bowtie" block 4.500" bore
454 Steel Crank stock stroke 4.00"
"Bowtie" (#-128)
Symmetrical Port
head chamber 68cc
Flatop forged piston
dome volume 0
Block deck height 0
Composition type head
gasket compressed thickness .038"
Head gasket bore diameter 4.5"
Total Cubic Inches 509
STATIC COMPRESSION RATIO 14.38-1

The above combination would need to run on high octane race fuels. The piston, with a very high lift drag race style cam, would also need some deep valve reliefs in the piston deck for valve clearance and this would reduce the static compression ratio slightly. You can see that if you give some thought to what you are going to build, flame travel can be improved, without losing the high compression values you might want. Now you do not need to build something with the very high ratio shown here. Other combinations are possible depending, in part, on bore, stroke, cam timing, piston and chamber volume of the heads used.

For some time race engine builders have had to spend a lot of time massaging the domes on pistons in order to remove sharp edges and generally smoothing out dome contours in order to achieve a piston dome shape they felt conducive to good flame travel and to reduce hot spots. Now however, many aftermarket piston makers deliver pistons with fully machined domes that greatly reduce the amount of hand finishing you may want to do. Pistons can also be lightened and balanced by the manufacturer eliminating another step.

Aftermarket piston makers supply just about any piston size and design you could ask for off their shelf. If they don't have what you want, they can and will "custom" make what you need. Call them and get their advice before you buy.

Remember that the piston pin is one of the highly stressed parts of the rotating assembly. Most pins that come with the new pistons you buy are reliable and strong enough for most applications. However, if you are building a race motor that is going to make a large amount of cylinder pressure say from very high compression levels or large amounts of nitrous or large amounts of blower boost or be used at extreme rpm levels, then you can look into upgrading the pin to a different design or total weight. There are some slight weight savings that can be made. Talk to your piston maker about what you need for your particular combination.

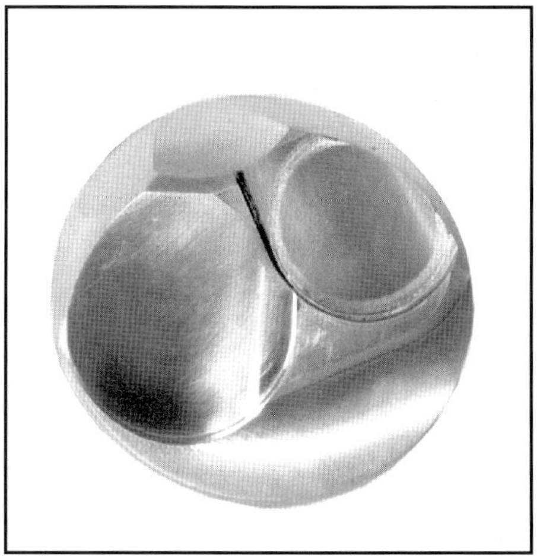

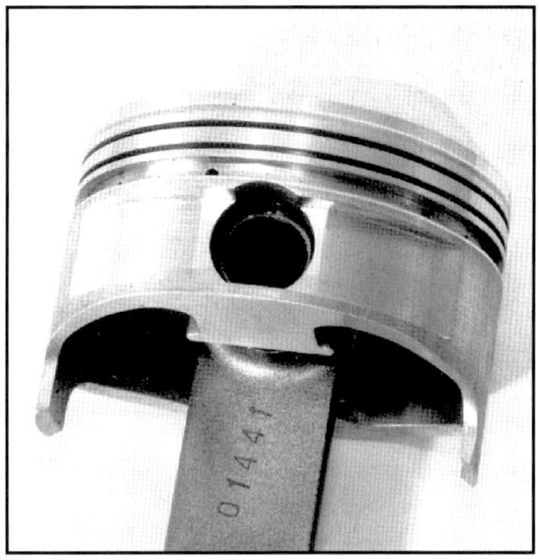

Ross lightweight race piston and 6.535" length Manley H-beam steel rod assembly makes for a strong rotating assembly that is easy to fit and clearance in a block.

Ross forged race piston used in this particular Bowtie tall deck application withstands alcohol and 14.7-1 static compression.

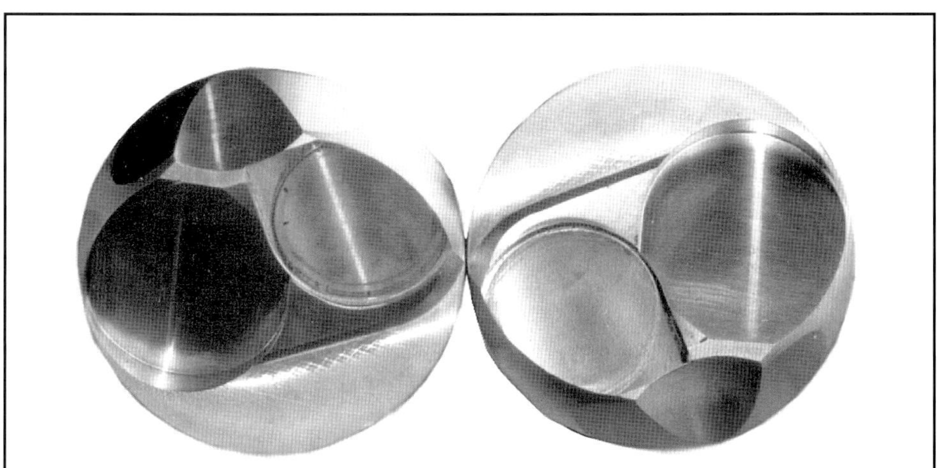

Ross forged race piston for 454 with open chamber heads has large, deep, valve reliefs and dome relief located near the spark plug.

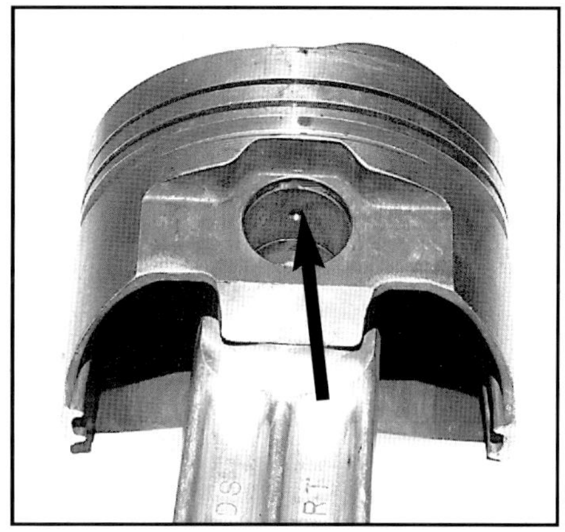

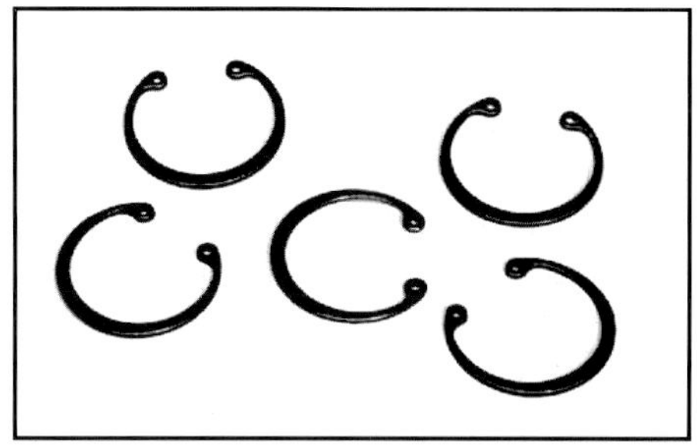

For a floating piston pin application you will need a piston with pin oilers and grooves cut for pin retainer "C-clips" or "Spirolocks". (lower photo)

PISTON RINGS

Piston ring grooves are cut for different ring widths for low performance and high performance piston rings. GM high performance and many aftermarket forged pistons use rings that have a top and second ring width of 1/16" with an oil ring width of 3/16". GM cast pistons use piston rings with a wider width of 5/64" for the top and second ring and 3/16" for the oil ring.

Piston top rings themselves are made from various materials plain cast iron, ductile iron, chrome-faced, plasma ceramic faced and moly-faced. Second rings can be made with various "zero-gap" features if you are looking for better cylinder sealing. Third or oil rings are also offered in low tension forms for race engines. Some ring sets are 0.005" over size and these require that you file the ring ends to make a custom ring end gap the size you want. There is some power to be gained here by getting the ring gap exactly right. However, if you reduce the gap by too much the ends of the ring can butt up against each other after they heat up and expand. Power will fall off and there may be some damage to the cylinder walls as a result.

BIG BLOCK Quick Facts & References

FORGED PISTONS FOR CHEVY BIG BLOCKS

PART #	ENGINE	COMP./CHAMBER RATIO/CC'S	DOME TYPE	PIN TYPE	USE
TRW-2383	396/402	8.95-1/106	CLOSED 13.9cc	P or F	
TRW-2240NF	396/402	9.5-1/106	CLOSED 21.0cc	P or F	
TRW-2328F	396/402	11.38-1/106	CLOSED 37.8cc	P or F	
TRW-2242F	396/402	11.6-1/106	CLOSED 39.5cc	P or F	
TRW-2287F SP-7021P	396/402	12.9-1/106	CLOSED 47.9cc	P or F	
TRW-2300	427 car	9.75-1/106	CLOSED 16.8	P or F	
SP-7048P	427 car	11.63-1/110	OPEN 38.5cc	P or F	
TRW-2268F	427 car	11.87-1/106	CLOSED 36.3cc	P or F	
TRW-2239NF SP-7011P	427 car	12.18-1/106	CLOSED 38.5cc	P or F	
GM-3959105	427 car	12.25-1/110	OPEN	FLOATING	L-88, ZL-1
GM-3909857	427 car	12.5/107	CLOSED	FLOATING	
TRW-2308AF SP-7041PA	427 car	13.3-1/110	OPEN 50cc	P or F	
TRW-2453 SP-7062P	454	7.81-1/118	DISH Blower -8cc	P or F	
TRW-2377	454	7.96-1/118	FLATTOP -4.9cc	P or F	
SP-7071	454	8.23-1/118	FLATTOP	P or F	
GM-14097018	454	8.75-1/118	OPEN	PRESSED	Gen.V HO 454
TRW-2399 SP-7060P	454	9.09-1/118	OPEN 13.8cc	P or F	
GM-6262976	454	9.0-1/106	CLOSED	PRESSED	LS-6 1971
TRW-2465	454	10.1-1/118	OPEN 27.1cc	P or F	
TRW-2349 SP-7053P	454	11.8-1/106	CLOSED 30.6cc	P or F	
GM-3976013	454	11.0-1/110	CLOSED	PRESSED	LS-6 1970
TRW-2372 SP-7039P	454	11.3-1/118	OPEN 38.5cc	P or F	
GM-3976014	454	12.25-1/110	OPEN	PRESSED	LS-7
TRW-2519F	454	12.35-1/118	OPEN 46.8cc	P or F	
TRW-2307 SP-7040PA	454	12.82-1/118	OPEN 50cc	P or F	
GM-10198977	502	8.75-1/118	FLATTOP	PRESSED	GEN.V HO 502
TRW-2513F	502+.030"	8.75-1/118	FLATTOP	P or F	

These forged piston compression ratio calculations assume a .008" deck height, the use of a .038" compressed thickness composition head gasket and a standard bore. Always calculate the static compression ratio for your particular parts combination. Many other pistons and piston types are available from aftermarket suppliers.

One rule of thumb for minimum top ring end gaps is 0.004" gap for every 1" of cylinder bore. For example, a 4.250" cylinder bore will require a minimum top ring end gap of 0.017" with gasoline. Be careful here. Any ring gap measurement you make will be when the piston and rings are cold. If you get the gap too small the ends may hit each other when the ring is heated up and expands. The top ring will be hotter than the second or third rings because it is so close to the piston top and combustion chamber heat. You can start out with a larger end gap and work your way down to smaller

PISTON PIN HEIGHTS

Pistons for each engine size had different pin heights:

Low–performance	Pin Ht.
396,402,427	1.770
454	1.640
366T (4–ring)	2.170
427T (4–ring)	2.160
High–performance	
396,427 1968 only	1.760
396,402,427	1.765
454,502	1.645

gaps when the motor is freshened. (This requires that you keep accurate records of the actual clearances given when the motor is first assembled.) If there is no sign that the rings gap ends have been rubbing or butting against each other, you can reduce the gap .001" from the previous clearance, but you have to be willing to risk the motor if you go too far. If cylinder temperatures are going to be high such as in a high compression or nitrous motor, blower engine, or with hypereutectic pistons, top ring end gaps will have to be larger. Use of alcohol or nitro fuels will also require changes in minimum ring end gap. Take a look at the accompanying chart from Speed-Pro concerning ring end gap clearances and always follow the manufacturer's recommendation for minimum ring end gaps.

For a street operated gasoline fueled engine you don't have to get fancy or expensive rings. A pre-gapped moly faced top ring, pre-gapped cast iron second ring and a standard tension oil ring assembly will do just fine and give excellent service. Minimum ring end gaps for a street driven engine can be larger and more conservative. Top ring gap should be 0.018"-0.022", second ring gaps should measure 0.012-.014" and the oil ring support rail gaps should be 0.012-.014".

If you are using an effective form of crankcase vacuum extraction system on a race motor, you can move to a low tension oil ring setup and free up a little more power by reducing oil ring friction on the cylinder walls.

BIG BLOCK Quick Facts & References

MAXIMUM AVERAGE PISTON SPEEDS

Here is a listing of average maximum piston speeds for various parts combinations. A conservative maximum average speed limit of 4500 feet per minute (fpm) or less should be used. On some short duration serious drag race motors, USING PREMIUM AFTERMARKET ROTATING PARTS that are lighter and stronger than stock parts, the maximum average piston speeds can reach 6000 fpm. Using the 4500 fpm figure will help you make other decisions concerning parts selections that will operate within the rpm range of your engine. Keep in mind that the maximum average piston speed has NOTHING to do with where in the rpm range your particular motor will make its best power. This is a calculation which will help you determine the reliability and longevity of a specific parts combination.

The formula for determining maximum average piston speed is

$$\frac{\text{CRANK STROKE x RPM}}{6}$$

The formula for determining the engine rpm at a maximum average piston speed of 4500 fpm is

$$\frac{27000}{\text{STROKE}}$$

The formula for determining the engine rpm at a maximum average piston speed of 5000 fpm is

$$\frac{30000}{\text{STROKE}}$$

To find the engine rpm at a maximum average piston speed of 5500 fpm the formula is

$$\frac{33000}{\text{STROKE}}$$

The formula for rpm at 6000 fpm maximum average piston speed is

$$\frac{36000}{\text{STROKE}}$$

	RPM WHERE PISTON SPEED WILL AVERAGE			
STROKE	4500fpm	5000fpm	5500fpm	6000fpm
3.31"	8157	9063	9970	10876
3.50"	7714	8571	9429	10286
3.60"	7500	8333	9167	10000
3.76"	7181	7979	8777	9674
4.00"	6750	7500	8250	9000
4.125"	6545	7273	8000	8727
4.25"	6353	7059	7765	8471
4.50"	6000	6667	7333	8000
4.60"	5870	6522	7174	7826
4.70"	5745	6383	7021	7660

Be cautious! Use the 4500 fpm average piston speed limit or less. This will help the rotating assembly to live longer and reduce the amount of stress on the parts. Going above 4500 fpm requires lighter, much stronger premium aftermarket parts that can take the amount of stress and force encountered at higher levels in a short duration race motor. The rotating assembly in a big block Chevy is relatively heavy compared to small block motors. Keep these piston speed limits in mind when you are making your parts selections. Also note that MAXIMUM AVERAGE PISTON SPEED is not the same as PEAK PISTON SPEED.

SETTING PISTON RING END GAPS

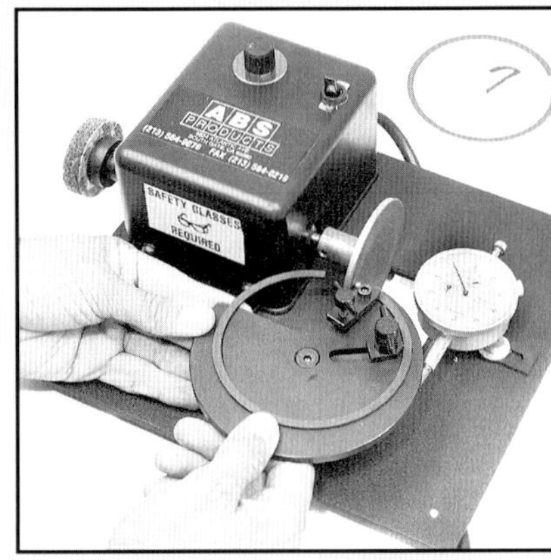

ABS Products ring gaping tool uses a rotating "table" and a dial indicator that makes gapping rings a precision process. Using this tool, it's easy to remove exactly the right amount of material.

Properly setting the end gap on oversize (gapable) ring sets can be a tedious, error-prone process. However, done precisely, it almost certainly will improve horsepower, reliability, fuel economy, and extend engine life. Done incorrectly, it can take the fine edge off an otherwise excellent engine or, in the worst case, result in severe ring damage and greatly shortened engine life.

Properly gapping rings places two demands on the engine builder: 1) The precise amount of material (exact to the thousandths of an inch) must be removed from the end of each ring (refer to the chart on page 69 for end-gap recommendations), and 2) the resulting gap must be parallel and have no burrs or sharp edges that can scratch the bore. Until recently there were two common ways to gap rings, both of them less than satisfactory. Filing while holding the ring in a vise is one common practice. So is using a simple cutting-wheel tool made by several manufacturers, including Perfect Circle®, K-D® and others. Both of these methods require substantial experience and manual dexterity. And even with experience, it's easy to ruin a ring or two (luckily most machine shops will sell individual rings at a reasonable price).

Thanks to ABS Products® (714-671-0798) there is another alternative. This tool uses a rotating "table" and a dial indicator that makes gapping rings the precision process it should be. In fact, using the ABS tool, it is possible for an experienced builder to gap all 16 top and second rings in about 15 to 30 minutes. And with only a little care, each ring gap will be exactly the right width and dead parallel.

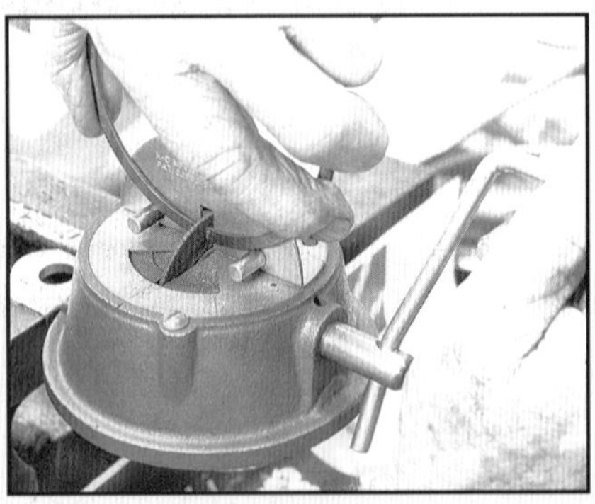

This simple cutting-wheel tool made by Perfect Circle, K-D and others requires a good deal of practice and is still good at "chewing up" rings if you aren't careful. Luckily there are other alternatives.

However, the ABS tool is rather expensive. Fortunately, nearly 500 machine and engine shops now use ABS gapping tools. We strongly recommend that you have your rings gapped by one of the shops that uses this tool. Since it takes them so little time to gap a set of rings, the cost should be reasonable.

If you are determined to do it yourself (and you don't want to buy the ABS tool), consider the Huggins® E-Z Gap® ring support and gapping tool as an alternative to the "K-D" type cutting wheel. Available from Taylor Engines (310-698-7231), the Huggins E-Z Gap is an inexpensive tool that includes a file, feeler gauge, and instructions. While it does require learning the proper "technique," you may have better luck with this tool than other low-cost methods.

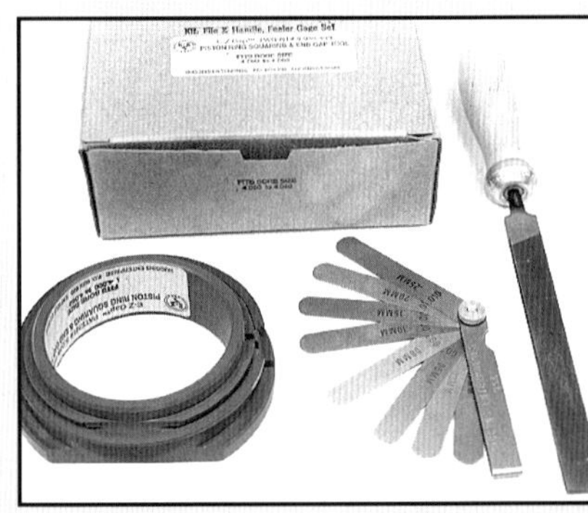

Consider the Huggins ring gapping tool. This inexpensive kit requires mastering the proper "technique," but some have had better luck with this tool than other basic methods.

OPTIMUM END GAP POSITION

Install all piston rings with end gaps staggered as shown here. With properly gapped rings and properly honed cylinder walls this procedure will provide the best cylinder sealing and most efficient oil control. Install rings on pistons with the proper expander tool whenever possible.

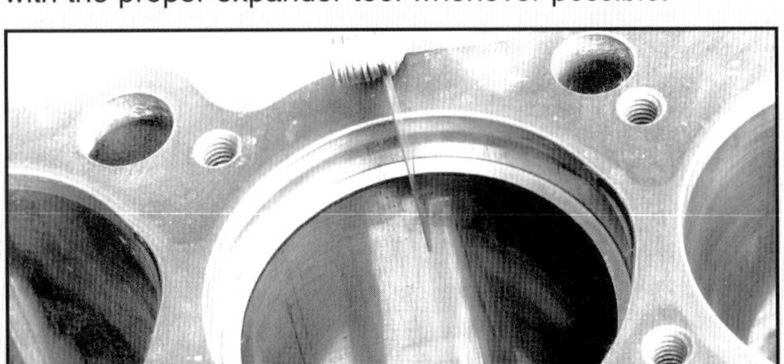

Position the ring down squarely in the cylinder bore and measure the ring gap with a feeler gauge.

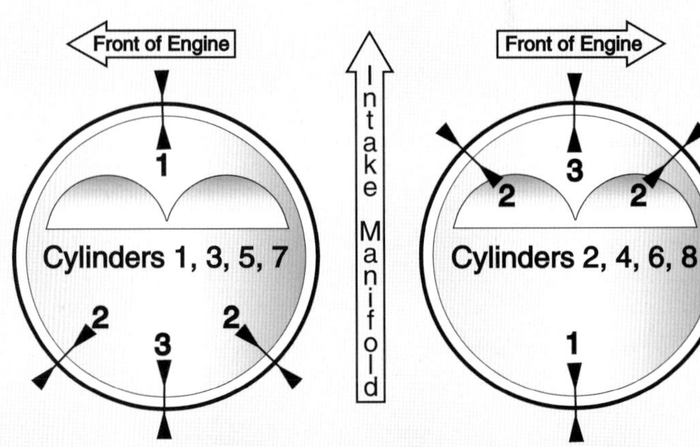

1—Oil Ring Expander And Top-Ring Gaps
2—Oil Ring Scraper-Rail Gaps
3—Second-Ring Gap

BIG BLOCK Quick Facts & References

Piston Skirt Clearances			
Piston Source	Piston Type	Street And Performance	Competition
Chevrolet	Cast Aluminum	.0015-.0025	Not Recommended
Chevrolet	Forged	.004-.006	.0055-.0065
Chevrolet	Hypereutectic	.0015-.0020	Not Recommended
TRW/Speed Pro	Hypereutectic	.0015-.0020	Not Recommended
TRW/Speed Pro	Forged	.003-.0055	.0055-.0065

Setting piston-to-cylinder wall (skirt) and end-gap clearances are critical steps in building a reliable big block. Use these charts as a starting point, and discuss your specific application with your machine shop.

Ring End-Gap Clearance Recommendations			
For 5/64 & 1/16 Ductile Iron Top Rings & .043 & .031 Pressure Back/Head Land Top Rings		For Regular (Plain) Iron Second Rings	
Using Special Fuels:			
Supercharged/injected nitro engines	.022/.024	Supercharged/injected nitro engines	.014/.016
Supercharged/injected alcohol engines	.018/.020	Supercharged/injected alcohol engines	.012/.014
Using Gasoline:			
Supercharged/injected gasoline engines	.022/.024	Supercharged/injected gasoline engines	.012/.014
Oval track rectangular rings	.018/.020	Oval track	.012/.014
Oval track head land rings	.024/.026		
Oval track pressure back rings	.020/.022		
Pro stock, comp eliminator, modified production, etc.	.018/.020	Pro stock, comp eliminator, modified production, etc.	.012/.014
Super stock	.018/.020	Super stock	.010/.012
Stock	.016/.018	Stock	.010/.012
Street engines	.016/.018	Street engines	.010/.012

Note: These Speed Pro end gap recommendations will be appropriate in most cases. They stress that this chart should be used as a guide to normal ring fitting, and that the ideal end gap may be different for specific combinations. If you're ever in doubt, check with other engine builders or add .002-inch to the recommended gap. Racers with plenty of experience often close up the gaps rom .002- to .004-inch, but you have to be willing to sacrifice a motor if you guess wrong.

BIG BLOCK Quick Facts & References

HIGH PERFORMANCE COATINGS

There are a number of high tech coatings available on the market for various parts of the motor. Pistons, combustion chambers, valves, valve springs, other rotating parts and exhaust headers can be coated with materials that reduce heat absorption or friction, and/or shed oil quickly. There has been some discussion of increased power, but these power gains seem to be marginal. The major benefit seems to be an increase in the durability and life of the coated parts. High tech coatings on engine bearings can help protect the bearing when oil pressures fall because of oil pump aeration. High temperatures in piston decks are reduced and so fatigue failures are reduced. Piston skirt coatings can reduce friction, so wear on the part is minimized. Valves faces and stems can be coated to reduce heat absorption and erosion. Coated valve stems have reduced wear and less tendency to seize and gall in the guides. Valve springs seem to last longer before the open pressures decline, since heat in the springs is reduced. Heat induced fatigue failures also decline. This is an area that will see a lot of advances over the years as new materials and application techniques are discovered and improved.

Dry film, liquid and sprayed on molten coatings are available. Not all of them penetrate into the metal itself. Some of the coatings are intended for non-ferrous aluminum parts such as pistons or aluminum heads, while others are used on ferrous materials like cast iron heads. Most form an extremely thin layer (some .0003-0005", others .001-.002") on the top surface of the coated material. The thickness of these layers may require that you take this into account when you are setting clearances on the motor. Check with the coating manufacturer for his recommendations.

BIG BLOCK Quick Facts & References

PISTOM DOMES & DISH VOLUMES

You can usually get information on the amount of piston dome and dish volumes from the piston makers. However, sometimes you don't or if the pistons have been machined for some reason (like a reduction in dome height or the deepening of valve reliefs) after they have been bought, you will need to measure their present volumes in order to determine the static compression ratio of your parts combination. It's easy. Put a piston with a top ring on it into a cylinder. Put a little petroleum jelly or white grease on the top ring so it will seal. Push the piston deck, (not the dome or dish) down in the cylinder exactly one-half inch or exactly one inch from the top of the block deck to the top of the piston deck. Make sure you wipe off any excess petroleum jelly or grease from the cylinder walls. Use a plastic plate to seal the top of the cylinder barrel. Now use your large volume syringe or burette and measure how many cc's of solvent or other liquid it takes to fill the cylinder barrel to the top of the block deck. When you finish and you know how much solvent was used, you can do the following calculations.

Determine how many cubic centimeters are in the cylinder without the dome or dish. You do this by taking the depth that you placed the piston deck in the bore, in this case either 1/2" or one inch down from the block deck and then multiplying the bore diameter and the decimal factor ".7854".

TOTAL VOLUME = Bore x Bore x Amount in Inches the Piston is placed down the bore x .7854
TOTAL VOLUME = 4.280" x 4.280" x 0.5" x .7854
TOTAL VOLUME = 7.1936356 CUBIC INCHES

Now convert that cubic inch sum to cubic centimeters by multiplying it by 16.387.
7.1936356 cubic inches x 16.387 = 117.8821 cubic centimeters (cc)
Now you can compare this calculated cc figure to the amount of volume in cc's you actually found from your cylinder measurement. The difference between the two figures is the volume amount of dome or dish or valve relief your piston has.
It takes less time to do if you use the "one/half inch down" measurement. But if you have an especially large, tall piston dome, it may be better to use the "1 inch down" amount.

DETONATION and PREIGNITION

One of the major things to avoid in any engine is detonation and pre-ignition. Maximum power is decreased and engine damage will happen if these problems occur, sometimes even for only a few seconds. When a gaseous mixture of air and fuel are compressed, the pressures increase and the temperature of the gaseous mixture will rise. In any performance engine, increasing these pressures are what you want to do since this has a direct effect on how much torque a motor will make. Best power is usually found at cylinder pressures and temperatures just below the detonation level. However, if pressures increase too much or too quickly or if the temperatures get too high, the gaseous mixture can be self ignited. The idea is to reach high cylinder pressures without detonation happening. Increasing the dynamic compression of any motor increases the mechanical efficiency of the engine. You want the mixture to be ignited by the spark plug at the optimum moment to produce the best power. In the effort to reach these maximum power goals race motors are often run right at the edge of detonation and sometimes the line is crossed. If you are lucky the motor just begins to miss and lies down on power. If you are unlucky, parts break or melt down and you get an expensive lesson in humility and the destructive power of the internal combustion engine.

Many things can cause detonation such as using too high a compression ratio without adequate fuel octane levels. An example would be someone who built a 12.5-1 static compression ratio motor and then tried to run the engine on 88 octane pump gasoline. Or someone who didn't match the camshaft to the dynamic compression ratio and built more cylinder pressure than the octane level of the fuel could handle. An inadequate cooling system that allows the coolant temperature to rise above normal, will allow the cylinder head temperature to rise because they can't be cooled quickly enough. The additional heat is not rejected from the heads to the cooling system and the increased temperatures prematurely ignite the air/fuel mix. Hot spots in the piston deck surfaces or dome or combustion chamber can get hot enough to ignite the mix. Carbon deposits or sharp edges on piston domes, valves or spark plug threads that protrude into the combustion chamber can get hot enough to glow red and set off the incoming fuel charge. Spark plugs that are the wrong heat range are not able to convey heat to the engine coolant quick enough so heat continues to build in the center electrode or the side ground wire. This can also lite off the fuel. An air/fuel ratio that is too lean will cause rapid cylinder temperature increases and once again the mixture can go off by itself. Ignition timing that is too far advanced for the particular engine combination can cause a preignition problem.

Improper piston dome shapes and/or too much piston deck to block head deck height can create small spaces that prevent the smooth travel of the flame front across the piston surface. Weird cylinder pressure spikes can be caused by the burning of some of the fuel in one part of the combustion chamber followed by a separate burning in another part of the chamber.

Oil that gets past piston rings or valve stem seals to contaminate the fuel mixture in the combustion chamber can also cause a detonation problem.

Detonation will probably not occur in every cylinder at one time. You may have one cylinder that does not receive the same amount of fuel as the other cylinders and thus run leaner, which is a common problem with big block Chevys because most factory production style heads have two long intake ports and two shorter port runners. Unequal fuel distribution can be a problem. Coolant circulation on high power motors can also be a problem if one or more cylinders run hotter than others and might be caused by a water pump that gives unequal or inadequate coolant distribution to each side of the motor.

Too much nitrous or improper nitrous/fuel/air ratios can cause cylinder pressures and temperatures to rise so much that detonation occurs. Too much blower boost will result in the same thing.

Remember also that the temperature of the air being ingested into the engine can have an effect on the motor temperature. Using hot air from inside the engine compartment should be avoided if possible. Feeding an engine cooler outside air either with scoops sealed to the top of the carb or ducting cooler outside air to the induction will help. Intercoolers on turbos or blowers will lower temperatures. Adding cooler induction air will have the added benefit of making more engine power. Nitrous oxide or methanol fuels can have a cooling effect on the incoming intake charge.

Aluminum heads offer some advantages over a cast iron head. One is that aluminum heads reject heat more quickly than a cast iron one. Head temperatures are usually lower and this reduces the chances of detonation. Static compression can be increased a bit before detonation again starts to be a concern when compared to the same motor with cast iron cylinder heads.

Some of the new high tech coatings applied to pistons, valves and combustion chambers can help reduce "hot spot" temperatures and keep detonation from happening in addition to reducing heat induced fatigue in parts.

PISTON DECK SURFACE AREA TO CRANK STROKE RATIO

PISTON DECK SURFACE AREA TO CRANK STROKE RATIO

One day I was sitting around and I started thinking about the relationship of bore size and the area of piston deck surface to crank stroke length. Why do some bore and stroke combinations seem to work out better than others? There is some relationship between the bore and stroke size and induction and exhaust systems that each combination requires and I was trying to figure it out. One of the things I did was to take the total surface area of the piston deck and divide it into the crank stroke length. I was trying to get a ratio for each and then I stared at them for a while trying to draw some conclusions.

ENGINE	BORE & STROKE	PISTON DECK SQ. INCHES	RATIO	RPMs AT 4500fpm AVERAGE PISTON SPEED
302 Chevy SB	4.00 x 3.00	12.5664	4.19-1	9000
327 Chevy SB	4.00 x 3.25	12.5664	3.87-1	8307
350 Chevy SB	4.00 x 3.48	12.5664	3.61-1	7758
377 Chevy SB	4.155 x 3.48	13.5591	3.90-1	7758
383 Chevy SB	4.030 x 3.76	12.7556	3.39-1	7180
408 Chevy SB	4.155 x 3.76	13.5591	3.61-1	7180
396 Chevy BB	4.096 x 3.76	13.1768	3.50-1	7180
402 Chevy BB	4.125 x 3.76	13.3640	3.55-1	7180
427 Chevy BB	4.250 x 3.76	14.1862	3.77-1	7180
447 Chevy BB	4.350 x 3.76	14.8617	3.95-1	7180
454 Chevy BB	4.250 x 4.00	14.1862	3.55-1	6750
472 Chevy BB	4.470 x 3.76	15.6929	4.17-1	7180
475 Chevy BB	4.350 x 4.00	14.8617	3.71-1	6750
499.9 Chevy BB	4.600 x 3.76	16.6190	4.42-1	7180
499.6 Chevy BB	4.700 x 3.60	17.3495	4.82-1	7500
502 Chevy BB	4.470 x 4.00	15.6929	3.92-1	6750
509 Chevy BB	4.500 x 4.00	15.9043	3.98-1	6750
526 Chevy BB	4.440 x 4.25	15.4830	3.64-1	6352
541 Chevy BB	4.500 x 4.25	15.9043	3.74-1	6352
557 Chevy BB	4.440 x 4.50	15.4830	3.44-1	6000
598 Chevy BB	4.600 x 4.50	16.6190	3.69-1	6000
612 Chevy BB	4.600 x 4.60	16.6190	3.61-1	5869

It would seem that if your intentions are to build a high rpm, high horsepower big block race motor, a large bore and short stroke seem to be the way to go. Some sacrifice in low and mid range torque would be made. On the other hand the long stroke motors give big jumps in torque, but have limited high rpm range.

There are many other factors to consider such as rod length, cylinder head flow, intake flow, exhaust flow, valve size, cam timing and their effect on the total engine package. I'm still trying to find that balance between all of the factors and then reduce it somehow to some sort of "global" equation or guideline. Maybe this will give you something to ponder.

Two things need to be added to the above. First, some of the above combinations can only be built using tall deck truck, Bowtie, Olds DRCE or aftermarket blocks with enough block deck height and bottom end clearance for large stroke motors. Second, the 4500fpm average piston speed limit requires high quality rotating parts. This average piston speed can be exceeded in some short duration drag race engines when using the strongest and lightest parts available. Certainly with high quality, premium aftermarket parts 6000 fpm average piston speeds can be tolerated for short periods. In a 499.90 cubic inch Pro Stock drag race engine with a 4.600" bore and 3.76" stroke, at 9574 rpm, the average piston speed is 6000 fpm. With a 3.60" stroke the Pro Stock piston hits an average of 6000 fpm at 10,000 rpm. You better believe they are using the best stuff they can find and cost is not a consideration. Peak piston speeds will always be higher than the average piston speed at any given rpm. Depending on the stroke and rod combination, peak piston speed will be reached somewhere between 75-90 crank degrees After Top Dead Center. Past that moment of peak piston speed the piston is actually slowing in speed even though it is still on its way to the bottom of the stroke.

Oil Lubrication

The oiling system in a Mark IV big block remained the same from 1965 thru 1990, however in 1991, the Gen.V motors and later Gen.VI HO blocks made changes in the path that oil takes as it circulates thru the engine. The main oil gallery in Gen.V and Gen.VI blocks runs along the camshaft tunnel, while the main oil gallery passage in the Mark IV blocks runs along the driver's side block oil pan rail. Keep this Mark IV main oil passage in mind if you are grinding on the oil pan rails for additional clearance for stroker cranks. The Mark IV rear main caps are different than the rear main caps of the Gen.V and VI which were changed to accommodate the new style one-piece rear main seal and crank rear main journal. The Gen.V and VI rear main cap also has different bolt boss heights, but the oil pump mounting surface on the rear cap is the same as that found on the Mark IV.

OIL PASSAGES AND PUMPS

Blocks made in 65' and 66' need an oiling groove in the rear camshaft journal and the rear cam bearing itself must also have a matching oiling groove. Blocks made in 1967 and later do not use this system.

Oil pumps and their intermediate drive shafts, from Chevrolet, are interchangeable between Mark IV and Gen.V motors. A number of performance pumps are available, LS-6 standard volume (PN-475908) with 1.1" pump gears, and LS-7 high volume pump (PN-3969870), and Gen.V high volume pump (PN-14097007) with taller 1.3" pump gears.

If you use a Mark IV pump on a Gen.V block and then use a stud to hold the oil pump on there may be a clearance problem that requires you to grind a little material off of the pump body.

Make sure that the oil pump and pickup screen you use will work with

Front of the Gen.V "182" block shows redesigned main oil passages and plugs and repositioned threaded holes for cam thrust plate.

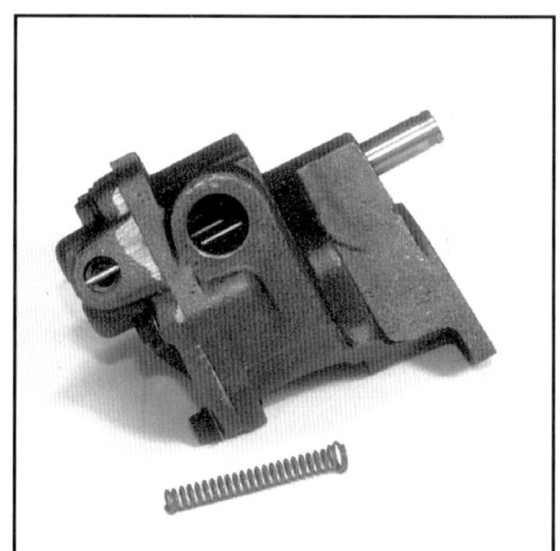

Stock big block pump should have the gear clearances checked and the pickup screen welded or bolted to the pump body. Stock oil pressure can be increased by using a stiffer pressure relief spring or by shimming the stock spring. Use an intermediate oil pump driveshaft with a metal coupler and make sure that whichever pickup screen you use that it matches the oil pan and sump depth.

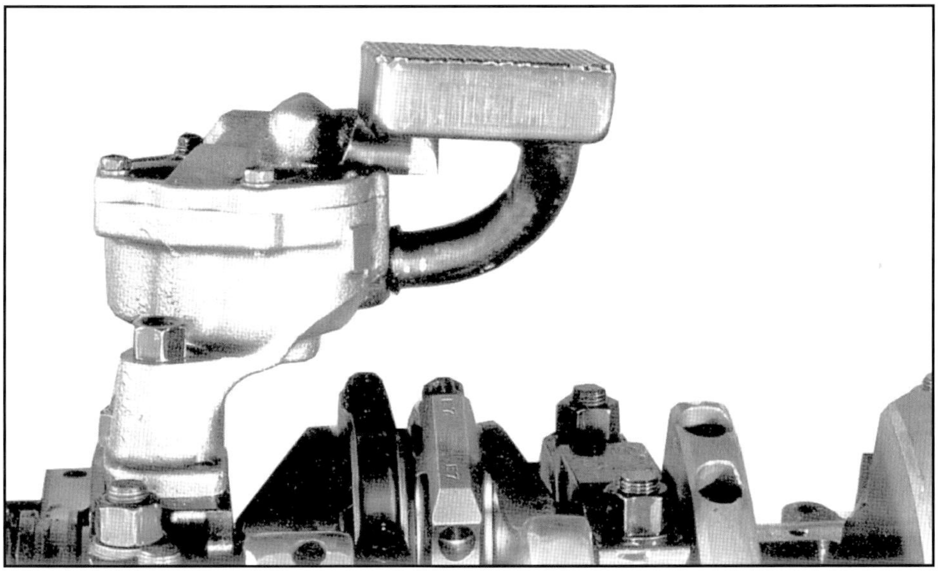

What kind of oil pump do you need on a 557 cubic inch Super Comp Bowtie? A stock pump will do just fine. Make sure the pump gear end clearances are correct and bolt it on. The aftermarket screen pickup bolts to the pump. Make sure you either weld the pickup to the pump or get one that bolts on.

the oil pan you have chosen. The pickup screen should be approximately 1/4"-1/2" off the pan floor. You can drive the screen and tube assembly on to the oil pump by using a 3/4" open end wrench or you can use a great tool made by Melling (PN-13075). In addition, the screen pickup tube should be tack-welded to the oil pump housing or use a pickup screen assembly that has a bracket that bolts to the pump cover or one of the rear main cap bolts. Make sure you take out the oil pressure relief spring before you weld the pump body to the screen tube.

The intermediate drive shaft for a small block Chevy V-8 is shorter than the drive shaft for a big block. Do not try to interchange these two shafts. Chevy has an intermediate oil pump drive shaft with a heavy duty steel coupler attached as (PN 3865886) and this shaft should be used any time you use a Mark IV style high volume oil pump. The intermediate oil pump drive shaft for Gen.V motors is (PN-3860365) and uses plastic sleeve coupler (PN-3764554). If you use a high volume pump make sure you use an intermediate shaft with a steel coupler.

Many aftermarket standard volume replacement oil pumps come with an additional pressure relief spring which is stiffer than the stock one. This will allow you to increase the pump pressure if you want to without having to use shims. Aftermarket pumps also can have additional features like anti-cavitation designs to improve very high rpm operation and adjustable pressure relief valves. If you are planning to use an especially deep pan, you may need an oil pump extension. This will move the pump lower in the pan without having to use a long screen pickup tube.

Don't bother putting a used oil pump back into a rebuilt motor. Prices for new stock oil pumps run from $20.00-35.00. A new one is cheap insurance.

Whichever oil pump you use check the end clearance between the gears in the pump cover. Try to get the end clearance to .025"- .0035". If the clearance is bigger than this, you can sand the pump body down. If the clearance is smaller than the recommendation, then sand down the pump gear ends. You can do this by putting down a sheet of 250 or 400-grit sand paper on a sheet of glass and moving the parts in a figure eight pattern over the paper.

Alright, so what is the oil pressure you should look for? One rule of thumb is that you should have 10 lbs. of hot oil pressure for every 1000 rpms. If the motor will be spun to 6000 rpms, then 60 lbs. of pressure will be fine. You should realize however, that this is dependant on other things beside the amount of pressure at the pump. If the bearing clearances have been opened up beyond the recommendations, pressure from a given pump may be lower than expected. Different oil viscosities will also change the pressure reading at the

You can plug the bypass valve in a oil filter adapter, but make sure that the oil filter you then use is large capacity and has a high flow capacity. This should only be considered on a race motor.

Fram "PH-373" two-quart spin on oil filter will give your oil system increased capacity. Many filter companies make similar two-quart truck style filters. These are helpful if you have blocked the oil bypass valve in the oil filter/block adaptor to insure full oil filtration. If you do block this bypass valve, don't rev the engine up, while the engine and oil are cold.

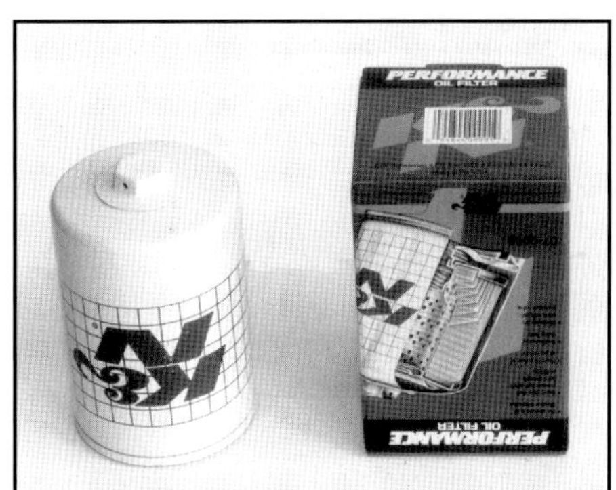

K & N oil filters will flow large amounts of oil, have high burst strength metal covers and filter out very small particles. Your motor will love them.

gauge. Also we are talking about hot oil pressure readings. When the motor is first fired up and the oil is cold and thick, the oil pressure may be much higher and maybe even peg your gauge. The pressure relief spring in the oil pump itself may open up at 65 lbs. and limit the total amount of pressure. Keep in mind also that the stock oil filter block adapters have a built in oil pressure relief valve that will allow oil to bypass the filter at pressures as low as 15 lbs. sending unfiltered oil back into the motor. Some folks plug or epoxy over the oil filter bypass valve to ensure that they get 100% filtration of the oil. If you do block this bypass valve make sure you run the biggest, highest flowing oil filter you can, (AC# PF-932 or Fram PH-373 are two-quart heavy duty truck spin on filters or one of K&N's excellent high flowing oil filters) and don't rev up the motor until the oil has warmed up. Otherwise you may blow the oil filter right off the motor. Moroso and K & N make high flow race oil filters with higher burst strength spin-on covers.

There is some advantage to heating the oil in the sump before starting a cold race motor. Moroso and a number of other companies offer a heat pad which can be stuck to the outside of the oil pan and runs off of electrical current from a remote battery or portable generator. Oil temperatures in the pan will be more consistent.

MINERAL OR SYNTHETIC BASED MOTORS OILS

There are a huge amount of companies selling motor oils these days. Each claims that theirs is best and has qualities not shared by their motor oil producing competition. It can all get pretty confusing to someone like me. I can tell you a few things. Mineral based oils have various additives that enable them to withstand most race situations. Synthetic oils also have a number of chemical additives which make them suitable for racing, and they have the ability to withstand higher operating oil temperatures without literally cooking. The synthetic oils will also withstand higher bearing pressures at lower viscosities, than mineral oils will. On the other hand the synthetics cost more, a lot more per quart than mineral oils. In an attempt at price compromise, there are some oil companies that are offering motor oils that are half synthetic/half mineral oil. There are some frictional losses in any engine and the synthetics seem to provide some help in reducing these losses. Being able to use a lower viscosity synthetic oil and still have the ability to stand up to race engine environments will make things easier on the oil pump. If the oil temperature happens to go over 240 degrees the synthetic will not break down and so there is a safety margin.

Performance motor oils are available as mineral based, a blend of mineral based and synthetic oils and 100% synthetic oil.

When all is said and done, I think the synthetic oils provide an extra measure of performance and wear protection, not found in mineral based oil. However, most name brand mineral oils will work in most street and bracket race situations just fine if you keep clean oil in the motor and change the oil filter routinely.

Don't use a synthetic oil when an engine is first fired up. Use a mineral based 30W oil. After the engine is broken in, change the oil and filter and use the oil you want to. Its a good idea to drain the oil out of the pan and thru a paper paint screener/strainer. Any objects that are found in the strainer, other than oil, can alert you to any problems. There are a number of other ways of looking for debris. You can cut open oil filters to look for anything unusual. There are little aftermarket screen filters being sold that mount between the oil filter canister and the block or other screen types can be placed in oil lines. Small magnets can be epoxied into the areas around oil drain holes in the rocker valley of heads and in the lifter valley of blocks. All of these things are early warning tools, but they must be checked frequently to be of any use to your engine and you.

WINDAGE

When the crank is spinning, oil is picked up by the crank and clings to it. The faster it spins the more oil gets entangled in it. You would think that oil gets thrown off the crank by centrifugal force, but in reality it doesn't. This is oil that is draining back to the sump from above the crank, but also includes oil bouncing off the pan walls back onto the crank. The oil clinging to the crank creates drag and results in measurable power losses at rpms generally over 6,000. Its like running along a beach. Its easier to run above the water line. If you try to run in the water, its harder. As you go deeper in the water and try to run, the effort is greater. If your motor is going to be spending time above 6000 rpms, then some additional power can be recovered if you try to reduce the amount of oil that can cling to the crank. There are a number of ways to do this.

One way is to lower the oil level in the pan in relation to the crank. A deeper pan or one with a wider, inverted T-shaped sump will move the oil surface further away. This may only give partial help because many front suspension crossmembers limit the front sump depth of the pan. Some of the crank throws may still come close to the pan in the forward half of the pan. Ground clearance may also be a problem if the pan sump you choose is too deep. You gain the most advantage in a situation where the pan sump is of equal depth from the front to the rear, but you may only be able to do this in a dragster or in a vehicle with an aftermarket tube frame and front suspension that is out of the way or in a boat. Front end suspension clearances may force you into a compromise situation. In any case, measurable amounts of power can be lost to windage above 6000 rpms. Anything you can do to reduce this will recover power lost to drag.

Pan baffling can help control the movement of oil in the pan while the vehicle is accelerating, decelerating or turning. Not only will baffling keep oil around the pickup screen, some types of baffling will keep sloshing oil from climbing the walls of the pan and getting picked up by the crank. Some pans have "kickouts" which are on the passenger side of the pan. The kickout catches oil scraped off of the spinning crank and forces it to slide down the side of the pan and into the sump area without bouncing back onto the crank. Crank scrapers help to do this job. Mounted on the passenger side of the pan or on the pan rails, scrapers literally scrape oil off of the spinning crank counterweights. Windage trays also help reduce the amount of oil on the crank. These trays can be solid with louvered slots, screen types or unidirectional screen types.

If you are using needle bearing roller rocker arms, which can get by with reduced lubrication, you can restrict the amount of oil going to the rocker arms with screw-in restrictor plugs in the oil passages or use pushrods or lifters that restrict the amount of oil. This in turn reduces the amount of oil returning to the sump and getting picked up by the crank as windage. Be careful here. If you restrict too much oil to the top there may not be enough to cool valve springs or the rocker arms. Check with your cam maker on which forms of oil restriction you should use. When you restrict the amount of oil to the roller rocker arms there is a reduction in the amount of oil available to cool the valve springs. Since the valve springs don't get as much oil splashed on them they may run hotter and their useful life span will diminish. Valve spring coatings can help the springs live longer. Chevy has reintroduced a short valve cover with oil drippers built into the roof of the cover. These help keep oil on the springs. If you need a tall cover, Competition Cams has introduced a "Oiler" valve cover which can be hooked up to a pressurized oil line. The inside of the cover has tubing and nozzles that direct a spray of oil onto the springs and help keep them cooler. A similar attempt has been made by T & D Machine which has a line of roller rocker arms which have a pressurized oil passage that exits and sprays oil directly onto the valve springs.

Work can also be done to the crank counterweights themselves usually called "knife-edging" or "wing contouring". The leading edges of the counterweights can be ground and cut in a way to not only reduce clinging oil but to direct the oil towards the main journals and away from the rod journals.

Another way to reduce windage is to keep the main and rod bearing clearances on the tight side rather than running them on the loose side and to have the proper rod side clearances.

WET SUMP OIL PANS

Different volume oil pans are offered by Chevy and from aftermarket suppliers. Be sure that the oil pickup tube and oil pump screen match the pan depth you choose. Also be aware that some pans will not fit in your particular vehicle because of clearance problems with the front crossmembers, steering boxes or steering arms, starters or exhaust pipes. This is especially true in early Chevelles and Camaros. Also if the ride height of your vehicle is very low, a "deep" pan may hit the road surface on bumps or while entering driveways or on a bumpy return road at the track. If you have a marine application, dragster or enough cross member/steering arm clearance, then an aftermarket full length sump pan can be used. Otherwise some compromise has to be made to fit the oil pan in your application. The sump can be deeper or be "T-shaped" in order to increase the depth and move the oil level further away from a spinning crank.

Oil pans for a Mark IV block will not interchange on to a stock production Gen.V or Gen.VI engine block and visa versa.

A four quart oil pan (PN-14081027) for the Mark IV 454 LS-6 engine assembly will fit early model Camaros and Chevelles, but can not use a factory windage tray. A five quart oil pan

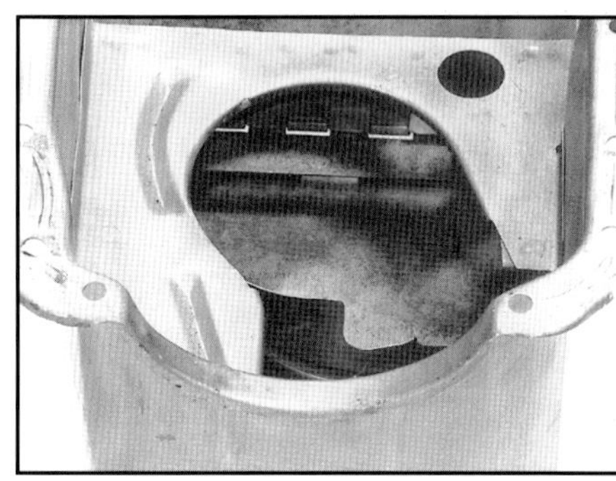

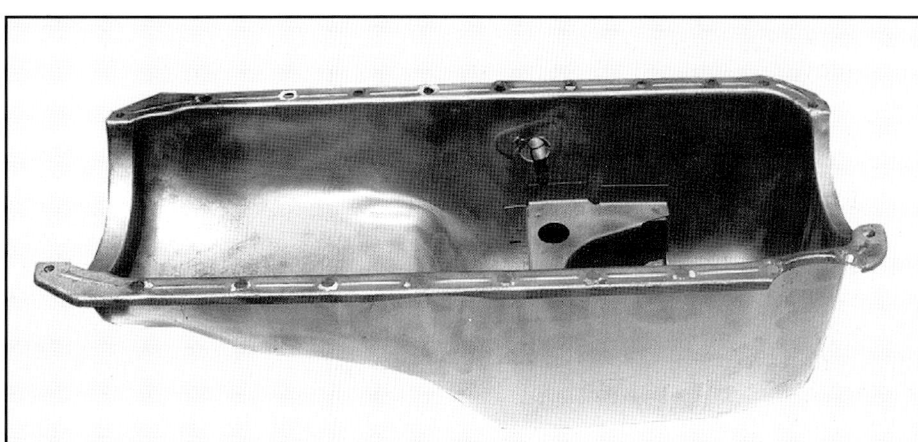

Mark IV LS-7 5-quart oil pan has good baffling and a trapdoor to keep oil around the screen pickup. You can fit a windage tray with this pan.

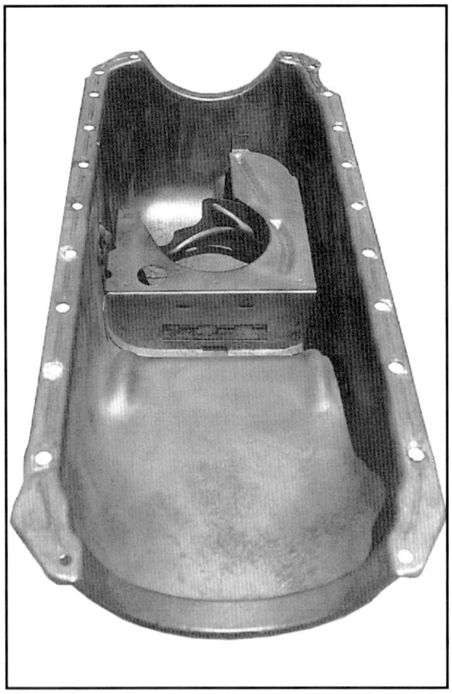

The Mark IV LS-7 6 quart pan was designed for high performance use.

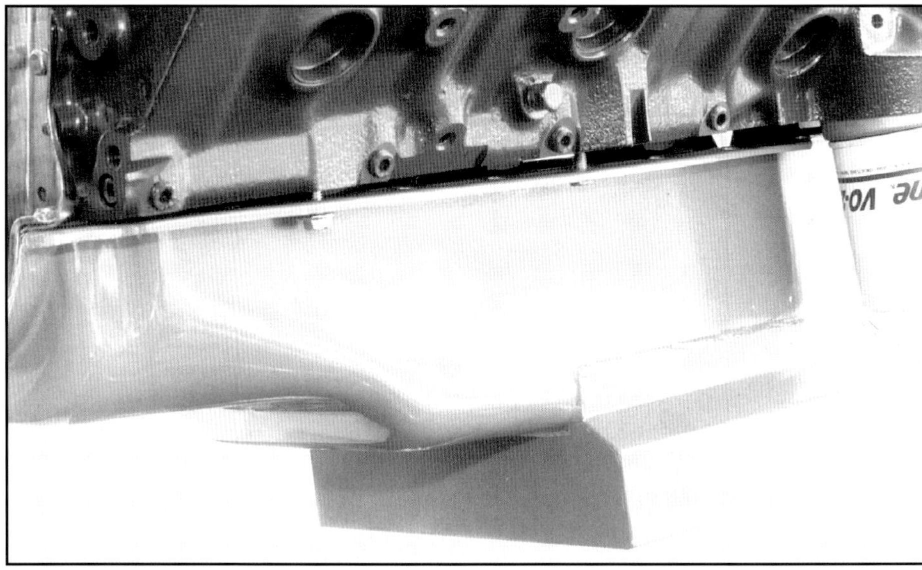

This type of oil pan increases sump volume and depth, but will still fit most vehicle applications.

that was used on the Mark IV LS-7 crate motor and big block Corvettes is available as (PN-14091356). This pan has a trap door baffle and a deeper and longer sump to aid in oil control. Six and seven quart oil pans were used on some Mark IV truck engines. Windage trays can be fit to the five quart and larger production oil pans.

Mark IV oil pans will fit Gen.V CNC "Race Prepared Bowtie" blocks, (PN-24502500) short-deck and (PN-24502502) tall-deck, which have splayed steel main bearing caps and use two-piece rear main oil seals. Gen.V "Bowtie" blocks, (PN-10185049) short-deck and (PN-10134367) tall-deck, with "straight" main bearing caps, use the stock production Gen.V-style oil pans for one-piece rear oil seals.

The Gen.V six quart truck oil pan, used on 454 HO and 502 HO engine assemblies, is (PN-10198992). This pan will not fit early Chevelles or Camaros. (PN-10185076) is a Gen.V four-quart oil pan by Chevrolet for the Gen.V and Gen.VI motors. This Gen.V pan is similar to the LS-6 style pan, but is used with one-piece rear main oil seals and this oil pan will fit in an early Camaro or Chevelle with a Gen.V or Gen.VI production engine. Aftermarket suppliers are also making different pans to fit Gen.V motors into various vehicle applications. Gen.V pans will fit Gen.VI HO blocks which also use a one-piece style rear main seal.

A magnetic oil drain plug (PN-2311420) can help spot any problems inside your motor and keep metal particles from circulating.

Aftermarket oil pan suppliers offer a wide variety of choices for the Mark IV and Gen.V engines. Wet-sump oil pans for circle tracks, marine, drag racing and restorations are available with all sorts of internal baffling, "scrapers", "kick-outs" and oil capacities. Some pan kickouts may require that you use one of the “mini-starters” in order for things to fit. Some aftermarket race blocks can mount the starter on the drivers side in order to reduce pan clearance problems.

WINDAGE TRAYS

Solid windage trays with louvered slots are available from GM for either Mark IV (PN-3967854) or Gen.V, VI engines (PN-14097040). These trays help keep splashing oil off of a spinning crankshaft and rods, which in turn, will help make more power at higher rpms. These trays will not fit four-quart, stock passenger car, oil pans. They can be used on five-quart Corvette or LS-7 oil pans or on six- quart Chevy or GMC truck pans. The trays require that you use special main cap bolts and studs (PN-3902885) to mount the baffle. Windage trays made with screen material are available from aftermarket suppliers such as Moroso or Milodon.

Crank "scrapers" can be mounted in between the block oil pan

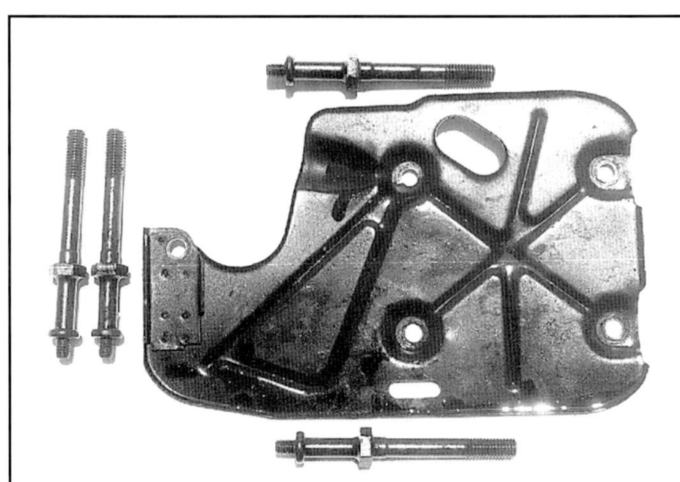

GM windage tray and the main cap bolts that are used to mount them. Other aftermarket trays feature louvers or screen material to help remove oil from the spinning crank and rods.

rails and the oil pan or mounted on the inside of the pan itself. Chevy does not offer these but aftermarket suppliers do and the "scrapers" also help keep oil from clinging to the spinning crankshaft in a race motor.

CRANKCASE EVACUATION SYSTEMS

When a piston is on its downward stroke in the engine it displaces the air that is beneath it in the crankcase. With seven other pistons doing the same thing and with the spinning crank and rods doing their thing, the air inside the crankcase, lifter valley and rocker arm covers can get very turbulent. Early engine design used a number of ways to help control these pressures, mainly by allowing excess air pressure to exit the crankcase spaces thru road draft vent tubes. Later Positive Crankcase Ventilation (PCV) systems allowed fresh filtered air to enter the crankcase thru a valve cover, while crankcase air was drawn out of the other valve cover by the use of intake manifold vacuum and then it was mixed in with the air/fuel in the carb and burned. Fresh air was always being drawn thru the crankcase, sludge build up was reduced so motors and oils stayed cleaner and lasted longer. Crankcase vapors and liquids were no longer being dumped on the road thru draft tubes. Oil leakage past various engine seals was reduced because the crankcase internal pressures were reduced.

However, some race engine builders realized that something else was going on. On the downward stroke of the piston enough positive pressure could build up so that the ring seal on the cylinder walls was being affected. In an attempt to create a negative pressure in the crankcase various crankcase vapor evacuation systems were designed. With these systems a negative air pressure can be created in the crankcase when the motor is running. Ring seal is improved and power goes up on the engine. Oil getting past the rings and into the combustion chambers is reduced or eliminated and a lower tension oil ring can be used effectively. Windage seems to be controlled a little better and with negative pressure oil seals on the crank seem to do a better job.

Most systems use hoses to connect the valve covers to the tube exhaust header collectors. When the rpms on the motor go up, the exhaust gases flow quickly thru the collector creating a lower pressure in the hose and this pulls crankcase pressures down. A one-way valve keeps the crankcase vapors moving in one direction and prevents an engine backfire from travelling up the hose to the crankcase. Oil baffles keep oil from being sucked out of the valve covers. A second connection can be made from one of these hoses to the intake manifold or carb PVC tube which will create a vacuum on the system when the engine is at idle or very low rpm speeds or when decelerating after a high speed run. An oil baffle should be used here also to prevent oil from getting into the intake manifold. Some racers have begun to use air pumps to create very low atmospheric pressures in the crankcase.

There is one thing to watch with these systems. If you install one, crankcase vacuum may get high enough to limit the amount of oil that gets past valve stem oil seals to the valve stem and valve guides. Some race engine builders don't use valve stem seals at all when using an effective crankcase vacuum system. Others put an oil seal on the intake valve only.

DRY SUMP OIL SYSTEMS

If you are bucks up and you are trying to get the last ounce of power from your engine and you have the clearance, dry sump oil systems may offer some advantages. Dry sump systems use external belt driven oil scavenge pumps and pressure pumps. A full length pan sump is used but it can be shallow which improves ground clearance, which in turn may allow the vehicle to sit lower. Removing the oil that normally sits in the pan sump and lifter valley greatly reduces the amount of windage found in other systems. Engine power goes up in a properly designed system. Reports of a 10% increase in power on a very high rpm motor, compared to a stock wet sump system, are not uncommon. On a 700 horsepower motor, for example, this can mean an additional 70 horsepower at very high rpms. However, these systems are expensive and may only be cost effective on very high rpm maximum effort motors where windage problems start to rob a lot of engine power. There have been a great deal of advances in wet sump oil system design and a good wet sump system can reduce windage losses somewhat without having to take out a second mortgage on your home in order to buy a dry sump system.

INTAKE OIL SHIELDS

An intake manifold oil shield (PN-14096974) will fit under any big block intake manifold and will keep hot oil from splashing on to the underside of the intake manifold. This will lower the temperature of the air/fuel mixture passing thru the intake ports resulting in slightly more power. On many modern design race intake manifolds the base of the intake acts as an intake oil shield. Intake runners and plenums are isolated from the manifold base by an air space.

OIL FILTER

Chevrolet and GMC/Truck used two different block oil filter adapters over the years. Car engines made from 1965 thru 1967 used an adapter that then allowed a metal oil filter canister and thru- bolt to hold a paper oil filter cartridge. In 1968 and later car motors, the inner block surface on the block oil filter pad was changed slightly and a different block oil filter adapter (PN- 3952301) was used to mount a spin-on type of filter. One-half quart oil filters (AC# PF-25), one quart filters (AC# PF-35) and two quart, truck style filters (AC# PF-932) and synthetic element (AC# PF-35L) which filters down to 12 microns, are available spin- on GM oil filters. K&N makes some of the freest flowing oil filters on the aftermarket and they come with high burst strength outside metal coverings that will withstand more pressure. Keep this in mind if you have

Adapters are available that enable you to remotely mount the oil filter. Header tubes or exhaust pipes may cause an interference problem with the normal block mounted oil filter location. Something like this with the appropriate hoses can solve the problem.

This block/oil filter adapter is used with the early metal canister and paper filter on '65-'68 blocks. You need to use this adapter when you convert to a spin on oil filter using another additional aftermarket adaptor.

This aftermarket spin on oil filter adapter fits below the old Mark IV canister filter type block adapter. You need BOTH when you make the conversion. I have seen people take off the old canister style adapter and replace it with one of these conversion adapters. The problem is that without this original oil canister adapter left in place, oil fills up the new spin on oil filter and then stops circulating thru it and then oil from the pan bypasses the filter 100% of the time.

blocked off the oil bypass valve in the stock GM oil filter block adaptor.

Some trucks and some police packaged cars continued to use the early paper cartridge, metal canister style filter assembly. You can change from one system to the other, but whichever one you use, you must then use the correct filter assembly that goes with it. Some people like the convenience of a spin-on filter, while others prefer the larger filtration capacity and lower cost of the cartridge style. Take your choice. Chevy makes a conversion kit (PN-5574538) to change a spin on filter set up to a canister style oil filter.

Aftermarket sources can provide adapters and lines to remotely locate the oil filter. These are helpful when you use an exhaust header system that comes too close to an engine mounted filter or routes header tubes right where the oil filter would be if left on the engine. They also make adapters to convert a cartridge style oil filter to a spin-on type filter or visa versa. Make sure that you install these conversion kits properly. One problem I have often seen concerns the conversion from a old style metal canister and filter cartridge to the spin on filters. A lot of folks take the old metal canister style block/oil filter adaptor and throw it away and then put on the new adaptor. They don't realize that the new adaptor uses the old block adaptor in addition to the new conversion adaptor. The problem is that if you don't use both pieces, the new spin on filter will fit and not leak oil, but the oil will bypass the filter after filling it up. (See the photos.)

OIL COOLERS

Chevrolet offers three types of oil coolers that can be used on 1968 and later Mark IV big blocks. (PN-997033) is a kit which uses an adaptor/spacer bolted to the oil filter pad on the block. It includes hoses, auxiliary radiator, adaptor/spacer and couplings. Your regular spin-on oil filter screws on to the adaptor/spacer. These kits can also be adapted to cool transmission fluid.

(PN-14084369) is an optional kit assembly on late-model Corvettes with small block 350 engines, that can be used on 1968 and later

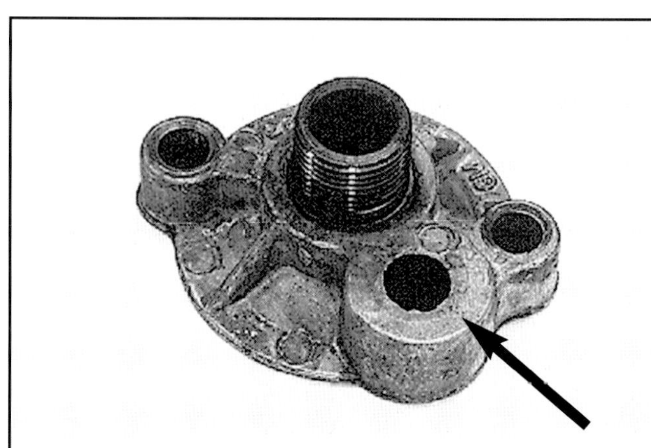

Mark IV blocks with a spin on oil filter use this type of adapter. It has a built in pressure relief valve. (arrow)

Mark IV big block motors. This heat exchanger uses an adaptor/spacer that allows 1/2" lines from the engine's water cooling system to cool and stabilize the temperature of the engine oil. The kit includes the heat exchanger adaptor/spacer, gaskets, seals and connectors, all of which can be bought separately if you wish. A standard spin-on oil filter will screw onto the heat exchanger.

Many four-bolt big blocks have two oil cooler threaded holes near the oil filter pad. External oil lines can be mounted to these holes and then run to whatever type of cooler you want to install. On Mark IV production and "Bowtie" four-bolt blocks, you must install an oil filter bypass valve (PN-5575416) in the rear hole to make this external oil cooler system operable.

If you are rebuilding a motor that previously used an oil cooler, it is a good practice to completely flush out the old oil cooler lines and the oil cooler radiator or replace the cooler and lines completely. You don't want any particles that may have built up in the old cooler contaminating your new motor.

Install an oil temperature gauge and try to maintain a hot oil temperature of 220 degrees or less. If your engine combination does this without an oil cooler, then leave it alone. An oil cooler that you don't need just adds weight and more fittings that can leak. If oil temperatures exceed 220 then get a cooler.

Dipsticks for the big blocks are all basically the same and are mounted to a passenger side exhaust bolt hole and enter the oil pan on the side. The dipstick tube is (PN-12551417) and the dipstick itself is (PN-10218765).

Always use an oil pressure gauge instead of relying on an stock "idiot" light. An oil pressure gauge will give you warning of a drop in pressure. An "idiot" light will come on when its too late.

Racers can also, in addition to the pressure gauge, install a

Gen.V and Gen.VI production blocks have a different spin on oil filter mounting boss and ports for an oil cooler.

warning light that will come on when oil pressure falls to a predetermined level. The pressure level that the light comes on can be set with a removable chip. Chips are available for various pressure settings. These warning lights are helpful if you don't have enough time in a race to watch a regular oil pressure gauge. The folks at Auto Meter make one. You can also set up the low oil pressure warning system to automatically cut electrical power to the ignition if the oil pressure drops below a certain point and possibly save an engine.

BIG BLOCK Quick Facts & References

OIL SYSTEM PREP CHECKLIST

1) **Always pre-oil your engine immediately before firing it up for the first time. Pre-oil it for sufficient time to ensure that oil is distributed to all components, including every rocker arm. This may take up to five minutes and may also require rotating the crank to make sure oil passages are aligned to feed oil through the entire engine.**

2) **Always verify oil pump clearances and deburr the pump gears, even if the pump is brand new.**

3) **Once proper pickup position is determined, tack weld or braze in place making sure to remove the bypass spring and valve before welding. Remove corrosion and make sure cover is flat before reinstalling.**

4) **Never use upper engine oiling restrictors with stock-type rocker arms or hydraulic lifters.**

5) **Use only the highest quality filter and ashless oil. Change oil and filter after initial fireup and frequently thereafter.**

Cylinder Heads

Photo courtesy of World Products

Mark IV cylinder heads, made by Chevrolet, will not interchange to a stock, production Gen.V Chevy block. There will be a water leak into the lifter valley because of a mismatch of water jacket core holes. Likewise, stock production Gen.V cylinder heads, made by Chevrolet, will not interchange onto Mark IV block cases for the same reason. There is one exception. You can put Mark IV cylinder heads on a Gen.V "Bowtie" block because the Gen.V "Bowtie" cases have smaller water jacket holes in the deck and will not leak when you use a Mark IV head. Some aluminum "Bowtie" heads, introduced in 1996, will work with any type of Chevy block. Stock head gaskets will not interchange between Mark IV and Gen.V stock production blocks.

All big block heads have screw in rocker arm studs or bolts and pushrod guideplates, however the Gen.V production heads use a non- adjustable rocker arm assembly that is different from the Mark IV heads.

Casting numbers on Chevrolet cylinder heads are found on the upper surface of the head casting beneath the valve cover. The last three digits of this number are repeated on the block side of an intake runner. There is also a head casting date located in the rocker arm valley under the valve cover. Heads have the words "TRUCK", "M.TRUCK", "HIPERF", or "PASS" cast into the rocker arm valley surface. Heads marked "HIPERF" usually have rectangular ports and large valves, but I have seen some late model, 1986-88 Mark IV GMC 427T truck heads (casting #-14092359) that were marked

Head casting dates and other identifiers (like "PASS" seen here) are cast into the rocker arm valley. "B 12 9" in this instance means this head was made on February 12, 1969.

Head casting numbers are found on the rocker arm valley surface and will help identify the type of head you have.

Casting of "HIPERF" found on cast iron square port "990" head. This is the current service replacement rectangular port, open chamber head for Mark IV engines.

Above is a small oval port head (#-236) and below is large oval port (#-950) head.

"HIPERF". These GMC 427T heads had large, oval ports and small valves! "TRUCK", "M.TRUCK", and "PASS" marked heads have small or large oval shaped intake ports and small valves. No rectangular port heads were factory installed on full size trucks. Hi-perf motors using square port heads were an option in early El Caminos.

Four different spark plug types have been used over the years. From 1965 to 1970, cast iron heads used 3/4" reach, 13/16" hex, gasketed plugs. From 1970 to 1994 heads used 1/2" reach, smaller 5/8" hex, tapered seat sparkplugs. Early aluminum heads used 13/16" hex, gasketed plugs, 3/4" reach. "Bowtie" aluminum symmetrical port, race heads require a 5/8" hex, .708" reach, tapered seat plug.

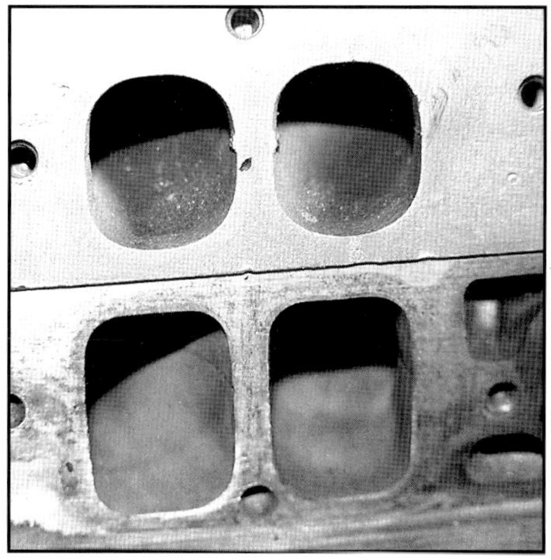

Large oval (#-049) port above a rectangular port found on head casting (#-990).

PORT TYPES

Production big block cylinder heads were made with three different intake port types, small oval (sometimes called "truck", "round" or "peanut" ports), large oval (found on cars and some trucks) and rectangular ports (also called square ports). These different intake port types were used to develop power in different rpm ranges. The small oval port opening dimensions are approximately 1.45" x 1.60". For large oval heads to opening is 1.82" x 1.60". Ports measuring approximately 2.37" x 1.71" are found on the stock production rectangular heads.

Most factory produced big block heads have "siamesed" intake ports. This results in a port runner arrangement that has two intake ports on each head that are slightly longer in length than the other two intake ports. In most situations this doesn't change the way the motor operates, but if you are building a motor for maximum performance this port length difference can cause some cylinders to operate differently from the other cylinders. The differences can be accommodated by stagger jetting the carb, intake manifold runner design changes, and different spark plugs heat ranges in different cylinders, all in an effort to equalize the power in each cylinder. Chevrolet and Pontiac have tried to improve the situation with symmetrical and spread port intake runner designs that are equal in length and these heads will be discussed later in this chapter, but first let's talk about the type of ports found on most factory production heads.

Mark IV, Gen.V and Gen.VI small oval port, cast iron, heads, commonly called "truck", "round" or "peanut" ports and made from 1976 on, make power and high torque at very low rpm's and are helpful when the motor is pulling heavy loads in trucks, trailers or RV's. The velocity of the air/fuel mixture in these heads is very high at low engine speeds, however they run out of breath at rpm's above 5000 and are not really suitable for street performance or race applications, although larger valves, bowl and other port work can improve the flow situation. Small oval port heads have intake valves that are 2.06" in size, with 1.72" exhaust valves. They have hardened valve seats. These heads can have closed or open combustion chambers. Most are found on trucks including late models, but there were some full size passenger cars and station wagons in the mid-70's that could be ordered with a 454 and may have the small oval port heads.

Mark IV large oval port heads make their power over a broader and higher rpm range. They are found on cars and trucks from 1965 thru 1990 and make excellent street performance heads. They can be used on bracket drag race motors, after bowl porting and installation of bigger valves, with great results because they make a large amount of torque and will breathe well to about 6500 rpm. They come with 2.06" intake and 1.72" exhaust valves. They have closed or open combustion chambers and depending on when they were made, have either soft or hardened valve seats. The open chamber style castings will flow more air than the closed chamber designs. A good starting place would be cast iron, large oval port, open chamber casting #'s 781, 049, 359, or 241. In 1996, Chevy introduced an over the counter aluminum open chamber head with large oval ports. This is the first factory big block aluminum head with large oval ports. This head will fit and seal with Mark IV, Gen.V and Gen.VI blocks.

Cast iron oval port heads, depending on which casting you use, have intake port volumes from 225cc to 262cc. Exhaust ports have volumes

CYLINDER HEAD CASTING NUMBERS

CASTING NUMBER	YEAR	PORT	CHAMBER	NOTES
330864	68-84	oval	OPEN	396, 402, 366T, 427T, 454 Truck
330865	68-84	oval	OPEN	396, 402, 366T, 427T, 454 Truck
330866	68-78	oval	OPEN	366T, 427T
330867	68-73	oval	OPEN	427T
336765	68-78	oval	OPEN	427T
336768	73-76	oval	OPEN	427T
336781	73-85	oval	OPEN	454
343771	68-86	oval	OPEN	366T, 427T, 454
343772	68-86	oval	OPEN	366T, 427T, 454
343783	70-76	oval	OPEN	454 Truck
346236	75-87	oval	OPEN	454
352625	70-76	oval	OPEN	454 Truck
353049	73-84	oval	OPEN	454
366765	74-78	oval	OPEN	427T
386781	73-79	oval	OPEN	454
473328	68-69	oval	OPEN	366T
3856206	65-66	oval	CLOSED	396
3856208	65	rect.	CLOSED	396/425hp 396/375hp "Z-16" Chevelle
3856213	66-82	oval	CLOSED	366T, 427T
3856260	68	oval	CLOSED	396 Truck
3872702	65-66	oval	CLOSED	396, 427
3873858	65-67	rect.	CLOSED	396, 427
3876875	77-85	oval	OPEN	427T
3904390	67	oval	CLOSED	396, 427
3904391	67	rect.	CLOSED	396, 427
3904392	67	rect.	CLOSED	427, aluminum, L88, L89
3904393	68-76	oval	OPEN	366T
3908952	67	oval	OPEN	427, "M.Truck", Marine
3909802	67	oval	CLOSED	396, 427
3917215	67-68	oval	CLOSED	396, 427
3917219	68-85	oval	OPEN	366T, 427T
3919840	67-69	rect.	CLOSED	396, 427
3919842	68-69	rect.	CLOSED	396, 427, aluminum L88, L89
3931063	68-69	oval	CLOSED	396, 402, 427
3933148	69-84	oval	OPEN	'69 396/265hp, 366T, 427T
3933149	68-76	oval	OPEN	427T
3935401	68-76	oval	OPEN	427T
3946074	69&71	rect.	OPEN	'69 427 L88, ZL1, aluminum '71 454 LS6, round exhaust port
3964290	69-70	oval	CLOSED	396, 402, 427, 454; large or small hex spark plugs used
3964291	69-70	rect.	CLOSED	396, 402, 427, 454; large or small hex spark plugs used
3965198	68-69	oval	CLOSED	396 Truck
3975950	68-70	oval	OPEN	396, 402 Truck, 366T, 427T
3986133	68-85	oval	OPEN	366T, 427T, 454 Truck
3986135	69-73	oval	OPEN	366T, 427T
3986136	68-84	oval	OPEN	366T, 427T, 454 Truck
3993820	71	oval	OPEN	402
3993820	71-84	oval	OPEN	454 Passenger and Trucks
3994026	71	rect.	OPEN	454 LS6
3999241	72	oval	OPEN	402, 454
6272292	70-78	oval	OPEN	402, 454
6272990	70-up	rect.	OPEN	454 service replacement, used on some Mk.IV crate engines
10045427		rect.		Pontiac/Chevy aluminum , large port, race
10049875		rect.		Pontiac/Chevy aluminum, small port, race
10051128		rect.		Bowtie, aluminum , 68cc chambers symmetrical ports
10051129		rect.		Bowtie, aluminum "raw"casting, symmetrical ports
10052902		oval	OPEN	MkIV, 427T
10101140	91-up	oval	CLOSED	Gen.V 366T, 427T
10114156	91-up	oval	OPEN	Gen.V & VI 454
10141279	96-up	oval	CLOSED	"Vortec 7400", 100cc chambers
10487052	77-90	oval	OPEN	366T
12363391	96-up	oval	OPEN	Bowtie, aluminum, Signature Series, fits Mk.IV, Gen.V, Gen.VI
12363401	96-up	rect.	OPEN	Bowtie, aluminum, Signature Series, fits Mk,IV, Gen.V, Gen.VI
14011077	69-up	rect.	OPEN	Aluminum, fits Mk.IV, "C-port" also sold as a "solid" head
14044861	84-96	rect.		Bowtie, aluminum, Mk.IV, raised runners", 1st design
14044861	96-up	rect.		Bowtie, aluminum, Signature Series, fits Mk.IV, Gen.V, Gen.VI raised runners", 2nd design
14044861	84-96	rect.		Bowtie, aluminum, Mk.IV "raw" casting of "861" MKIV 1st design
14081045	78-87	oval	OPEN	454
14081052	85-87	oval	OPEN	366T, 427T, 454 Truck, Marked "HiPerf"
14092359	86-90	oval	OPEN	366T, 427T, Marked "HiPerf"
14092360	86-88	oval	OPEN	454 Truck
14096188	70-71	rect.	OPEN	454 service replacement, also used on later Mark IV LS-6, LS-7 and 454 HO crate motors
14097088	91-up	rect.	OPEN	Gen.V 454/502 HO
24502585	97	rect.		Olds/Chevy, DRCE 2, Pro Stock aluminum, for blocks w/4.900" bore centers

T = Tall deck

Oval includes both "peanut" (also called "round") ports and large oval ports

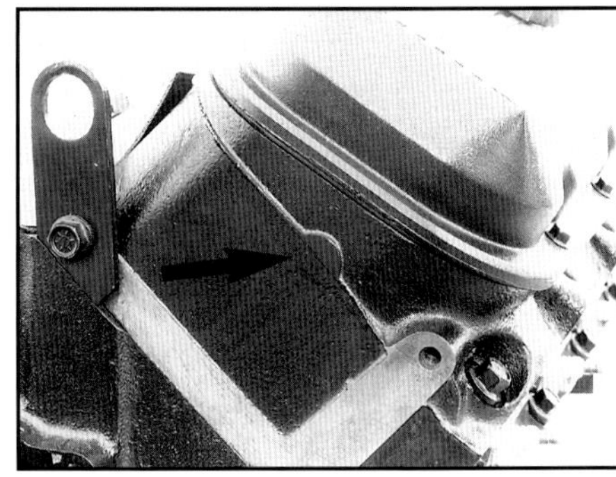

Hi-perf rectangular port production heads have a "bump" identifier on the ends. Oval port heads do not have this "bump".

from 110cc to 119cc. For you folks interested in NHRA stock class drag racing, the largest "as cast", cast iron oval intake port volumes are found in closed chamber head casting #-3964290, with the 5/8" hex small spark plug. In 1970 GM changed from the large hex old style spark plugs to the small 5/8" hex plugs, so look for the small plug "290" castings made in 1970 thru 1976. The "as cast" intake port runner volume on the "290" small plug head castings measure a NHRA maximum of 262cc.

The largest "as cast", cast iron, large oval port exhaust runner seems to be found on the casting #-353049 head made from 1970 to 1984. It measures a NHRA maximum of 119cc. This head was used in 396, 402 and 454 passenger and truck motors.

Rectangular or "square" port, cast iron and aluminum cylinder heads are strictly for high-performance use. Because the intake port dimensions are so large, air/fuel velocities are at a low speed in the ports until the rpm's of the motor get high. Then the power comes on like gangbusters. Applications include high performance street use and racing where large volumes of air and fuel are required at high rpms. Square port heads came with larger 2.19" intake valves and 1.88" exhausts. Some early closed chamber heads had 1.84" or 1.72" exhaust valves. Open or closed chambers were available with soft or hard valve seats. Once again, the open chamber designs on production heads will flow more air than the closed chamber production variety. If your engine is going to operate in the 7000+ rpm range or if you are

LET IT BREATHE

If we built an engine and made incremental changes to it, we could show the improvements in power made by those parts. The base motor could look something like this:

Cubic Inches 427
Bore 4.250"
Stroke 3.76"
Large oval port, cast iron "781" heads with 2.06"/1.72" stock valves
9.5-1 CR
Split plenum intake with a 700 cfm carb
Street headers with mufflers
Hydraulic flat tappet cam, seat to seat timing of 15-53-61-19, lift-.453"/.484", LSA-110

	BASELINE #1		#-2		#-3		#-4	
RPM	HP	TQ	HP	TQ	HP	TQ	HP	TQ
2000	187	491*	188	494*	188	494	151	396
2500	232	486	235	493	236	496*	196	412
3000	273	478	279	488	283	494	241	422
3500	311	466	321	482	328	491	352	462
4000	340	446	358	470	369	484	399	465*
4500	348*	406	379*	443	398	465	436	457
5000	332	349	378	397	408*	428	451*	430
5500	299	286	358	342	398	380	449	393
6000	250	218	315	276	363	318	428	346

In test #-1 this short duration, moderate lift cam provides a large amount of torque off idle and the torque curve is fairly flat. Because of the short breathing provided by the cam, valve sizes and stock heads, HP peaks at 4500 and then starts to decline. As street engines go it would provide good service and reliability in the rpm ranges where most street motors are driven.

In engine #-2 everything stays the same but, you break out your grinding burrs and sanding rolls to "pocket port" the heads. Just knock off the casting edges, machining ledges and casting seams below the valve seats, smooth out and slightly widen the short side radius and clean up the bowls. This helps the air in the port runners make the turn into the cylinder and power goes up. It doesn't cost anything except your time. It hasn't changed the operating range of the motor much, but because of the better breathing, it will pull to 5000.

Bigger valves (2.19"/1.88") are added to the pocket ported heads and you can see the results in test #-3. Breathing capacity and valve ratio is improved, power is up across the board and peak power levels go up 500 rpms. So far, you have the cost of new larger valves, valve seat grinding and your labor on the pocket porting. The gain is more torque, an even flatter torque curve and 60 more horsepower at a modest 5000 rpm.

In test-4 its time to change the cam for more duration and lift. Cost will now include the cam and lifters and a new set of matching springs. This hydraulic cam has a duration at .050" of 236/236 degrees and lift is .556" on both sides. LSA is 110, but the cam is ground with 4 degrees advance giving a 106 degree intake lobe centerline. Horsepower takes a jump to 451, but the torque from the motor is down and the torque peak has moved from 2500 to 4500. We have to get the cylinder pressure and torque back, so by optimizing the static compression, we can get the dynamic compression of the motor back up to work with the intake valve closing point. This cam requires a compression ratio of 11.0-1, so in test -5 we will leave the heads as they are and use a different piston to get the desired static compression.

	#-5		#-6	
RPM	HP	TQ	HP	TQ
2000	159	417	159	418
2500	206	433	207	434
3000	254	444	255	446
3500	313	469	314	472
4000	369	485	372	489
4500	420	490*	425	495*
5000	459	482	466	490
5500	478	456	487	465
6000	480*	420	491*	4300
6500	463	374	476	385
7000	433	325	449	337

Test #-5 getting the compression right for the new cam gained 29 more horsepower at 6000 rpm and the torque on the motor came back up to 490 ft./lbs. If you were rebuilding a stock motor, depending on which parts you chose, there could be a 100 horsepower difference in engine output and about the same torque figure. However, peak torque would occur 2500 rpms higher and this would effect the driveability of the motor a bit. A vehicle with a stick transmission would not drive that much different, but an automatic car would need a higher than stock stall converter to operate best. A lower rear gear with engine #-5 would also give better response. Still from an economic standpoint, the engine rebuild cost would be about the same.
Test-6 makes one final change. Since the engine now breathes better and will run to a higher rpm range we can substitute an 850 cfm four barrel carb for the original 700 cfm model. The split plenum manifold allows a larger carb size and will still give good throttle response. Power and torque levels picked up a bit more above 2500 rpm. There is a 143 HP increase from test #-1, while total peak torque climbed to 495, but peak torque occurred at a much higher rpm when compared to the baseline combination in test #-1.

building a large cubic inch race motor, then the rectangular port heads will make the power you need, but at a higher rpm.

Rectangular port heads, depending on which casting you use, have intake port volumes from 290cc to 327cc, while the exhaust ports have volumes from 103cc to 127cc. Aluminum "Bowtie" casting have 315cc to 404cc intakes and 126cc to 158cc exhaust port volumes. Valve sizes for production square port heads are 2.19" for the intakes and either 1.72", 1.84" or 1.88" for the exhaust depending on the casting. Pontiac Motorsport aluminum heads for the big block Chevy blocks have intake port volumes from 365cc to 460cc, with "D-shaped" exhaust ports on the head having volumes of 145cc.

The largest "as cast" rectangular intake port runner volumes for factory installed heads seem to be 1970 LS-6 castings #-3964291 which the NHRA lists as having a maximum of 327cc's. These head castings also came on the 1970 402/375 motors in Novas, Chevelles and Camaros. The largest "as cast" exhaust port runner volumes for factory installed heads are casting #'s-3994026, 14096188, and 6272990 which all have an NHRA maximum of 127cc's.

Cast iron or aluminum square port heads are available. Aluminum heads weigh from 27 to 31.5 lbs. per head compared to cast iron versions which weigh approximately 63 lbs. each. As you can see, there is a substantial weight reduction if you move to aluminum heads and the weight (60-65 lbs.), comes off the front end of the vehicle. Aluminum is also easy to repair and has one other important characteristic. Because aluminum sheds heat quicker than a cast iron piece, the use of aluminum heads allows a slightly higher compression ratio (generally one-half to one full compression point). Since the aluminum heads run cooler, heat caused detonation is reduced and may let you use slightly more compression than you could with a cast iron head combination.

The larger 2.19" (PN-3969815) intake and 1.88" exhaust (PN-3946077) valves can be put into your Mark IV, Gen.V or Gen.VI small valve, oval port heads by any competent automotive machine shop and your engine performance will be enhanced. Chevy also makes a 2.30" intake valve (PN-3879618), however, you need a very high flowing cylinder head to make good use of this valve size. Most production heads will do just fine with the 2.19" intake and 1.88" valves. Valve stem stock diameters are 3/8" and smaller valve stem sizes are available from aftermarket sources. If you do add larger diameter valves remember to unshroud the chamber walls adjacent to the valve edges so that you take advantage of the increased flow. Adding larger valves puts the edge of the valves closer to the existing chamber walls. If you don't move the chamber walls away from the valve edge, the heads will probably flow less air than if you had left the stock diameter valves in place. Remember that if you unshroud the valves the combustion chamber volume will be increased and this will have an effect on the compression ratio.

Gen.V rectangular port heads (PN-14097088), with a non-adjustable valve train, use 2.19" (PN-14097045) and 1.88" (PN-14097049) valves. These heads should not be used with production Mark IV blocks.

If you have an early set of heads that have soft valve seats, any competent auto machine shop can install hardened valve seats so that you can run the engine on unleaded pump gasoline.

Also keep in mind the valve weight. There is quite a difference between manufacturers and some of the valves are heavier than others. Take a look at various brands and you will notice that valve margins are thicker on one valve compared to a different maker or model. The valve face may be flat or dished. The stems may be undercut or not and may be a smaller diameter. Some of the valves may have more meat (and weight) on them because they were intended for a street driven engine with a long life expectancy. Others are intended for race motors and designed with weight reduction in mind, along with the ability to withstand higher operating temperatures and increased valve spring pressures. Manley, for instance, has basically three different levels in valves, street, race and severe duty. Each type is made differently for its intended use and there is a big difference in weights and materials used.

High tech valve coatings are available from valve manufacturers or you can get your valves coated by companies specializing in applying these coatings. These coating can reduce the heat absorbed by the valve and add to its life span by reducing stem wear and reducing heat induced metal fatigue.

INTAKE PORT VOLUMES AND CUBIC INCHES

A drag racing acquaintance had a very strong running 427 motor with the #-077 aluminum heads on it. The stock intake ports on these heads have a stock port volume of 295cc. They have the smallest volume of any factory square port head. A friend of his did some work on the ports and got the volume up to about 315cc and he also increased the total flow. He increased the intake valve size to 2.25". The flowbench changes cost the racer about $1,000.00. Then the racer later decided to move up to a tall-deck "Bowtie" block and increased the cubic inches on the new motor to 557. He again used the same aluminum #-077 head castings from his 427. The new motor was quicker, but not up to the racer's expectations. We started talking about it and he realized that the port sizes were too small for the size of motor he was now running. He also realized that even after spending money for the aluminum heads and then additional money getting them to flow more air, they didn't flow any more air than a stock cast iron square port head with a larger port volume of 325cc.

What's the point of this story? Well, these days you can spend your money trying to get a set of heads flowbenched and modified to work on a larger motor or you can spend the money on a set of heads that already have the flow numbers you need. Manufacturers now have various intake and exhaust port sizes to match and adequately feed some very large or high rpm engines. Some of these heads initially cost more than the stock factory heads, but when you add up the cost of inadequate heads, paying someone to flowbench and modify the heads, milling the head deck to decrease chamber volume, add bigger valves, widen the spring pockets, install longer stem valves, guides, etc., the costs seem to equalize.

Now don't go out and buy the biggest port volume head you can find. You need to match the intake port size to the size and intended use of the engine. For a stock size motor that is going to spend its time at low and middle rpm ranges use a large oval port head. In a race motor, try to match the heads to the size and rpm limit of motor you build. If you are building a 427-454 bracket race engine with a maximum rpm limit of about 6500, then some of the large oval port, open chamber head castings will do just fine in a steel bodied car. The smaller port volume (295cc) square port aluminum heads (casting #-077) or aluminum large oval port head with 290cc (PN-12363399 made available over the counter in 1996), will also work. These heads will provide large amounts of mid-range torque and often will out perform the large port volume square port heads in a heavier car or boat in these rpm ranges. The velocities in these heads will stay high up to about 6000-6800 rpms in a production size motor.

If the rpms are going to go over 6500 rpms, then move up to the square port heads with a port volume of at least 315-325cc. In larger race motors of 500 cubic inches, 325-380cc heads will do the job. In 550 and bigger cubic motors, 380-460cc heads will move enough air for a high rpm motor. Likewise increases in valve head diameters can also be made. Intake valve sizes of up to 2.50" and exhaust up to 1.95" are possible depending on which head casting you choose.

COMBUSTION CHAMBER VOLUMES

Big block cylinder heads have had many different combustion chamber volumes. Combustion chamber shapes were changed regularly over the years from 1965 to 1994. The first heads had bathtub-shaped "closed" chambers which had volumes as low as 96cc. Quench area, on the spark plug side and from the chamber walls, was removed little by little over the years, resulting in larger "closed" chamber volumes of 98.4cc, 101cc, 103cc, 107cc, 108cc and 109cc. "Open" chamber heads, first introduced on the 1969 L-88 and ZL-1 motors, had even more quench area removed over the years and had chamber volumes of 110cc, 112cc,114cc, 116cc, 118cc or 124cc. These volumes and the resulting chamber shapes, lowered compression ratios and exhaust emissions after 1970, but they also flowed much more air and fuel than the closed chamber heads. "Bowtie" aluminum heads have "semi-open" and "open" chambers with volumes of 68cc, 72cc, 105cc, 110cc or 118cc depending on which casting you choose.

Note that combustion chamber volume has a direct effect on the static compression ratio of a motor. Motors run on low-octane "pump" unleaded gasoline should not have a static compression ratio greater than 9.5 to 1. Higher static compression ratios might result in detonation if run on pump gas.

PORTING HEADS

You may be thinking that you want to "port and polish" your cylinder heads for improved performance. If you want to send the heads out to a performance or race cylinder head specialist there are a number of fine companies who can do the work for you. Be warned however, that this kind of work is an art form and also requires technical skills and equipment which are expensive and time consuming.

Pontiac Super Duty "875" aluminum head has a heart shaped 83cc chamber. These "spread port" heads also require special piston dome shapes to fit this chamber.

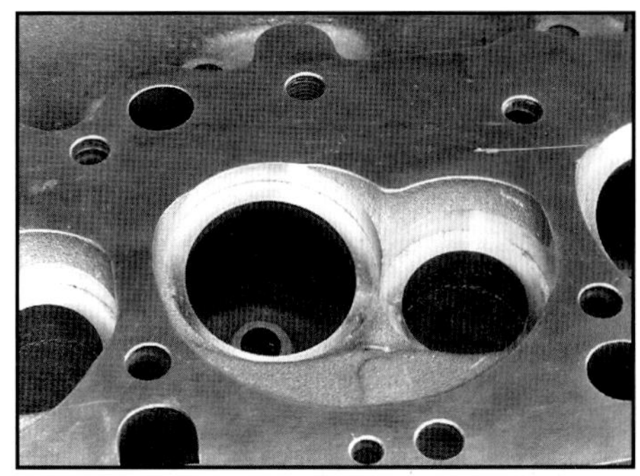

"128" symmetrical "Bowtie" aluminum heads have heart shaped chamber volumes of 68cc. The quench area on the spark plug side is laid back slightly on the raised square port heads. Piston domes used with these heads must accommodate the "beak" of quench area between the valves. This chamber does not have any valve seat inserts installed.

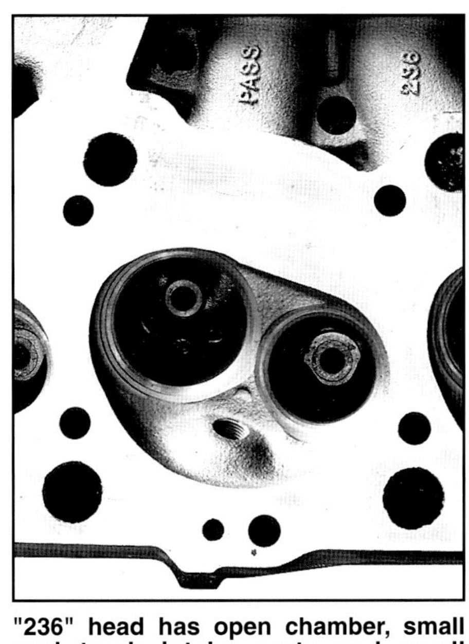

"236" head has open chamber, small oval truck intake ports and small 2.06"/1.72" valves. Spark plug boss protrudes into the chamber.

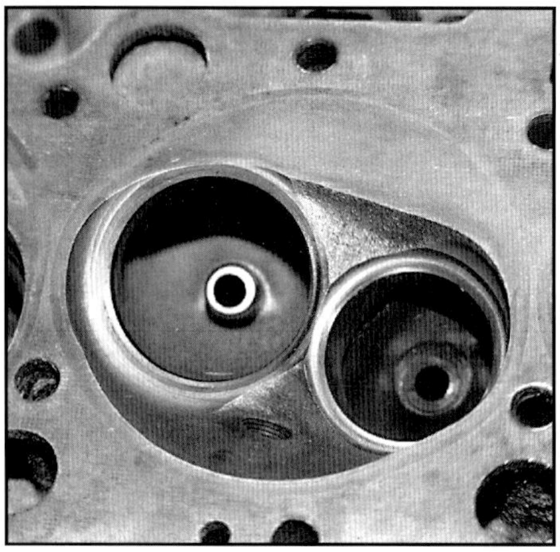

Rectangular port, cast iron "990" and "188" heads have open chambers, came with 2.19"/1.88" valves and had 118cc chamber volume. Aluminum open chamber heads like the "077" head are similar, however the "077" has a "C" shaped exhaust port.

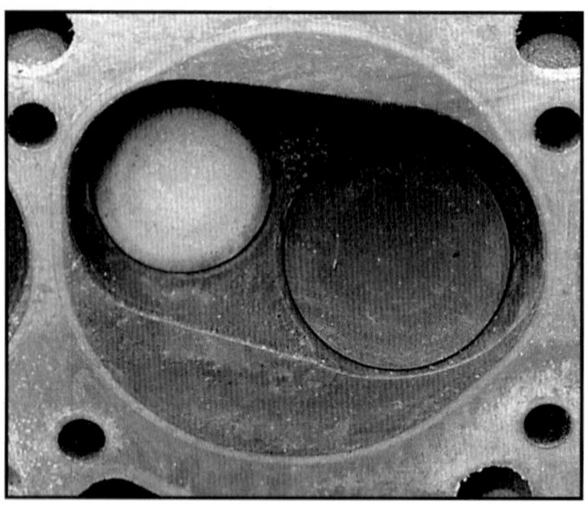

Closed chamber, rectangular port "840" head had 2.19"/1.72" valves and approximately 108cc volume. This is also a "bathtub" type chamber, but it has more volume that the oval port, closed chamber variety. Early aluminum closed chamber heads, like the "842" casting, have a similar chamber volume (107cc), but used 2.19"/1.84" valves. The "842" castings had round exhaust ports.

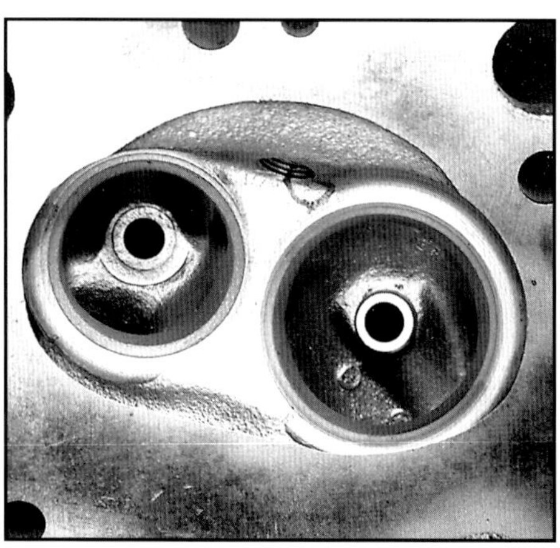

This "063" closed chamber head has a portion of the quench area on the spark plug side laid back giving a chamber volume of about 101cc. These heads had 2.06"/1.72" valves and large oval ports.

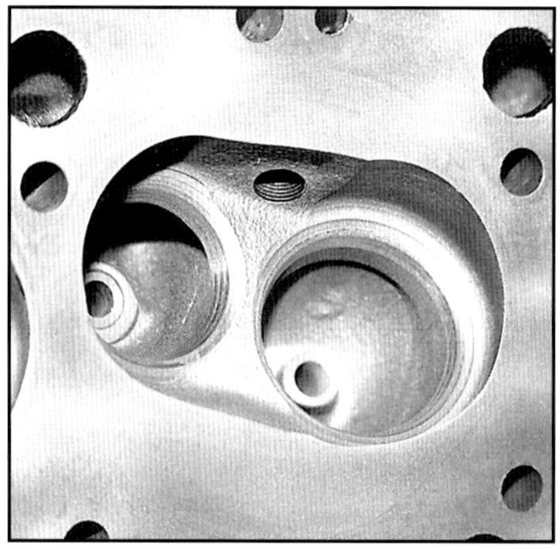

Early closed chamber, large oval port heads, like the "390" version had 2.06"/1.72" valves and have a "bathtub" shaped chamber volume of about 98cc.

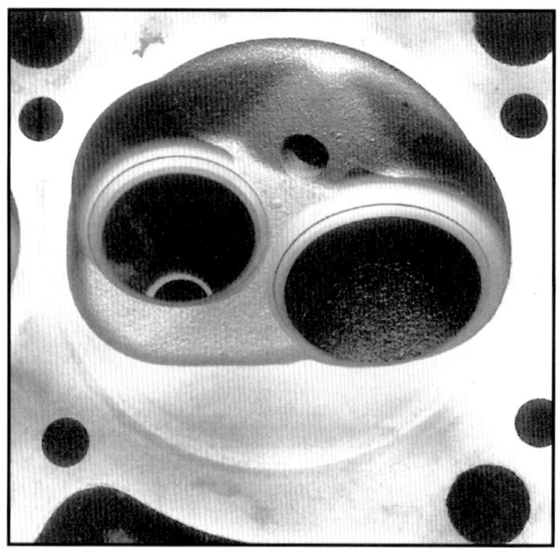

This open chamber shape is found on many oval port truck heads, this one is casting #-950. The quench area around the exhaust valve has been partially removed and the spark plug boss in the chamber bulges into the chamber quite a bit. This head used 2.06"/1.72" valves. Open chamber type truck castings have volumes that run from 112cc to 124cc depending on the casting.

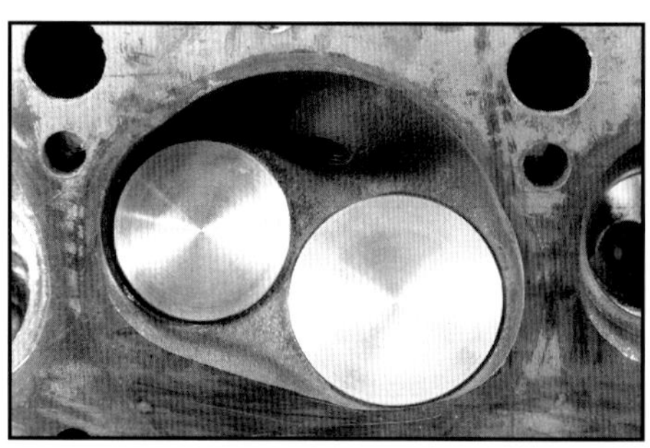

"049" open chamber, cast iron head has large oval ports and originally came with 2.06"/1.72" valves. "781" castings are very similar. Both of these cast iron heads are good choices for bracket drag racing with bigger 2.19"/1.88" valves added. With their large oval ports they keep flow velocities high and will help make large amounts of torque. Chamber volume runs from 114 to 118cc. Make sure you "CC" any head you are going to use.

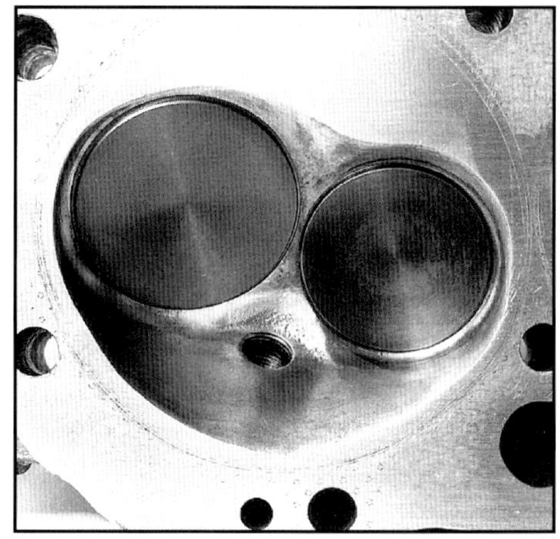

Aluminum Bowtie "861" raised port heads have 105cc open chambers and are set up to use 2.19"/1.88" valves, but larger valve sizes can be used. They have vanes in the exhaust ports that give them their "W" port shape.

Casting #-360 Mark IV iron head has a open chamber and "peanut" style small oval intake ports. This 1988 casting came with 2.06"/1.72" valves.

After checking the prices for some of this work, you may decide to try to do the work yourself. While I have no problem with that there are a few things to keep in mind.

First, you should know that in most cylinder head ports the major flow restriction is found at the valve seat, valve head, and the bowl area directly underneath the valve seat when viewed from the combustion chamber. This can account for more than two-thirds of the total restriction found in the intake port runner. The runner portion from the intake manifold/head mating surface to the bowl section of the runner causes comparatively little restriction. Therefore, you will get the most benefit from removing excess casting material from the bowls directly beneath the valve seat and from making sure that the heads receive an accurate four-angle valve job.

Valves with undercut and smaller diameter stems (11/32") are available from aftermarket suppliers and these reduce the restriction further. Backcutting the valve 30 degrees will help airflow. Removing material from the combustion chamber walls near the edges of the valve head will help the airflow get past. If you add larger diameter valves these chamber walls may become a problem and actually restrict the flow of air. The edge of the larger valve actually comes closer to the existing chamber walls. You need to unshroud the combustion chamber walls appropriately, otherwise you may get little or no benefit from the increased valve size. You can take the head you are going to use and bolt it to the bare block. Reach inside each cylinder bore and scribe a line on the head indicating the circumference of the cylinder bore. Take the head off and you can now remove material from the chamber walls to the scribed gasket line in an effort to unshroud the valve edges. This will increase the chamber volume slightly so it will have an effect on compression ratios. You can live with that or you can reduce the volume by decking the head.

Adding larger valve head diameters can also improve the valve seat position on a used head that has sunken valve seats. The larger valve seat diameter will be on previously uncut material on the chamber roof and will lift the new bigger valve higher on its seat. Lifting the valve surface up in relation to the chamber roof will also have the effect of reducing chamber volume which in turn has the effect of raising static compression ratio. Remember if you add larger valves you will need to unshroud the chamber walls adjacent to the valve edges.

Do not "polish" the intake port runner to a mirror-like finish. Just knock off any excess casting material, machining ridges or casting seams. Leave the surface of the intake runner slightly rough. This type of finish helps keep the air/fuel mixture in vapor suspension. You can also remove any excess casting flash or seams from the exhaust port runner and you can polish and smooth this

BIG BLOCK Quick Facts & References

CC'ing CYLINDER HEADS

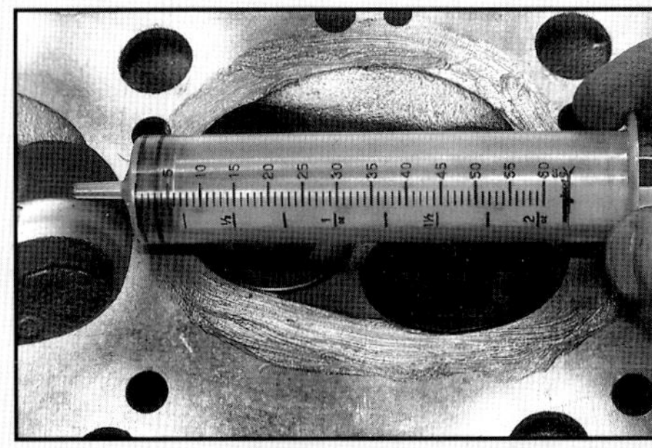

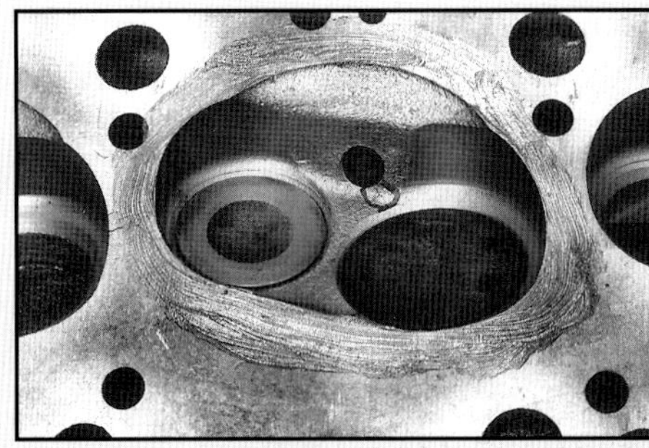

You can check the amount of chamber volume on any set of heads very easily. Get a plastic syringe from your local pet and veterinary supply store. These usually have syringe volumes of 60cc and they are cheap and accurate. You don't need the syringe needle. Get a piece of clear Lexan, plexiglass or plastic sheet at least 1/8" thick, which is also cheap. Get an intake valve, an exhaust valve, a spark plug of the type you are going to use, some petroleum jelly and enough water or solvent in a cup to fill the syringe twice.

Now place the cylinder head on a flat surface with the chambers facing up. Put a little petroleum jelly around the edge of the valves to seal them and then put the valves and spark plug in the head. Spread a thin film of petroleum jelly around the edge of the combustion chamber to seal the clear plastic sheet to the head. If you haven't done it already, drill a small hole in the plastic sheet so you can add the water or solvent thru it into the chamber.

Press the plastic sheet to the head with the hole near the edge of the chamber wall. Fill up the syringe, being careful to completely fill it and remove all of the air in the syringe. You can flick your finger against the side of the syringe to shake loose any air bubbles. Then slowly depress the slide on the syringe until you have exactly 60cc of liquid in the syringe and no air.

Fill the chamber with fluid, slowly. As the fluid fills the chamber, air in the chamber will exit via the small hole in the plastic sheet. You may need to incline the head slightly to get the air to leave the chamber completely. If your syringe empties before the chamber is completely filled up, refill the syringe again and continue. Look in the ports to see if you have any fluid leaks. If you do, reseal the valves and start over.

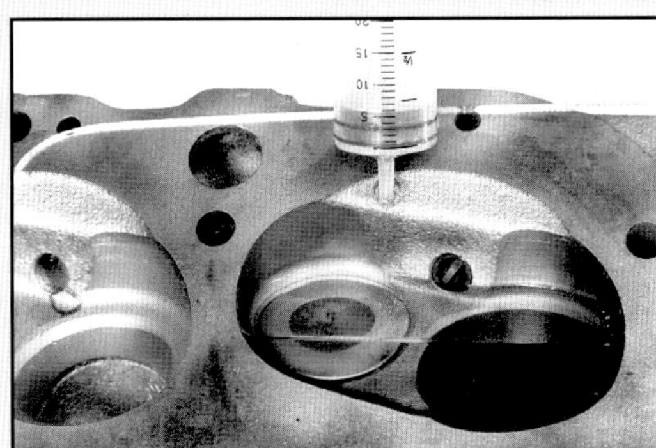

When the chamber is full to the bottom edge of the small hole drilled into the plastic sheet, you're done. You will now know exactly how much chamber volume your heads have.

You can check each chamber in turn, which is a good idea. Casting tolerances at the factory allow minor differences between similar castings. Also if the heads have been "decked" previously at a machine shop, the chamber volume will have been reduced from normal. If someone previously was grinding metal away from the chamber, the volume may be more than usual. If the heads have had many valve jobs, the valves may have been "sunk" into the head and increased the chamber volume. Some used heads have been around for almost thirty years. During that time many things could have happened to change the combustion chamber volume from its stock volume dimension. That's why its better to check.

Intake and exhaust port volumes and piston dish and dome volumes can also be determined using the above "CC'ing" equipment.

runner since there is no fuel in suspension to worry about.

Be careful when you are grinding away on heads. There is a temptation to take too much material off. Just smooth out and blend in areas under the valve seat in the bowl and runners. It is all too easy to cut into a water jacket or to make different cuts in different runners resulting in unequal port runner volumes. Remove any sharp edges and smooth out any sharp casting contours or edges on the short side radius. Take your time. Don't try to change the shape of the port, leave that to someone with a flow bench. You can practice and get the feel of the grinder by first cutting on a junk head. Be careful around the valve seat area. A high speed carbide cutter will work on the heavy stuff and sanding rolls of various textures will do the finish work.

Another area which you can work on is the opening on the intake manifold runner where it meets the intake port opening on the head. Sometimes there is a mismatch of these two openings which results in a ledge on the head side that can restrict the flow down the runners. You can use an intake manifold gasket to scribe lines on the head and intake port openings. Once the lines are drawn you can use your grinder to carefully open up the port openings so that they match. Most folks will extend this dimension about one inch down the port runner, where it then smoothly joins the existing port runner walls. Remember, you are not trying to make the opening as large as physically possible or to match the largest gasket opening you can buy. All you are trying to do is have the same port opening on the manifold and on the head with no ledges on the head port opening to impede flow.

Manley Race Flo stainless steel 2.19" intake valve has an under cut 11/32" stem to help flow past valve and it removes a few grams of total valve weight.

Stock production cast iron heads can use a relatively inexpensive exhaust port plate. These port plates fit between the head and the flange of an exhaust tube header. The plate has a port vane attached to the plate and this vane extends down into the port opening and floor. It acts the same way that the cast-in vane does on the factory "C-port" and "W-port" aluminum heads.

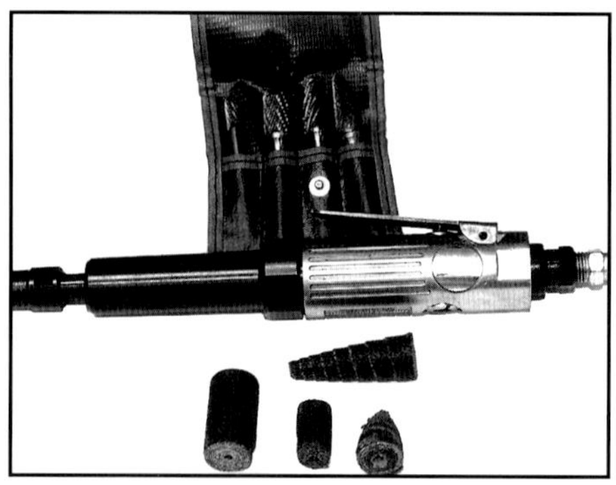

If you are doing any bowl porting, gasket matching or deburring a block, you will need a high speed electric or air driven tool along with a selection of carbide burrs and sanding rolls with different shapes and safety glasses. Special burrs are available to grind aluminum that won't clog when you use them.

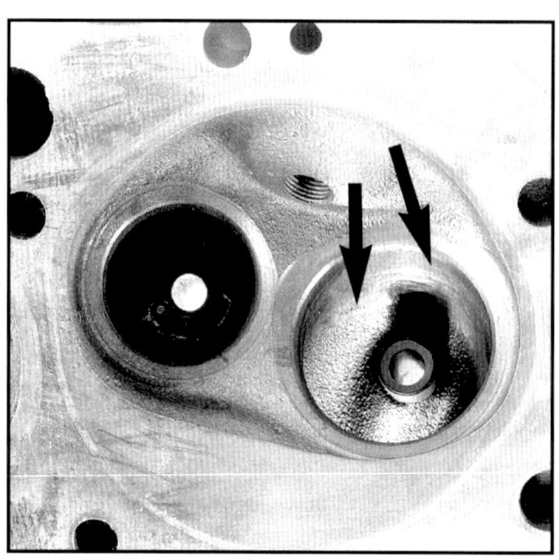

Open chamber found on #-049 cast iron, large oval port head. The intake port bowl rough cuts (arrow) have been made to remove the ledges found beneath the valve seats on the long wall. The as cast surface can still be seen around the valve guide boss.

It increases velocity and flow in the port and helps solve the problem of reversion in the stock exhaust ports on cast iron heads. The port plates can be used in oval or square port heads. MPG Head Service makes these and they are a "bolt in" item that works.

If you do get your heads flow-benched, one thing you want to do is get the exhaust port to flow approximately 80% of the total flow found on the intake ports. The other thing to look for is velocity thru the port. I think this is even more important than total flow. High velocities will fill the cylinder quickly especially at high rpms and can take advantage of inertia effects necessary for good cylinder filling.

One of the major changes in technology in recent years has been the widespread use of CNC milling machines to port cylinder heads. Before this, cylinder head work was largely done by experienced hand and expert eye. Cylinder head flow

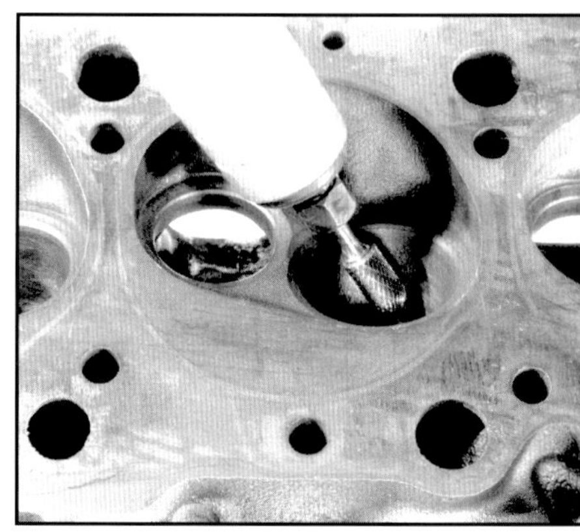

An electric or air driven grinder with a carbide burr will remove material very quickly. Finish with sanding rolls.

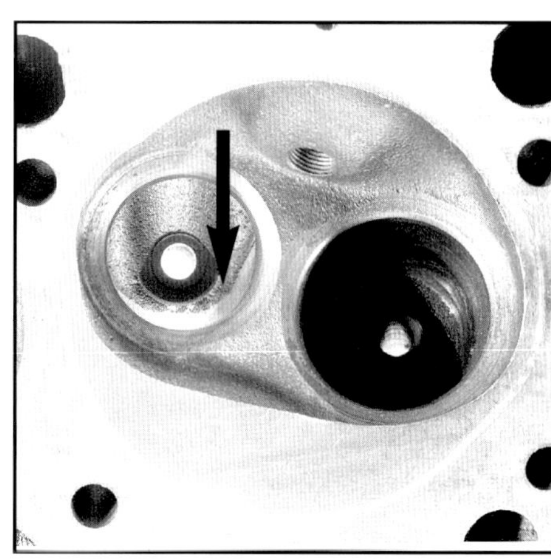

Here is the exhaust bowl on the #-049 head with the rough cut on the long wall begun.

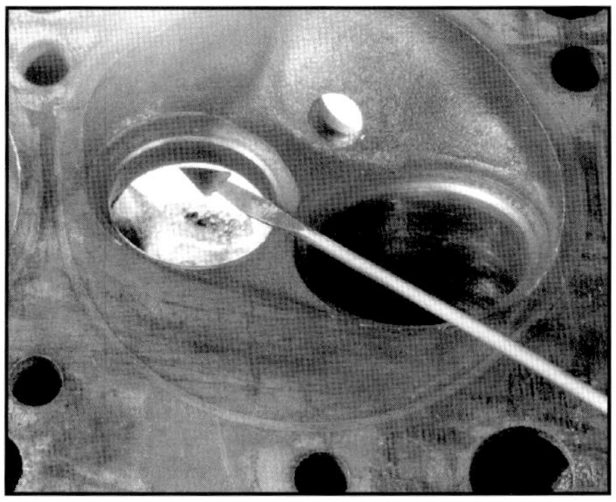

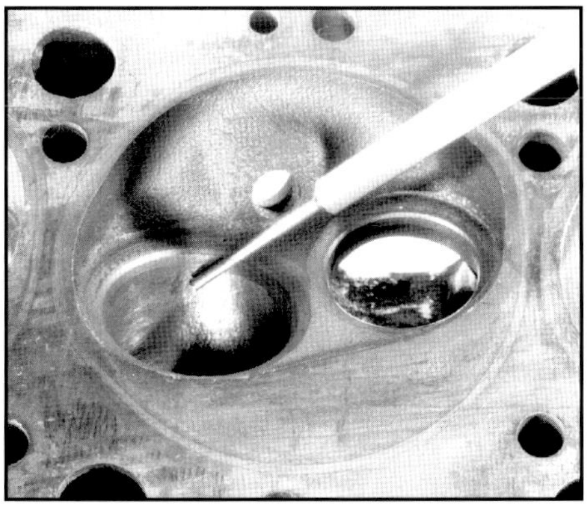

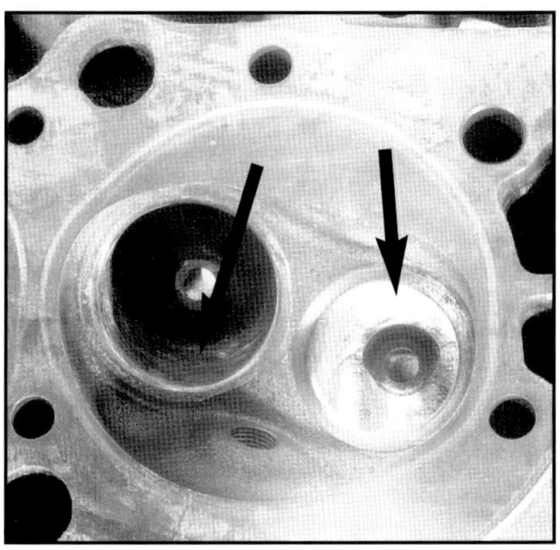

Arrows show area below valve seat where material can be removed from the long wall and short side.

Exhaust port view showing material removed from the valve seat bowl side. The port itself can be smoothed out and any casting seams or slag removed from the port walls and exit opening.

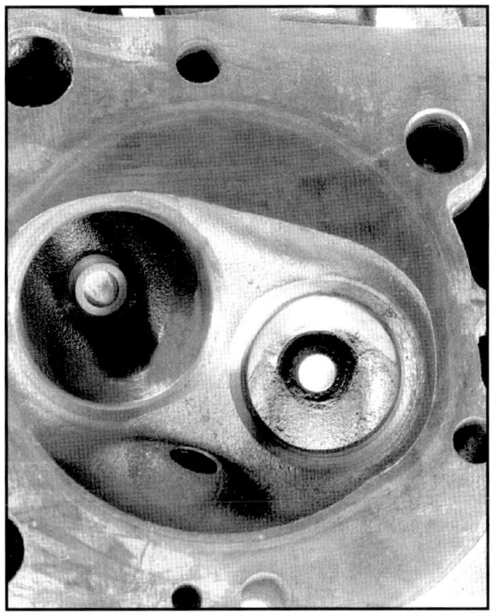

Large oval port "049" head ports can be improved by removing casting ledges, smoothing out the bowl area and smoothing out runner walls.

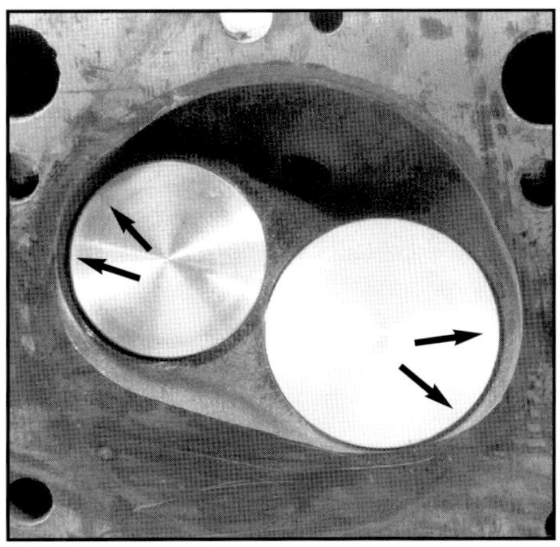

Adding larger valves to any head moves the valve edges closer to the existing chamber walls. Make sure that you relieve the chamber walls so that flow is not impeded.

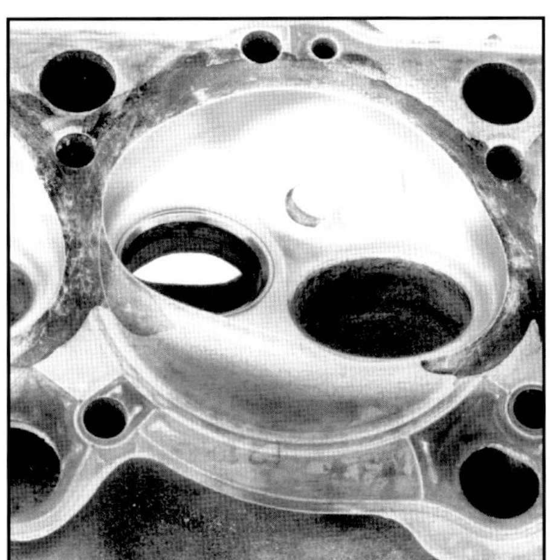

Chamber walls have been relieved to the gasket line and the rest of the chamber smoothed out. Machinist blueing has been used to make the gasket scribe lines stand out.

Using an intake gasket, you can scribe a line on the intake port openings to guide you. A little machinist blueing will help the scribed line stand out. Remember the main reason for this "gasket matching" is to make sure that there is no ledge existing when the intake port makes the transition from the intake manifold section of the runner and into the head portion of the runner.

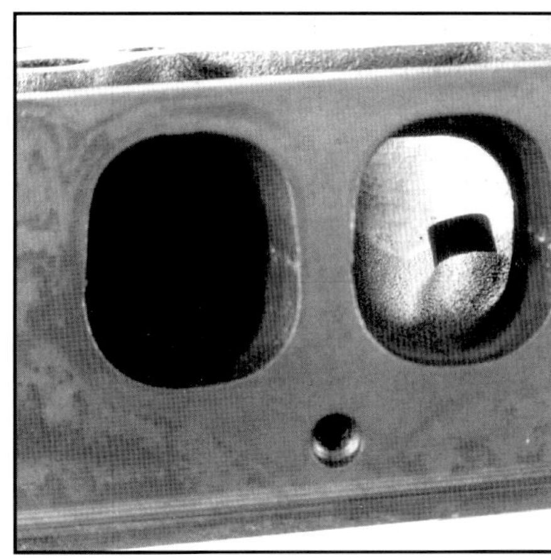

View of intake port on Large oval port casting #-049 shows rough grinding of the bowl section and the "as cast" finish of the guide boss area which needs to be smoothed out with sanding rolls without changing its shape or contour unless you have access to a flow bench.

Big Block Quick Facts & References

VALVE GUIDES & VALVE SEAT ANGLES

VALVE GUIDES

Valve guides are made from a number of different materials and configurations. Cast iron, phosphor bronze or silicon bronze guide material is used and the guide itself may be solid or wall inserts or spiral inserts. The inserts are used by drilling and honing the original guide to size and then the inserts are fitted inside the original guide. Some inserts are sleeves and others are thread in types. When solid guides are used the old original guide is driven out and replaced with an entirely new guide.

Bronze guides seem to be able to endure better with less chance of seizure occurring to the valve stem and can be run with slightly tighter guide to valve stem clearances. Silicon bronze guides have exceptionally long life.

Big block guides, if they are driven out, must be driven outward thru the top of the head (the rocker arm valley) and inserted from the rocker valley and driven into place. The guides are not driven out or inserted thru the combustion chamber side of the head.

The most important thing is to get the guide to valve stem clearance correct. With bronze guides .001" clearance for the intake and .0015" for the exhaust clearance should be maintained. With cast iron guides the clearance should be a little larger, intake .0012-.0015 and exhaust .0018-.0025". Valve stems with .001" or less taper should be used. When a guide is replaced you will have to recut the valve seat angles on the head, since the seats must be accurately concentric with the new guide. If you are going to use the positive type oil seals with your heads, the rocker arm valley side of the guide outside diameter must be machined down to create a stepped shoulder to accept these seals. Also let your machinist know the total cam lift at the valve so that the top of the guide can also be reduced to provide a minimum clearance of .050" between the bottom of the spring retainer and the top of the guide and oil seal when the valve is completely open.

Once again make sure that the valvetrain geometry is correct in order to prevent the premature wearing of the guides and valve stem because of excessive side loading. Use of a roller tipped rocker arm will help reduce the side loading of the guides.

VALVE SEAT ANGLES

Most "four angle" valve seats on a performance head are cut at the following angles.

The first top cut is a 30' cut to lay back and smooth out the transition from the existing combustion chamber surface to the valve seat.

The next cut is a 45' cut that the valve face will actually sit on and seal to.

The third cut is at 60' to begin the transition on the other side of the valve seat.

The final lowest cut is at 70' to help blend in the angles to the existing port surface.

Now there are a lot of variations on this theme, but this is the basic series of cuts and will work well in most situations. "Radiused" valve seats can be cut but it is more time consuming, unless done by CNC machining. There may be an increase of flow, but it is at the expense of the longevity of the valve seal. Likewise seats that are cut with very narrow seat angle widths may allow more flow but since narrower seats do not transfer heat as well as wide seats, long term durability of the narrow seats will be diminished.

The valve head itself should have sufficient margin so that it doesn't warp when it is hot (especially the exhaust valve). The 45'degree seat can be cut on the valve and then followed by a 30' backcut to aid flow around the valve edge.

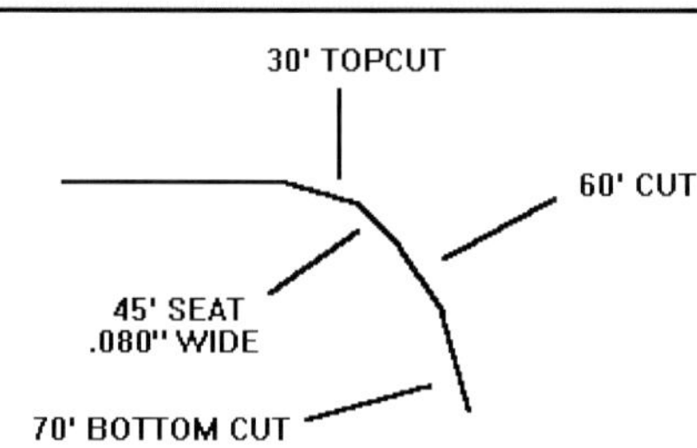

FOUR ANGLE VALVE JOB

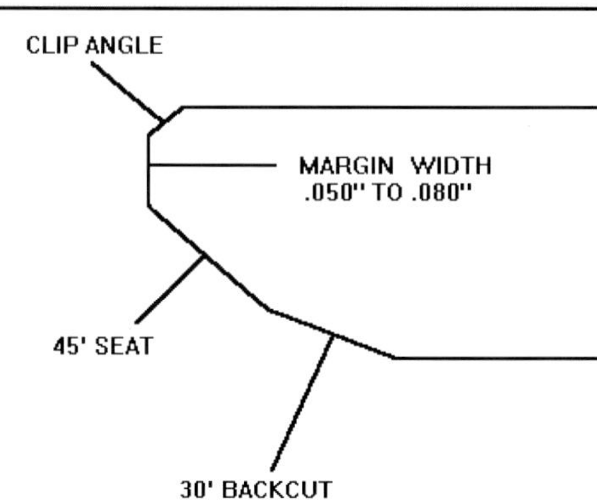

bench experimentation showed where flow and velocity could be increased, but it took an experienced, steady hand to actually remove material or in some cases build up port walls or floors. CNC equipment has changed much of this. Once a particularly good port design has been found on the flow bench, a tracer can be used to record the shape and dimensions of the finished port. The port design is recorded on computer, which in turn controls the CNC milling machine. The CNC port can then be duplicated over and over in exactly the same way. This eliminates much of the hand grinding work formerly involved and it also makes the work extremely accurate and equal from port to port. Some head castings themselves have been redesigned to add additional material to decks, port runner walls, spring seats, and rocker arm stud bosses. Water jackets were moved away from areas that formerly were too close to areas being worked on. These changes corrected weaknesses in the original stock production designs and allow for more extensive port modification.

If you are trying to build a maximum effort race head and money is no object, send them off to a race head specialist. He will have the CNC milling machines, seat grinding equipment, airflow bench and most importantly the skills and experience to do the job right.

GM and aftermarket sources make a wide variety of heads. If you plan on modifying a stock production type head you might consider buying a different set of heads that already have the features you are seeking. By the time you finish adding up the time and cost of modifying a set of stock heads, (machine work, decking, valve job, new guides, crack checking, cleaning, new larger valves, hard seats, bowl porting, valve unshrouding, labor and the cost of the used heads themselves), the price of a new set of race style heads, with the features you want, may not seem too high.

CHEVROLET SERVICE REPLACEMENT CYLINDER HEADS

CAST IRON

(PN-10140633) (Casting #-3964291) has a 108cc closed chamber, rectangular intake port with a square exhaust port exit. It uses 2.19" and 1.72" valves. Intake runner volume is 325cc. Use with closed chamber type pistons. This head should not be used on a Gen.V production block.

(PN-6260482) (Casting #-6272990 or 14096188) has a 118cc open style chamber, with rectangular intake ports with a square exhaust port exit. It uses 2.19" intake valves and 1.88" exhausts. Intake runner volume is 325cc. Open or closed chamber type pistons can be used with this head. This head should not be used on a Gen.V production block.

(PN-14096802) (Casting #-14097088) is a Gen.V rectangular intake port head with intake runner volumes of 325cc. It has 118cc open style combustion chambers and is set up for 2.19" and 1.88" valves. This head is a fully machined "bare" casting. This head should not be used on a Mark IV style block. This head is set up for a non- adjustable Gen.V valvetrain. It can be converted to a Mark IV style adjustable valvetrain by retapping the rocker stud bosses from 3/8" to 7/16" and then installing the appropriate Mark IV adjustable rocker arm assembly and pushrods.

(PN-14096801) (Casting#-14097088) is a Gen.V head exactly like (PN-14096802), but comes with springs, retainers, locks and 2.19"/1.88" valves. Remember that Gen.V production heads are originally intended to be used with Gen.V non-adjustable valvetrains and Gen.V blocks. This head should not be used on Mark IV style blocks.

(PN-10101137) is a Gen.V 427T head with 100cc closed combustion chambers and small oval intake ports. It comes with 2.06" and 1.72" valves. It uses a non-adjustable valvetrain and should not be used with Mark IV style blocks. Use with closed chamber style pistons only.

(PN-10213533) is a Gen.V 454/502 head with 118cc open chambers. Small oval round intake ports are used and it comes with a Gen.V non-adjustable valvetrain with 2.06"/1.72" valves. This head should only be used with Gen.V style blocks.

(PN-12552888)(Casting #-10114156) This cast iron, small round port Gen.VI head has a non-adjustable valvetrain with 2.07"1.72" valves. The open chamber volume is 118cc and comes with valves and springs.

ALUMINUM

(PN-3919838) (Casting #-3919842) is a Mark IV closed chamber head. Chamber volume is 107cc and the rectangular port intake runner volume is 325cc. Valve sizes are 2.19" and 1.84". The exhaust port exits are round. Do not use this head with stock production Gen.V blocks or with open chamber style dome pistons. This is considered the first design aluminum head and was used on L-88 427 motors.

(PN-14011076) (Casting #-14011077) is a Mark IV open chamber cylinder head with 2.19" and 1.88" valves. The chamber volume is 118cc and the rectangular port intake runner volume is 295cc. This is the "smallest" intake port volume found on a factory square port head. The exhaust port has a vane cast into the port floor and these heads are referred to as "C-ports". It is the second design L-88, ZL-1 427 head. Do not use this head with Gen.V stock production blocks.

Remember that when you use an aluminum head you need a composition style head gasket and you need to use hardened steel flat washers under each cylinder head bolt. You also need to use a steel shim or spring cup under the valve springs so that the spring does not cut into the aluminum head's valve spring pocket.

CHEVROLET "BOWTIE" ALUMINUM HEADS

"Bowtie" aluminum heads offer a substantial power improvement over production cast iron and aluminum heads when used with other correct engine components in a large displacement, high rpm racing engine.

(PN-14044861 & 14044862) (CASTING #-14044861), 1st design. These "Bowtie" heads are Mark IV style aluminum heads and as such can be only used on any Mark IV block and Gen.V "Bowtie" blocks. They have raised intake and exhaust port runners, and 105cc, open style combustion chambers. The exhaust port has a vane cast into the bottom floor of the port runner to enhance flow. These heads are sometimes referred to as "W-ports". Mark IV "861" heads require that you use four longer than stock head bolts (PN-12337919) or studs in the holes adjacent to the four exhaust ports and they utilize valves which are .200" longer than the stock stem length. An "861" 2nd design Bowtie head was introduced in 1996 in the "Signature Series."

This 1st design head was introduced in 1984 as (PN-14044862) and is fully machined. (PN-14044861) is fully machined but, lacks seats and guides. The intake ports entrances are raised .100" over a stock intake port and the intake runner is raised .100" again inside the port for a total of .200" difference from a stock port. The shortside radius of the intake port was also raised for improved flow. Intake runner volume was increased to 380 cc. Stock intake manifold ports will not match up to the head intake ports because of the raised runners. You will have to modify the port openings on the intake manifold to match the raised ports on this head.

The exhaust port exits are raised .750" over stock exhaust port exits and have a rectangular shape. The port floor has a vane cast in it and this reduces turbulence and backflow in the port. With a high shortside radius this exhaust port provides excellent flow. Exhaust port volume is 126 cc.

Round exhaust port found on early aluminum closed chamber head (#-842).

The open combustion chamber volume was reduced from stock open chamber designs by adding quench area between the intake and exhaust valves and by not relieving the chamber walls around the two valves. This results in a open style chamber volume of 105cc.

Rocker arm stud bosses are reinforced with larger ribs to withstand high valve spring pressures and the rocker cover rails have been raised 1/4" to improve sealing. Stock pushrod guideplates do not fit these heads because of the larger rocker stud reinforcement ribs. You will have to grind the ribs to fit the guideplates or use a hardened flat washer under the rocker arm stud to act as a shim to raise the guideplates above the reinforcement ribs. These "861" 1st design heads can be used on any Mark IV production block and on any "Bowtie" block.

(SYMMETRICAL PORT "BOWTIE" PN-10051128 & 10051129) (CASTING #'s 10051128 & 10051129) Symmetrical port Mark IV "Bowtie" heads, have four equally spaced intake runners and the exhaust ports are raised even higher than the "Bowtie" 861 and 862 heads. These aluminum alloy heads have 68 (raw castings) or 72cc semi-open chambers. These heads can be used on any Mark IV production block or "Bowtie" block, however, because of major differences in construction special pistons, spark plugs, exhaust headers, head bolt fasteners, gaskets, valvetrain parts and intake manifolds are required from aftermarket suppliers when you use these heads. These head castings need valves which are .800" longer than the stock stem length.

Introduced in 1987 and revised in 1992, this aluminum race head casting was intended for use in the professional Pro Stock drag race category which has a 500 cubic inch engine size limit. The casting is unique in that the head has four equally spaced, equal length intake runners, unlike previous siamesed runner head castings for the big block. This change necessitates the use of unique pistons, valvetrain, custom fabricated intake manifolds, cylinder head fasteners, gaskets, and exhaust headers to match the raised exhaust ports. The head is intended for use by individuals who wish to maximize engine power and have the knowledge and skills to fully utilize the head's features and to customize and fabricate whatever related parts they require.

The symmetrical intake ports equalize port volumes to all cylinders and in order to use these runners and still use the head bolt pattern on existing Chevy blocks, access holes are located above two of the intake port runners. This allows fasteners to be screwed into place, the heads of which then sit slightly below the intake port floor surface. The access holes are then plugged. Cylinder head fasteners for this head are not offered by Chevrolet and must be obtained from aftermarket sources or fabricated by the engine builder.

The intake runners are .750" higher than the intake runners in other

"Bowtie" head castings. Intake runner volume is 400 cc, but can go larger since the ports can be extensively modified by engine builders using this head. The port floors have been moved upwards and the port walls are thick allowing for extensive possible modifications by builders. These heads require valve stem lengths that are 5.85" and 6.00" and utilize a shaft mounted rocker arm ratio of 1.8-1, although other custom ratios are available. Custom length pushrods are required and should be approximately 9.10" for intakes and 10.0" for exhausts on a short deck block.

The exhaust ports were raised .400" higher than other "Bowtie" castings and have round exits. The exhaust port exits have the same bolt pattern as other big block exhaust exits, but since they sit higher custom headers are required. "As cast" exhaust runner volume is 158cc.

The combustion chamber, on the "128" head, was reduced to 68cc by adding quench area between the intake and exhaust valves and on the spark plug side of the chamber, resulting in a heart-shaped shallow chamber. Deck surfaces and wall thicknesses are .600" allowing for extensive modifications. The heads use .708" reach, 5/8" hex spark plugs.

The intake valve angle were reduced from the stock 26 degrees to 18 degrees and the exhaust valve angle was reduced from 17 degrees to 10 degrees. The compound angle of relationship between the two valves was reduced from 4 degrees to 2.5 degrees. These valve angle changes unshroud the valves and permit more airflow.

(PN-10051128) is fully machined but has no seats or guides installed and is made of 356-T6 aluminum alloy.

(PN-10051129) is a "raw" unmachined casting and allows an engine builder to custom machine the heads. This head comes with combustion chamber volumes of 72cc and is made from a different aluminum alloy 355-T6.

Chevrolet does not make pushrods, rocker arm assemblies, intake manifolds or gaskets for these heads and therefore they must be fabricated or

Notice that the "861" 1st design Mark IV head has round water jacket holes in the deck to match the jacket core holes in Mark IV production blocks and Mark IV and Gen.V "Bowtie" blocks. Arrows point to the two additional bolt holes found in two intake port runners on these heads. You can use them for additional head bolts or plug them if they go unused. "861" 2nd design heads will work with any Chevy big block.

Bowtie "861" first design have 105cc chambers. This chamber has been smoothed out and the valves unshrouded. When the owner was done, the chambers measured 118cc.

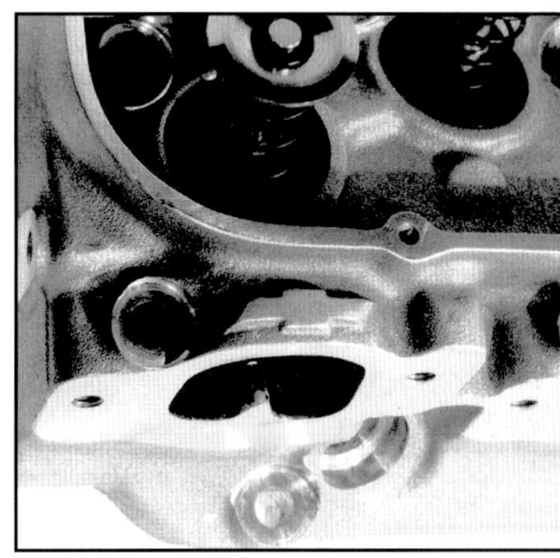

Bowtie logo found on "861" first design head is above the exhaust port with the port vane found on these "W" port heads.

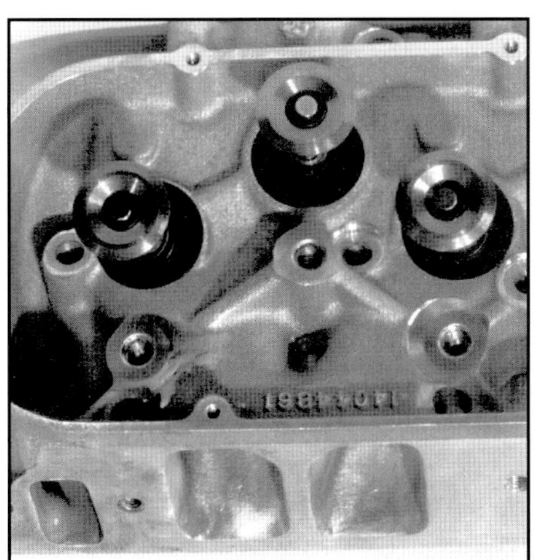

"861" first design aluminum head has bigger ribs around the rocker stud bosses and raised ports with large 380cc volume intake ports.

Arrow points to a hole in the Bowtie "861" intake port floor. An additional allen head bolt can be used here for additional head clamping to a block with a matching hole. If you decide not to use an additional head bolt, make sure that these port floor holes are plugged.

The exhaust port with a vane on the port floor is found on the Mark IV Bowtie aluminum "861" head. The port is raised .750" above the stock port position.

Bowtie "861" casting has large ribs to strengthen the rocker arm stud bosses. Stock pushrod guideplates do not fit unless the ribs are modified or you use a flat hardened washer under longer rocker arm studs.

obtained from aftermarket suppliers.

"SIGNATURE SERIES" BOWTIE ALUMINUM BIG BLOCK CHEVY HEADS

In 1996, GM/Performance Parts introduced three new cylinder heads for the big block Chevy. These three heads have been cast so that they will fit on any Mark IV, Gen.V or Gen.VI 396 thru 502 engine block without worries of water leaks past the head gasket into the lifter valley.

(PN-12363399, Casting #-12363391) is a CNC machined aluminum large oval port head. The head deck is 9/16" thick for better gasket sealing. This head has 290cc/110cc port volumes with hand blended bowls and a 110cc semi-open combustion chamber. It is set up for 2.19"/1.88" valves and comes with interlocking hardened seats. The bronze guides are semi-finished and can be finished to 11/32" or 3/8" inside diameters. The head is setup to use 3/8" pushrods. Adjustable 7/16" rocker studs are used as are 1.55" valve spring diameters. The rocker stud holes and exhaust bolt holes are helicoiled.

(PN-12363400, Casting #-12363401) is a rectangular port head with the ports in the standard locations. The head is CNC machined with hand blended bowls and port volumes are 315 & 300cc/110cc with 118cc "open" combustion chambers. It uses 2.19"/1.88" valves and has interlocking hardened seats. The bronze guides are semifinished and can be final honed to 11/32" or 3/8" diameters. 7/16" rocker stud holes are helicoiled as are the exhaust bolt holes. The decks are 9/16" thick and the spring pockets are set for 1.55" diameter valve springs.

Use the following parts with the two aluminum "Signature Series" heads listed above.

Part #	Description
12337919	7/16"x14x5.25" head bolt
10141204	7/16"x14x4.06" head bolt
10141205	7/16"x14x2.08" head bolt
14011040	.45"x.78x12 head bolt washer
12363414	head gasket, use with '65-'90 MkIV block std. bore to 4.370" bore
12363413	head gasket, use with '65-'90 MkIV block with 4.47"- 4.54" bores
12363412	head gasket, use with '91-'96 Gen.V block std. bore to 4.370" bore
12363411	head gasket, use with '91-'96 Gen.V block with 4.47"-4.54" bores

(PN-12363425, Casting #-14044861 and 12363425, 2nd Design). This is a "raised runner" "W-port" aluminum head. This head is the "2nd design" "861" Bowtie head and like all "Signature Series" heads will fit any MkIV, Gen.V or Gen.VI block. The first design "861" head is still being made, but only as a "raw" casting under (PN-14044876) and it will not fit Gen.V or Gen.VI production blocks. The 2nd design "861" head has 380cc raised intake runners which are .750" higher than the standard intake port location. The raised exhaust port has a vane cast into the port floor to increase flow. 115cc semi-open combustion chambers are used. Semi-finished bronze guides and seat inserts are included for 2.19"/1.88 valves. There are two extra head bolt holes in these heads for bolts to go into the lifter valley and provides each cylinder bore with six head bolts around the barrels for increased sealing. You will need

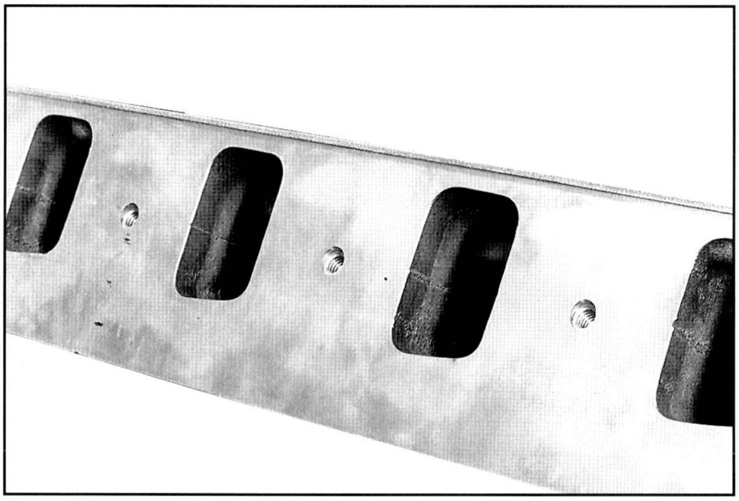

Casting #-128 symmetrical port Bowtie aluminum race heads have non-siamesed port runners. These heads require special intakes, valvetrain parts and head fasteners.

Symmetrical port "128" head requires some special head bolts that pass thru threaded opening in rocker arm valley and fit into the port floor. Threaded opening in the rocker valley is then plugged.

The raised exhaust port on the symmetrical port #128 aluminum head.

Chevrolet's new "Signataure Series" Bowtie aluminum cylinder heads are the latest high performance offering from the Chevy RaceShop

valve stems which are .200" longer than stock and four extra long head bolts (PN-12337919) near the raised exhaust ports. Use 3/8" pushrod guideplate (PN-3860038) or 7/16" (PN-3879620) and remember to grind enough clearance in the guideplate to clear the rocker arm stud boss reinforcements to fit the guideplate. You will need an aftermarket intake manifold and exhaust system that will match the raised intake and exhaust ports found on "861" heads.

PONTIAC SUPER DUTY CYLINDER HEADS FOR CHEVROLET BIG BLOCKS

Pontiac Motorsports has developed six different aluminum race cylinder heads to be used with Mark IV Chevrolet and all Chevrolet "Bowtie" engine blocks. These heads are offered for the individual seeking a maximum effort race engine. These heads are not intended for street use and some are sold only as "raw" castings which require extensive machine work before use. All of these heads feature "spread port" equal length, intake ports, raised runners, and thick decks. They will require aftermarket or custom fabricated intake manifolds, rocker stands and other valvetrain parts to use them.

(PN-10045427)(Casting#-10045427) Pontiac Super Duty "Large Port" aluminum race cylinder head has "spread port" intake ports and runners of equal length. The raised intake port volume is 460cc and the runner walls have a minimum thickness of .450" to allow extensive port runner modifications, however these heads do not require much in the way of porting. They were designed to incorporate much of the latest development in race head design. The exhaust ports are "D-shaped" and have a port volume of 145cc. Port volumes in these heads are kept within 1% of each other.

The valve angle on the intake is 18' and the exhaust is 9'. The open style combustion chamber volume is 91cc and will accept up a 2.40" intake valve and a 1.90" exhaust.

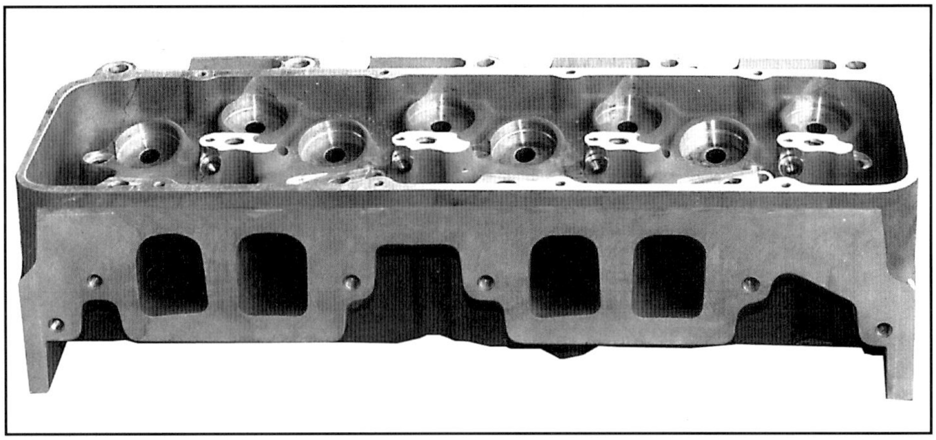

Pontiac "Super Duty" casting #-875 aluminum race head has "spread ports" and requires custom intake, water crossover manifold, valvetrain parts, piston domes and head studs.

Pontiac Super Duty heads require a special water crossover manifold that connects the heads through the outlet shown here.

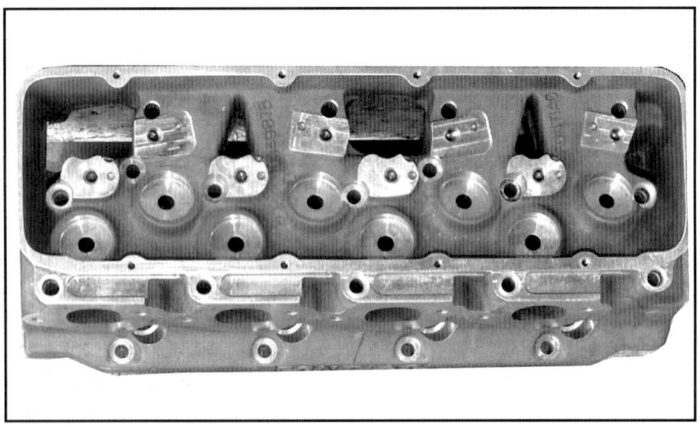

Aftermarket rocker arm stands are used with the Pontiac "Super Duty" heads.

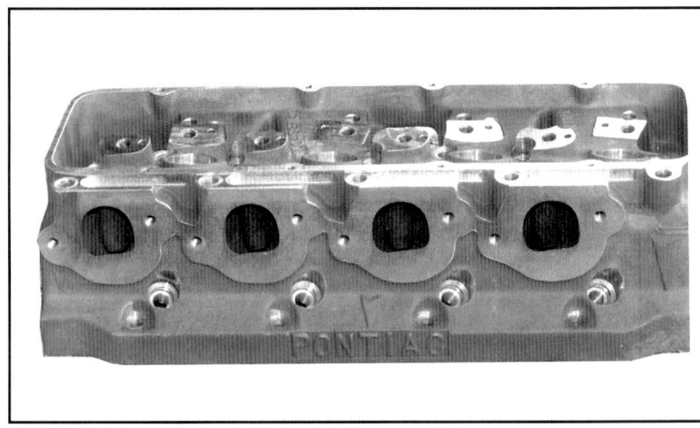

Raised exhaust ports on the Pontiac #875 "Super Duty" aluminum head are "D" shaped.

These cylinder heads were designed to be used on very large cubic inch motors from 450 up to 600 ci. Aftermarket custom intake manifolds, pistons, exhaust headers, extra long head bolts and valve stems and aftermarket valvetrains are required with these heads. Pontiac cast aluminum valve covers (PN-12341643) are required. 3/4" reach spark plugs are used. Phosphorous bronze valve guides are supplied with the heads but are not installed.

(PN-10049875)(Casting#-10049875) is the Pontiac Super Duty "Small Port" aluminum race head. It also has "spread port" intake runners. It is similar to the "Large Port" casting but has some differences. This head was designed for use on race motors up to 450 cubic inches and as such, the raised intake port volumes are 376cc and the "D-shaped" exhaust port volume is 145cc.

The exhaust valve centerline is moved .080" closer to the intake valve to reduce airflow restrictions between the exhaust valve and the cylinder wall on smaller displacement engines. The heart- shaped combustion chamber has a volume of 83cc that take a 2.25" intake and 1.88" exhaust valve. Extra long cylinder bolts and valve stems are required. Custom intake manifolds, exhaust headers, pistons and valvetrains will have to be obtained. Pontiac valve covers are also necessary. Valve guides are included, but not installed. A bare head weighs approximately 47 lbs.

(PN-10093386) is another Pontiac "spread port" aluminum race head. However, in this head the intake runner roof is rounded to improve airflow. The intake port runner volume is 365cc and will feed a 450 cubic inch race motor. The exhaust port is "D-shaped".

The valve angles are set at 15 degrees and the heart-shaped combustion chamber volume is 61.4cc. 2.40" and 1.90" valves can be used.

(PN-10093385) is a "raw" unmachined version of the 10093386 aluminum head, but has a port volume of 420cc's and a heart-shaped chamber volume of 60.4cc's.

(PN-10093378) is a machined Pontiac Super Duty "spread port" aluminum head designed for race motors up to 600 cubic inches. It also has a rounded intake runner roof but with a port volume of 420cc. The "heart shaped" chamber volume is 60.4cc and will take valve sizes of 2.40" and 1.90". The exhaust port is "D-shaped" like all other Pontiac Super Duty heads for the big block.

(PN-10093383) is a "raw" unmachined version of the 10093378 aluminum head.

Pontiac offers a coolant manifold (PN-10045450) for the above heads which connects to the front of each head. Coolant is not circulated thru an intake manifold.

EDELBROCK ALUMINUM HEADS

Edelbrock "Performer" aluminum oval port heads are available either bare (PN-6046) or complete (PN-6047) and they are "50-state" legal for emissions regulated vehicles. These heads will work with either Mark IV, Gen.V or Gen.VI blocks. The chamber volume is 110cc and the intake port volume is 290cc. Stainless steel valve sizes are 2.19" and 1.88" with 11/32" stems. ARP rocker arm studs are used. The heads do have the heat riser passage drilled.

Edelbrock "Performer RPM" 454-O complete heads, (PN-6045), have large oval ports with CNC matched 290cc/110cc volume runners. They will fit on either Mark IV, Gen.V or Gen.VI blocks. The valve sizes are 2.19"/1.88" with 11/32" stem diameters and the semi-open style combustion chambers measure 110cc. The heads are machined for 1.55" stock style valve spring diameters and take ARP 7/16" screw- in rocker arm studs and an adjustable Mark IV rocker arm assembly. Bronze guides and hardened interlocking seats are used. The decks are thicker than stock. Stock large oval port intake manifolds, exhausts manifolds and accessory brackets will all bolt right up to these aluminum heads. You can buy them complete (PN-6045) or bare (PN-6044). These heads are for non-emission regulated vehicles.

Remember to use an early style large oval port intake manifold with these heads and a composition type head gasket. You also need to use a hardened steel flat washer under all of the head bolts and use a hardened spring cup or shim under the valve springs when using aluminum heads.

Edelbrock "Performer RPM" 454-R (PN-6055) are aluminum rectangular port complete heads. Intake port volumes are 315cc on the long runners and 300cc on the short intake runners, with 110cc exhaust port runners. The open style combustion chamber is 118cc and uses 2.19"/1.88" valves with 11/32" stem diameters with undercut stems. The ports are CNC matched and the bowls are hand blended. Port openings are in the stock location and have stock accessory mounting holes and bosses. With the complete heads you get stainless steel, one-piece, swirl polished valves with undercut 11/32" stems, PC-style oil seals, springs, guideplates, rocker studs, retainers and keepers. These heads will fit either Mark IV, Gen.V or Gen.VI blocks. Bare Performer RPM 454-R heads are (PN-6054).

The Edelbrock aluminum heads feature interlocking hardened valve seats so they can be used with unleaded fuels and the port runners and bowls are hand blended and CNC port matched. The decks are thicker and the bolt bosses for the rocker studs and exhaust manifolds are heli-coiled.

Edelbrock "Victor" aluminum big block race heads (PN-7760) are bare square port heads with 380cc intake port volumes. The intake port is raised .100" and exhaust ports are raised .750" from the stock position. They have 105cc semi-open combustion chambers and require .400" long valve stems. The machined valve cover sealing rails have been raised .200" to prevent leaks while retaining the stock valve cover bolt pattern. The rocker arm stud bosses are reinforced with ribs. Two extra cylinder head bolt holes are provided for additional clamping power. The heads need to be port matched to your intake and the bowls beneath the valve seats need to be blended. These heads will fit either Mark IV, Gen.V or Gen.VI blocks and should be used on high rpm (over 7000) or large motors.

If you use these heads on a Gen.V block remember that the stock production pushrods for the non-adjustable Gen.V valvetrain are shorter than those used with adjustable Mark IV heads. When using an adjustable valvetrain you need to switch to the Mark IV style pushrod and rocker arm assembly or to an even longer pushrod with the "raised port" heads.

MERLIN BIG BLOCK HEADS

Merlin cast iron cylinder heads are available from World Products with either large oval ports or rectangular intake ports. The oval port heads have intake runners that measure 262cc or 295cc and the rectangular intake runners are 315cc or 330cc. Merlin cast iron heads all have equalized intake runner volumes. The decks are all thick and the combustion chambers are 119cc in volume. The chambers have hardened seats and are machined for 2.19"/1.88" valves, but can accept up to 2.30"/1.90". The valve spring pads are machined for 1.550" diameter springs but can be opened up to 1.625". The valve train is the same as the stock Mark IV Chevy adjustable valve train. The heat riser passage is left undrilled, but you can drill it out if you need it. Extra material is in the port runner walls to

Photo courtesy of World Products

World Products 49 state legal Merlin replacement heads are a popular chioce for street performance engines and even tow vehicles. The oval port version of these heads makes a ton of torque for street and bracket applications.

allow additional porting. Stock intake manifolds, exhaust manifolds and accessory brackets will bolt to these heads. The casting has been reinforced in critical areas to prevent cracking sometimes found on some production Chevy heads.

These heads will work with either Mark IV, Gen.V or Gen.VI blocks and are "49-state", emissions legal (California pending) replacements. World Products recommends that you use a Gen.V head gasket with these heads, when using a Gen.V block, to insure that the proper water core holes are blocked and that there is sufficient gasket surface around block water passages. World Products also recommends that you install tube exhaust headers when you install Merlin Heads to aid in cylinder head cooling and to take advantage of increased breathing. You need to use eight extra long cylinder head bolts, with hardened washers, in the holes near the exhaust port exits because the bolt bosses were moved up to improve exhaust port flow. The required extra long head bolts are available as a kit (#7301) from World Products. The heads can be ordered bare or complete.

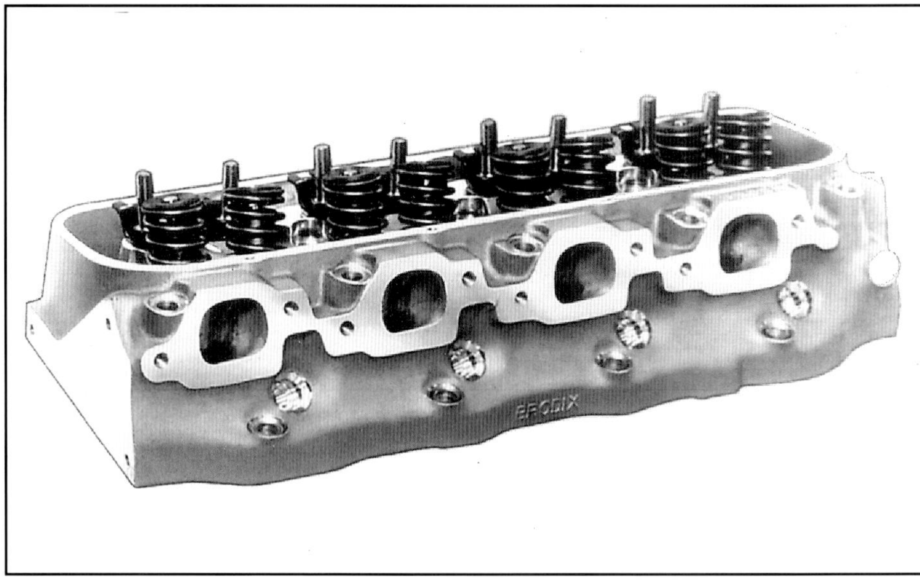
Brodix makes aluminum heads for the big block Chevy (BB-2 shown) in various port volume sizes and valve sizes. They have different heads for the street enthusiast thru all levels of competition, including some for 700+ cubic inch "megamotors". (Photo courtesy of Brodix)

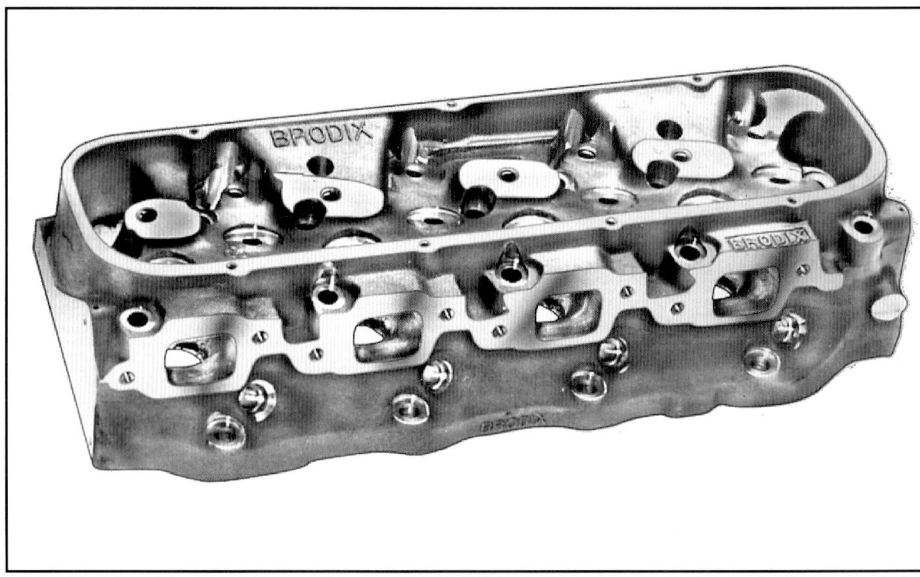

Brodix "BB-4" aluminum head is available for serious big block racers. This rectangular raised port head is recommended for 468-572 cubic inch race motors. It has 360cc intake ports that "out of the box" flow 361 cfm at .700 lift and will work well above 7000 rpms. (Photo courtesy of Brodix)

BRODIX

Brodix offers a wide range of aluminum big block Chevy heads. Their heads will fit either Mark IV, Gen.V or Gen.VI blocks.

All complete heads come with bronze guides, ductile iron seats, 10 degree steel retainers and locks, springs seals, valve bowls blended, 3-angle competition valve, 7/16" rocker studs that are .250" longer on top and intake gasket port matching. Valve stem diameters of 3/8", 11/32" and 5/16" are available.

BB-1 OEFI Oval Port - This aluminum head has 119cc open chambers and intake port volumes of 265cc, with the exhaust port raised .600". These heads can be ordered bare or complete with 7/16" studs that are .250" long on top, the valve springs of your choice, 10 degree steel retainers and locks, 3/8" guideplates, valve bowls blended and 2.25" and 1.88" valves and oil seals. Spring pocket diameters can be cut to 1.625", but they should not be deepened. Extra long head (6.0") bolts or studs are needed next to the raised exhaust ports. Valve stem length is 5.218"/5.394". Out of the box these heads will flow 272 cfm at .600"-.700" lift on the intakes and 225 cfm on the exhaust side at .700".

BB-1 Rectangular Port - The BB-1 is aluminum head with 119cc open chambers, 280cc intake port volume and this head also has exhaust ports which are raised .600" over the stock height. Angle milling to 108cc and angle machining to 98cc chambers are possible. Spring pocket diameters can be cut to 1.625", but do not deepen them. The same valvetrain parts are used on this head and the BB-1 OEFI to complete it. This head is recommended for 427-454 engines with a flat tappet cam. Holes and bosses to fit standard header flanges, intakes, accessory brackets, pushrods and valve covers are provided. 10 longer head bolts or studs are required. Valve stem length is 5.218" and 5.394". Out of the box these heads flow 302 cfm intake and 226 cfm on the exhaust at .700" lift.

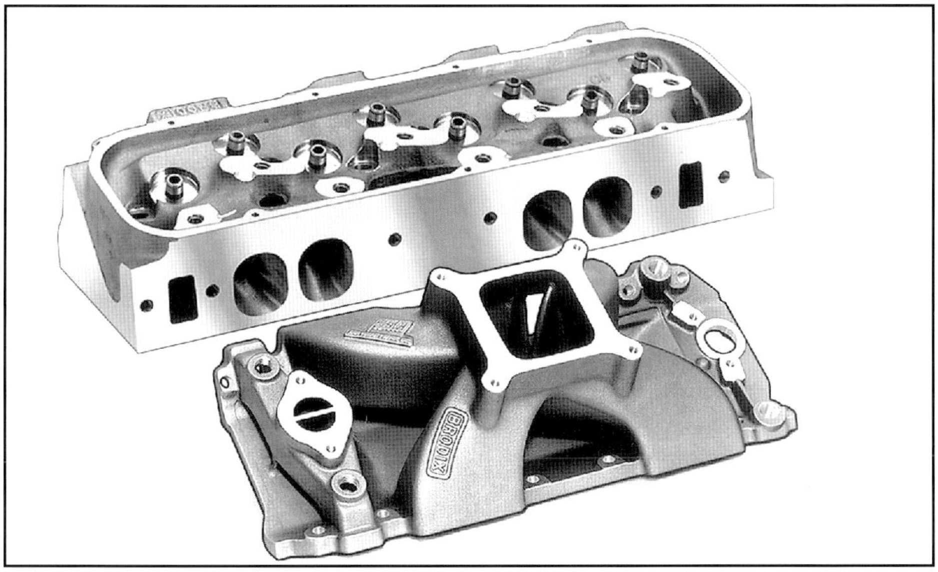

If you are looking for a large oval port aluminum head and intake, the Brodix "BB-1 OEFI" heads and "HVBB-O" intake manifold will make an excellent high flow velocity combination. (Photo courtesy of Brodix)

BB-2 - This head has 119cc open chambers and 300cc intake port volumes. It is recommended for 427-509 cubic inch motors with at least 12.0-1 static compression. The bronze guides are replaceable and the seats are ductile iron. Rocker arm bosses are reinforced and may be milled for Jesel shaft rockers. Valves are 2.25" and 1.88" and stem lengths are 5.218"/5.394". The exhaust ports are also raised .600" requiring 10 longer head bolts or studs. The heads can be angle milled to 108cc chambers or if angle machined to 98cc. Out of the box performance on the flowbench is 319 cfm and 225 cfm at .700" lift.

BB-2X - 119cc open chambers and larger 340cc intake port volumes come on this head. It is recommended for 482-572 cubic inch engines. Valve diameters are 2.30"/1.88" and the intake stem length is .250" longer at 4.468", while the exhaust stays at 5.394". Replaceable bronze guides and ductile iron seats are provided. Spring pocket diameters can be cut to 1.625", but no deeper. 108cc chambers are possible with angle milling and 98cc with angle machining. Exhaust ports are raised .600" higher than stock. Standard rocker arms, valve covers, intakes and pushrods can be used. The flow on this head out of the box is 323 cfm/226 cfm at .700" lift and can be fully race ported to flow 408 cfm on the intake side.

BB-3 - 370cc intake port volumes and 119cc open chambers are recommended for 540-605 cubic inch big blocks and this is the largest conventional head Brodix offers. Exhaust ports are raised .600" over stock. Longer head bolts or studs are required near the exhaust ports. This head requires a .250" long intake valve and valve sizes are 2.30"/1.88". On this head the exhaust valve has been moved an additional .030" from the intake position. Since the exhaust valve has been moved slightly, you need to make sure the corresponding valve reliefs in the piston are properly located. Also because of the exhaust valve placement, you will need a special size stud girdle with these heads. Bronze guides and ductile iron seats are provided. The head can be angle milled to 108cc, but should not be angle machined. It is not recommended that you use a Jesel type rocker arm system with this particular head. Spring pocket diameters can be cut to 1.625". This head out of the box flows 371 cfm/228 cfm and race ported can flow 395/280 cfm.

BB-4 - 134cc open chambers and 360cc intake port volumes are found on this head recommended for 468-572 inch big block screamers with at least 13.0-1 static compression and which operate at 7000 rpms or above. Both the intake and exhaust ports have been raised. Valve sizes are 2,30"/1.90" and the intake stem is .250" extra long, 5.468" overall, while the exhaust uses a .100 longer stem for a total of 5.494". The BB-4's, if used on a short deck case, will require the use of a wider tall deck intake manifold or the use of spacers. If used on a tall deck block, you will need the wider tall deck style intake plus the spacers. The valve positions have been "rolled over" to 24/4 on the intake and 15/4 degrees on the exhaust. A pedestal type rocker arm system from T & D Machine or Jesel is required. Like other Brodix heads the spring pocket diameters can be opened to 1.625", but don't cut them deeper. Depending on what type of combination you plan to use make sure that the piston dome and/or valve reliefs will work with these "rolled over" heads. As delivered these heads will flow 361 cfm/251 cfm and can be ported to flow more than 400 cfm on the intake side.

In addition, in conjunction with WELD TECH, Brodix heads can be 100% CNC prepared.

BRODIX "EB" Series

Brodix also offers an "EB" line of professional level drag race or endurance race heads which feature various large port volumes for big block race motors of 400-800 cubic inches. The "EB" series are Brodix castings that are designed by Engine Prototype Development (EPD). Marine versions and solid heads are available in the "EB" line. These heads can be furnished "as cast" or fully CNC race ported and can be hand finished to your specifications and then given a valve job, cc'ed, surfaced, flow benched, rocker stands installed and pressed in head stud washers installed. Steel guides with bronze liners are used. These are serious heads.

Drag Race Applications "EB" series

EB-420 - Recommended for 400-475 inch race motors and slightly larger motors with automatic trannies, these heads need a minimum of 4.375" cylinder bores and will work well with 4.440" bores. 2.375"/1.88" valves are used.

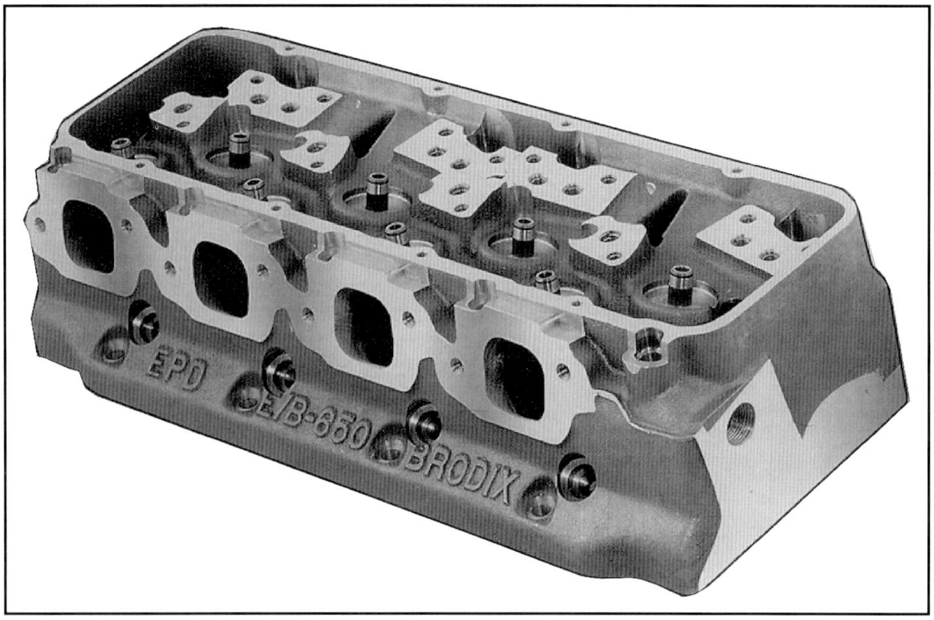

The EB series of aluminum race heads from Brodix were designed in conjunction with Engine Prototype Development (EPD) and come in various port volume sizes for serious racers looking for the most power. The CNC machine ported race heads can be custom prepared and then flowed for use in big block engines from 400 to 800+ cubic inches. (Photo courtesy of Brodix)

EB-510 - For 475 to 575 inch engines with minimum 4.500" cylinder bores and up to 4.600" bores. 2.420"/1.88" valves are used.

EB-600 - This head is recommended for 575-632 monster motors and comes with 2.45" intakes and 1.90" exhaust valves.

EB-650 - Head comes with 2.475"/1.92" valves and is made for 632- 700 cubic inch big blocks with a minimum bore size of 4.560" and up to 4.600". Brodix says this head will flow over 500 cfm on the intake and 345 cfm on the exhaust.

EB-700 - For the guys who think that no Chevy is too big and have 700+ cubic inch motors with 4.560-4.625" cylinder bores. Valves sizes are 2.475" and 1.92" and this head has larger port volumes than the EB-650 head.

EB-800 - This head is the largest in the "EB" series line and is for Chevy motors from another planet who bring 700-800 cubic inch engines to the pavement wars. Minimum bore size of 4.600" is needed because this head uses 2.50"/1.93" valves. The new DRCE-2 or other aftermarket 4.900" bore spacing blocks with 4.700" bores, are the foundation for these heads.

Endurance Race Applications "EB" Series

These heads supply more flow with lower lift cams in the mid rpm ranges for marine, oval track or off road race motors.

EB-498 - For 480-560 size motors with 4.375" to 4.440", it comes with 2.375"/1.88" valves.

EB-598 - Should be used with 560 to 650 cubic inch engines. Valve sizes are 2.40"/1.90" and require 4.500" minimum cylinder bores and up to 4.600".

WHICH HEADS SHOULD I USE?

I guess that one answer would be to buy the best ones you can afford and which match your operational requirements. Also, keep in mind that you can use a head with a particular port volume and then have additional work done to improve the flow on those heads or you can buy a head with a better port design or larger port volume and run them almost "as is" in terms of improved flow. Make sure that the heads you use will work with the block you intend to use. The water jacket core holes in the block deck must seal with the water core holes in the head deck. If you are building a motor for street use or bracket drag races or a boat, with a stock production cubic inch engine size and a maximum rpm limit of 6,600, large oval port heads will provide more torque and mid rpm range punch than square port heads.

If the rpms on a race motor will go over 6600 or if you are building a large cubic inch engine, then use a square port head. Large cubic inch motors over 460 cubic inches can move up to the larger square port intake volumes of 300cc or more depending on the size of the motor, rpm range and intended use.

Cast iron heads can make a great deal of power. Aluminum heads offer a number of advantages. First is obviously weight reduction. It reduces the total weight of the vehicle and except for a rear engine dragster or boat, the weight comes off a high position in the front end. Second because of the quicker heat dissipation rate of aluminum, a slightly higher compression ratio may be used with aluminum heads because the point where detonation occurs will be slightly higher. Some folks feel that cast iron heads actually offer some advantage in this respect by keeping more heat and energy in the cylinder which can be turned into power. Also keep in mind that aluminum heads are easier to repair if something goes wrong. The main concern I think for most people is cost. Aluminum heads always cost more than the cast iron variety. Maybe some aftermarket head producer will come up with a manufacturing process that allows a price that is comparable with cast iron heads. Perhaps someone will cut their per unit profit margins in return for increased unit sales.

BIG BLOCK Quick Facts & References

DART CYLINDER HEADS

PN	Intake Runner CC's	Chamber CC's	Valve Sizes	Notes
	265	119	2.19/1.88	oval port
6000	320	139	2.25-2.30/1.88	square port
6001	360	139	2.25-2.30/1.88	square port
6002	320	119	2.25-2.30/1.88	square port
6003	360	119	2.25-2.30/1.88	square port

The following Dart/Pontiac "Big Chief" 355-T6 alloy aluminum heads are maximum effort, professional level Pro Stock style heads. They feature raised, "spread ports", very large port volumes, small heart-shaped combustion chambers, revised valve angles and require aftermarket rocker arm assemblies. CNC machine programs can provide custom valve spacing and valve diameters.

PN	Intake Runner CC's	Chamber CC's	Valve Sizes	Notes
9000	395	85	2.375/1.90	18/9 degree valve angles
9001	395	74	2.375/1.90	14/7 degree valve angles
9002	395	74	2.375/1.90	16/7 degree valve angles

DART HEADS

Dart Machinery also makes high performance and race aluminum heads for big Chevys. Some of their big block heads come with larger combustion chamber volumes of 123, 139, or 143cc. These chamber sizes can be left as is for supercharged applications or the decks can be cut down to reduce chamber volume and increase static compression in normally aspirated motors. The heads are made in siamesed or spread port configurations. Solid versions are available for alcohol fueled motors.

These Dart heads will require valve stem lengths that are longer than stock because of the raised ports and will also require head bolts or studs longer than stock to fit the raised head bolt columns. Some Dart heads will require aftermarket custom intake manifolds and aftermarket valvetrain assemblies.

BIG BLOCK Quick Facts & References

CYLINDER HEAD PREP CHECKLIST

1. Degrease and perform a careful visual inspection. Look for any signs of damage.
2. Magnaflux and pressure test all used heads to locate cracks.
3. Inspect valves and guides for wear. Inspect seats for cracks, erosion.
4. Recondition or replace valve guides if required.
5. Cut spring seats for large diameter springs if necessary. Note: Some heads have very little material between the spring seat and the water jackets; proceed with caution.
6. Machine the tops of the guides for press-on seals if used.
7. Match intake ports to heads and exhaust ports to manifolds or headers.
8. Smooth port contours, especially in the bowl areas (pocket port ing), as desired.
9. Perform a quality three-angle or four-angle valve job with proper seat widths and angles.
10. Install valvesprings to proper height and check spring pressures.
11. Check for adequate spring, retainer to guide and seal, and rocker clearances.
12. Check for uniform valvestem height.
13. Measure and equalize combustion chamber volumes.
14. Install studs and guideplates if required.
15. Thoroughly clean heads with soap and water, blow dry with com pressed air, and store in plastic bags until ready for assembly.

Valvetrain

TIMING SETS

Timing chain covers are not interchangeable between Mark IV (PN-330850) and Gen.V (PN-10191648 or 12551544)) motors, even though they have the same bolt pattern, because oil pan gasket mating surfaces for each type of engine are different. The timing chain cover found on the Gen.VI 454HO and 502HO is a different composite material design that uses only six bolts to hold it on and uses a notch in the cover, instead of a timing tab, to indicate ignition timing. The Gen.VI timing cover works with the one-piece oil pan gasket found on these motors.

Very early Mark IV blocks did not have two dowel pins to locate the timing chain cover. Later, two dowels were added. If you need them the block dowel pins for the timing chain cover are (PN-10105879).

If you need the little camshaft dowel pin, it is (PN-10105879). A .015" thick shim (PN-3975949) is available to fit behind the cam sprocket. Camshaft bolts (PN-9424877) are 5/16"-18 x 1". A cam thrust button is listed as (PN-12364086).

The bolt on timing pointer tab for an eight inch diameter harmonic balancer is (PN-3992073) and can be used on Mark IV or Gen.V blocks.

Many of the timing sets available today come with a crank gear that has multiple keyways. These different keyways allow you to advance or retard the cam timing in relation to the crank. Generally speaking advancing

When setting and checking cam timing it is helpful to use a large diameter degree wheel. Small changes are easier to see. An easily adjustable Jesel timing belt set up makes cam dowel offset bushings unnecessary for cam timing changes.

the cam timing will increase engine power at the low end of the rpm scale, while retarding the cam timing will increase the top end power. There are some aftermarket timing chain systems that have adjustable cam gears that allow for cam timing changes without resorting to offset dowel pins. However, I suggest that if you are unfamiliar with these types of cam timing changes and what will result to your particular engine and cam combination, that you use the stock "0" setting on the crank keyway.

Keep in mind that some cams are ground with a certain amount of advance when they are made. Other cams are ground "straight up" but may require that you advance the cam by means of offset bushings or other means. Check with the manufacturer and the cam timing card you receive when you purchase the cam. Any changes you do make can be made when the cam is "degreed in". Some of the engine dyno simulator computer programs can help you determine whether cam advance or retard will make more power in your situation. Specific cam timing can be entered into these programs and then you can advance or retard the cam timing from the manufacturer's specs on the computer. When you actually build the engine make sure that there is adequate clearance between the valve and piston deck. Changes to the cam timing can effect the point when the valve comes closest to the piston as it approaches and passes over Top Dead Center.

Five different types of timing sets are available. Chevrolet has used a stock, production "silent" link chain with metal gears that have plastic teeth on the large cam gear and metal teeth on the crank gear. These timing sets are quiet, inexpensive and dampen crank vibrations transmitted to the cam gear. However, the chains stretch very quickly with use and cam timing quickly changes because of chain slack. Many racers use these chains but they must be replaced frequently because of stretch. A GM "silent", single roller chain is (PN-10114177). Use cam sprocket (PN-10106429) and crank sprocket (PN-10106430).

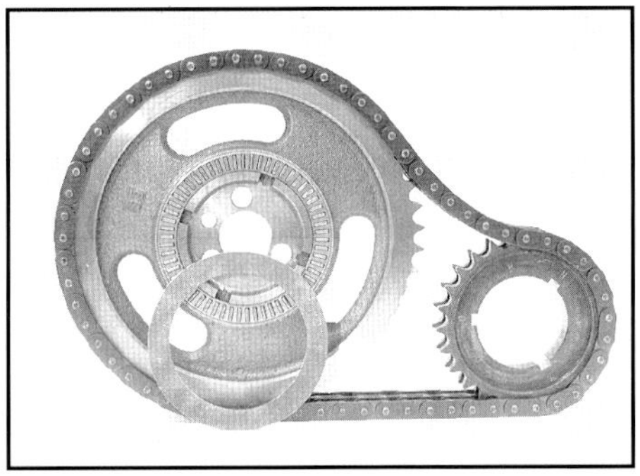

Roller timing chain will work in most any application. They are relatively inexpensive and can be replaced when they show signs of stretch and chain slack.

"Double roller" chains offer more strength, have metal teeth on the gears, are a little noisier, but still stretch with use. (PN-3891519) is a "double" roller chain design and uses cam sprocket (PN-3891517) and crank sprocket (PN-3891518).

"True roller" chains have increased strength and have larger solid bushing rollers that roll as they pass over the gear teeth. They do not stretch as much and will give accurate cam timing for thousands of miles.

Gear drive timing sets were offered by Chevrolet for early reverse- rotation marine cam applications and on some early tall deck truck engines. These sets used two gears, but no chain to turn special reverse-rotation camshafts on these motors. The reverse-rotation cams also required that the distributor drive gear be reverse- rotation. So if you change a reverse-rotation gear drive over to a normal rotation chain driven setup, make sure you change the distributor drive gear also.

Aftermarket suppliers make a number of different variations of three and four gear drives for regular rotation camshafts. These gear drives are noisy and transmit crank vibrations to the cam. However, since there is no chain to stretch, cam timing stays relatively accurate.

Edelbrock "Accu-Drive" has a four gear timing drive for big blocks that also allows easy adjustment of cam timing.

No matter which type of timing set you use make sure that the cam gear does not rub against anything on the front face of the block that will interfere with the rotation of the cam gear. You can use a brass shim (PN-3875949) under the cam gear for clearance or have the block cam thrust surface or cam timing gear machined for clearance. Cam thrust bearings can be used to provide clearance and protect the cam thrust surface on the block.

Roller and some flat tappet cams will need some sort of device to keep the cam from moving forward in the block. Cam button spacers or bumpers from Chevrolet (PN-12364086) and other aftermarket designs are available.

Adjustments in cam timing, whether it is to correct a manufacturing problem or to try advancing or retarding the cam placement for power reasons, can be accomplished a number of ways. All of these changes are made with the timing gears. For a long time folks would get out their degree wheels and piston stops (GM PN-12364087) to check the accuracy of the cam as delivered and make whatever adjustments they felt necessary. If a change in cam placement was to be done, the cam dowel pin hole on the cam timing gear was slightly enlarged and then an offset bushing that gave the appropriate increment of change was placed in the hole. Offset crank keys were also offered to do the same thing. It is a cumbersome process. Now many timing chain sets are available that have multiple keyways that allow you to change cam timing by choosing the appropriate keyway to get the advance or retard that you may want. On many belt timing sets and some timing chain sets, very precise changes can be made quickly and easily thru adjustments made on the cam gear itself. It is one of the advantages of these timing systems. You still need to check the cam timing with a degree wheel to assure yourself that the cam you have purchased is in fact doing what it should and that the valve events occur when they are supposed to.

If you have had your block align bored, the distance from the crank centerline to the camshaft bore centerline may have been decreased. Cloyes or Competition Cams make a timing set that is slightly shorter and will make up for the decrease. Likewise longer timing sets are available for the aftermarket blocks that feature a raised cam bore. Timing chain sets are available that are already "pre- stretched".

CAMSHAFTS

Hydraulic and mechanical flat tappet cams have been used in the big blocks from Chevrolet/GMC. The following chart gives information on camshafts available from the factory and aftermarket suppliers. They are interchangeable between Mark IV and Gen.V blocks. In 1996, Gen.V "Vortech 7400" production 454's began the use of factory installed hydraulic roller cams as did the Gen.VI 454HO and 502HO.

Aftermarket cam makers make an almost endless and sometimes confusing variety of cams for the big block. It is impossible to list them all. The best advice I can give you is to talk with the suppliers. They will help you make the best choice for your engine. Don't chose a camshaft just because it worked well in somebody else's engine combination. The camshaft is the "brain" of the engine and has an enormous influence on how the engine performs, how it idles and what other assorted parts it needs to operate correctly.

Cams used in a engine block made in 1965 or 1966, (check the block casting date), need an oiling groove in the rear cam journal and in the rear cam bearing itself, in order for the oiling system to work correctly. This groove in the rear cam journal is 3/16" wide and 7/64" deep and any machine shop can cut the rear cam journal groove if you need it. 1967 and later blocks do not use a groove in the cam or the cam bearing.

You have to be careful when you make cam comparisons because cam

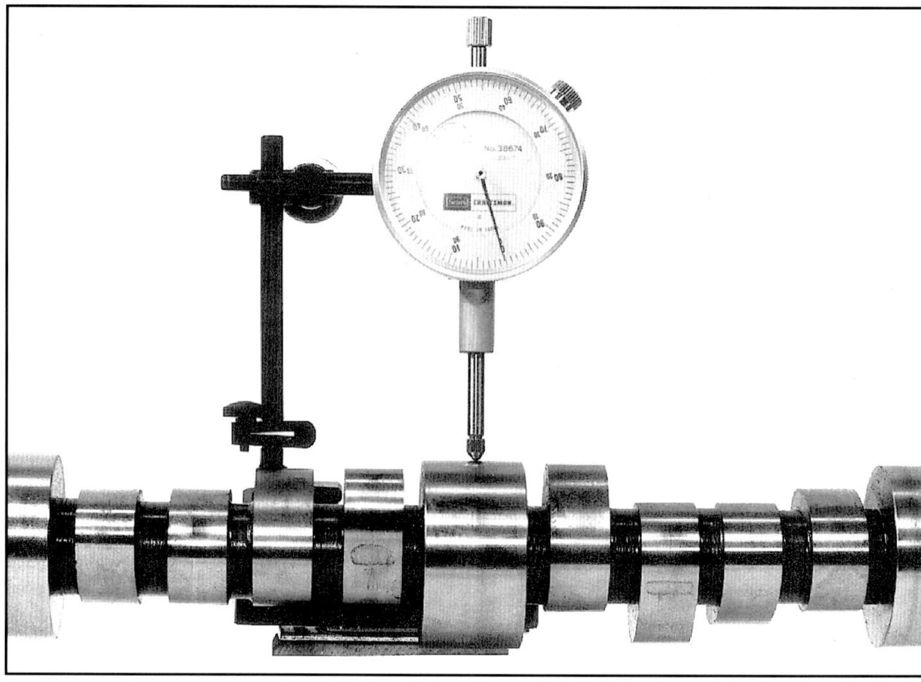

You can set up your cam in a set of V-blocks and check for runout and straightness. Lobe lift and wear can also be checked.

timing may be calculated as "seat to seat" timing, "SAE timing" or at ".050" of "tappet lift" timing. In addition, with a solid lifter cam that requires additional lash made at the rocker arm, this lash has to be subtracted right off the top when comparing total valve lift figures. Finally, different cam companies use different methods and starting points when calculating cam specifications and dimensions. This complicates your ability to make comparisons between camshafts made by two or more different cam makers.

Cam duration is comparatively measured by different cam makers at .050" of actual valve tappet lift by the cam. It is expressed as "duration @.050" and is not the same as "advertised duration" measurements. This "duration @ .050" is an important figure to look for when you compare camshafts. A "short" duration figure of say 195-225 degrees at .050" of tappet lift, will work with a stock torque converter in an automatic car, have a decent idle and will generally show good street manners in terms of low and medium rpm power. Above 225 degrees of "duration at .050", you will need to change to a higher stall torque converter with an automatic transmission, the cam will tend to make more power in a higher rpm range and may not have a smooth idle. Manual transmission vehicles or larger cubic motors are more forgiving of cam duration increases than an automatic trannie or a smaller cubic inch motor.

The lobe separation angle (LSA) is not to be confused with intake lobe centerline angle (LCA). The lobe separation angle is the angle of differ-

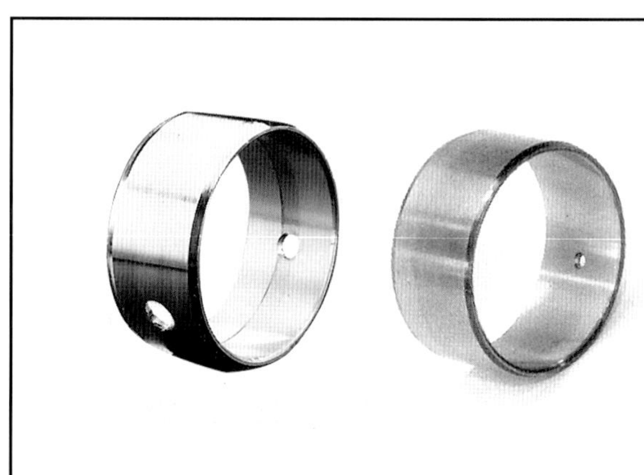

Photo shows the two types of rear cam bearings. Bearing on the left with groove in the inside circumference is used in '65 and '66 Mark IV blocks. These blocks also require the use of camshafts that have a matching groove in the rear cam journal. Non-grooved bearing is used in '67 and later big blocks and uses cams with a non-grooved rear journal.

BIG BLOCK Quick Facts & References

BIG BLOCK CHEVROLET CAMSHAFTS

PN	LIFTER TYPE	DUR. at .050"	LIFT at VALVE	LASH	LOBE CNTR.	SPRING P/N
3904365 3874872	HYD.	196/196	.398/.398	–0–	113	3859911
Mild street cam, 396/325hp, 8.5–1 to 10.25–1 compression Used in most low performance, stock big blocks						
12364055 * 3904359 3883986	HYD.	214/218	.461/.480	–0–	115	3859911
Mild street performance cam, 396/350hp, 454/365hp, 427/390 hp, LS–5, 8.5–1 to 10.5–1 compression						
12364056 * 3963544	HYD.	192/203	.398/.430	–0–	110.5	3859911
Mild street performance cam, 402/300hp, 454/345hp, LS–3, LS–4, 427 tall deck truck, 454 light truck, 8.5–1 to 10.5–1 compression						
353040	HYD.	228/228	.440/.440	–0–		3859911
Mild street cam, 454/215, 245hp, 8.5–1 compression to 10.5–1 compression						
12353920	HYD.	228/228	.540/.540	–0–	114	3989354 3916164
GM/Crane Cams grind, street and marine performance, off road, needs 9.5–1 to 10.75–1 compression						
12353921	HYD.	208/218	.475/.500	–0–	112	3970627
GM/Crane Cams grind, street and marine, towing, needs 8.0–1 to 10.5–1 compression						
12353922	HYD.	218/228	.500/.500	–0–	114	3970627
GM/Crane Cams grind, street and marine performance, needs 8.75–1 to 10.5–1 compression						
12361314	HYD.	202/210	.468/.485	–0–	110	3859911
GM/Crane Cams grind, Truck Torque design, 8.25–1 or higher compression						
12361316	HYD.	210/218	.485/.490	–0–	115	3859911
GM/Crane Cams grind, 90–93 SS454 truck, 8.25–1 compression						
10185060	HYD.	224/224	.510/510	–0–	116	3970627
Street and marine performance, 454 Gen.V HO, advanced 5 degrees, 8.75–1 compression						
14096209	HYD.	220/220	.500/.500	–0–	115	3970627
Street and marine performance, 502 Gen.V HO, 8.75–1 compression						
12364057 * 3904362 3863148	MECH.	242/242	.520/520	.024/.028	114	3970627
Hi–perf, LS–6, 396/375hp, 427/425hp, L–71, L–78 needs 11.0–1 or more compression						
12364058 * 3925535	MECH.	264/270	.560/.580	.024/.026	112	3916164
Hi–perf, L–88, 427/435hp, needs 12.5–1 or more compression						
3959180	MECH.	262/273	.556/.594	.024/.026	110	3916164
Hi–perf, ZL–1, LS–7, needs 12.5–1 or more compression						
340282	MECH.	262/262	.560/.560	.024/.026		3916164
Off road, short track competition, needs 12.5–1 compression						
3925533	MECH.	264/270	.540/.560	.024/.026	112	3916164
GM gear drive, reverse rotation, 427 L–88 off road competition, needs 12.5–1 or more compression						
3994094	MECH.		.600/.600	.024/.026		3989354
Drag, Road race, Large cubic inch, fuel injected or multiple carb, used in Can–Am road racing, needs 12.5–1 or more compression						
6272989	HYD.		.500/.500	–0–		3970627
Street and marine performance, needs 8.5–1 to 10.25–1 compression						
12552296	HYD Roller	204/209	.480/.483	–0–		
'96 Gen.VI 502 Truck conversion crate motor						

Note: Some of the above camshafts are no longer available over the counter from Chevrolet. However, they are available from aftermarket cam makers as "Factory Blueprint" reproduction cams. * = current GM part number

BIG BLOCK Quick Facts & References

CAM LOBE SEPARATION ANGLES

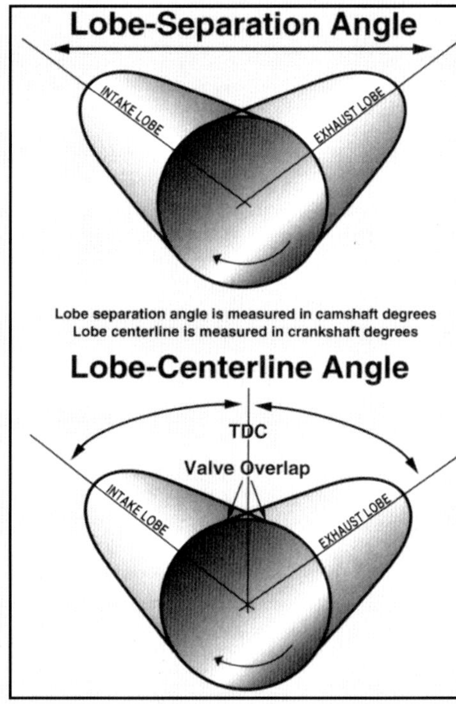

Cam lobe separation angle is often ignored or misunderstood. The first thing many people look at is total valve lift or duration figures when they are making cam selections and comparisons. Part of the problem is the confusing number of terms used when folks talk about cams. So many of the words or terms used sound similar, so a person thinks they understand, but really doesn't. The lobe separation angle or LSA for short, is the amount of camshaft rotation degrees measured from the centerline of the intake lobe to the centerline of the exhaust lobe. This amount is manufactured into the cam when it is made and can not be changed by the end user unless they have the cam lobes reground. Various LSA's can be built into a cam and the amount of degrees can generally vary from 100 to 120 degrees. If the lobe separation is "tighter", say from 100-110 degrees, the amount of time that both the intake and exhaust valves are open at the same time is increased. With a wider LSA this "overlap" period is decreased.

Now there are a few general things that one can say about Lobe Separation Angles. Keep in mind that this and other cam and engine factors will have an effect on your particular motor. Wide LSA's, (112-120) because of the shorter "overlap" period will build more low rpm torque because more air is trapped in the cylinder at low and mid engine level speeds, give a better vacuum signal in the motor at low engine speeds, idle well at low rpms without cough and sputter and will work with vacuum powered accessories like power brakes. Since overlap is less, less raw fuel is pulled out the exhaust valve and fuel mileage is improved. Low and mid range speed "driveability" are enhanced and the motor generally has a wide power band. However, at high rpms these wide LSA's and valve timing may inhibit the breathing characteristics of the motor.

Narrow LSA's will generally cause opposite effects. Idle quality may be rough and lopey, power brakes may not receive enough engine vacuum to work, the torque peak on the motor will move upwards in the rpm band and low speed power will be less. However, the mid and high rpm speed breathing will improve and power will increase, but over a more narrow power band.

Some of these effects can be offset by changing valve timing, duration, induction systems, use of a manual transmission or cubic inch displacement increases.

Talk to your cam manufacturer and get his recommendation. Make sure he understands what you intend to do with your engine. Different grinds are available for different types of street performance and racing and even different race classes in order to take into account class rules and engine limitations. If you are going to use nitrous or a blower of some sort, there are appropriate cam LSA's for these types of induction.

ence, measured in degrees of cam rotation, between the centerline of the intake lobe and the centerline of the exhaust lobe. Narrow LSA's (108-112) will generally make a big block cam that tends to have a "lopey" idle, that makes more power at higher rpms and creates a lower engine vacuum. Wider LSA's (112-116) tend to have a smoother idle, higher engine vacuum and generally operate well at low and medium rpm levels. If you were to take two cams, that had the same lift and duration, but made one with a LSA of 108 and the other had an LSA of 116, and then put them in the same engine, one after the other, the engine would exhibit completely different ways that it idled, ran and where it made power.

Cam duration, total lift and lobe separation angles affect the volume of air/fuel that the engine breathes in and out. The amount of time a valve is open (duration), how much it opens (lift), the time both intake and exhaust valves are open at the same time (LSA) and how quickly a valve opens and closes (determined by the shape of the cam lobe) have a direct effect on the way the engine works. If you want an engine that idles smoothly, has enough engine vacuum to run vacuum operated accessories like power brakes, will work with a stock stall speed torque converter and makes it's power at low and medium rpm levels for street driving, then use a cam that has a short duration and a wide LSA.

If you are looking for high performance street use or you want to go racing, then look at high lift, longer duration, narrow LSA cams. But be warned! You may have to increase the engine's static compression ratio, use a lower gear ratio in your rear end, change the stall speed of your torque converter, upgrade your valvetrain and use a separate vacuum reservoir container to make your power brakes and other vacuum assisted accessories work correctly. Blower or turbo and Pro Stock drag racing motors can use wider LSA's to help prevent incoming fuel and air from being pushed right out the exhaust port.

GM in 1996, began to use hydraulic roller cams in big block 454/502 motors. Aftermarket roller cams and roller lifters, both hydraulic and solid, are available that will reduce internal friction and open and close valves quicker in the motor. Roller cams can provide large performance gains, but they are more expensive than solid or hydraulic flat tappet valvetrains and cams. You have to purchase roller lifters, different pushrods, matching springs, retainers, roller rockers, a cam thrust bumper, and a bronze distributor gear to install a roller cam. Depending on how radical a roller cam you purchase, you may also need a rev kit, stud girdle and tall valve covers. However, because the roller lifter can follow a cam lobe more closely, much more aggressive lobe opening and closing ramp and flank profiles can be used. Valve opening and closing can be made to happen more quickly when compared with a cam using a flat tappet. The valves stay open longer in the cam rotation. Because the roller lifter must stay directly on the cam lobe, the springs used on a roller cam are much stiffer. If the roller on the cam follower loses contact with the cam lobe, the lifter will quickly self destruct. These stiffer valve springs may be taller or larger in diameter than a stock spring, so additional changes may be necessary. These changes can include widening the valve spring pocket on the head to accommodate a wider diameter spring and/or increasing the valve stem length to fit a spring with a taller installed height. Keep in mind that most factory stock distributor drive gears are compatible with the stock cast iron camshaft gear. If you change to a steel roller cam you will need to change the distributor gear to a bronze gear that will live longer. Some roller camshaft makers are making steel roller camshafts with a cam /distributor drive gear that is made with a different material that is compatible with the stock gear found on most distributors.

As with any cam change, you will need to check the rocker arm geometry to see if any changes in pushrod length are also required. It is a little more involved and certainly more expensive, but changing to a well thought out roller cam can provide a substantial increase in power when compared to a flat tappet style camshaft and lifters.

Stock production Gen.V engines come with a non-adjustable valvetrain. If you change cams and substitute a cam with less or more lift, it may not work with the non-adjustable valvetrain. See the section on ROCKER ARM ASSEMBLY for ways to install an "adjustable" valvetrain on stock production Gen.V heads.

Keep something else in mind. A number of the Chevrolet factory cams for the big block are designs that are 25-30 years old. Now this does not mean that the cams are unsatisfactory. In the right combination they will do just fine. You may even be required to use them under certain class race rules. The point is that many changes in camshaft design have taken place in the last 25 years. Chevy and Crane Cams have teamed up to offer a number of new camshafts for the big block and they are available over the counter as GM parts. Manufacturing techniques and tooling, materials, ramp designs and the use of computer assisted design and mathematics have made significant advances possible. This means more camshaft for your dollar. You may ask yourself, why didn't Chevy offer these types of perfor-

BIG BLOCK Quick Facts & References

CAMSHAFT LOBE "INTENSITY"

There is an aspect concerning cam comparisons and selection that you might find helpful. When you look at the "advertised duration" of a cam and compare that to the "duration @-.050" tappet rise figures given for the same cam, they are obviously different. There are some clues to cam profile and performance that can be made by analyzing these two figures. Some "advertised duration" measurements are generally made at the point where the lifter rises off of the base circle from .004-.006" and begins its travel up the ramp. Some cam types use a different measurement starting point (.020" lifter rise) for their "advertised duration" measurement further up in lift. Solid flat tappet and roller cams also require a certain amount of valve lash and this has to be taken into account when initial tappet rise is considered. In an attempt to make comparisons somewhat easier, most cam makers also give a "duration @-.050" figure which begins its duration calculation when the tappet has risen .050" in the tappet bore.

If you subtract the "duration @ .050" figure from the "advertised duration" you can get some idea of the "intensity" of the ramp profile of that particular cam. A cam with a more aggressive and quicker valve opening ramp will have a lower intensity figure than one with a larger figure when you do the subtraction. Let's do an example. If we have two cams and one cam has an "advertised duration" of 280 degrees and a "duration @ .050" of 212 degrees, the "intensity" of the lobe ramp will be: 280-212 = 68. If we have another cam and its "advertised duration" is also 280, but its "duration @ .050" is 225 degrees we get: 280-225 = 55. The lower the "intensity" number (55), the quicker the valve starts to open or close. The higher the number (68 in this example), the more slowly a valve will be opened or closed because the lobe opening and closing ramps have a more gentle profile compared to the other cam.

Intensity figures around 50-55 or lower, will give you a cam that quickly opens a valve and this will add more power to the motor. Some roller cams, with very aggressive opening ramp designs have an intensity figure of 34-40. On the other hand, with a higher intensity figure, it may be argued that since the valve is opened at a slower rate, the life of the valvetrain is longer because the stresses made on the valvetrain are not so acute. The valvetrain is also quieter. However, modern camshaft design, manufacturing techniques and materials, especially when used with a roller cam design, enable cam makers to produce cams with intensity figures of 34-60 that will live just fine in a race motor if you follow the correct cam pre-lubrication and break in techniques and match the rest of the valvetrain to the cam requirements. It is possible that if you compare two cams with the same lift and duration figures, but have different "intensity" in the opening and closing ramps, the one with the lower intensity figure and therefore the quicker opening and closing ramps, will put more lift "under the curve" than the slower opening camshaft.

Roller cams have very quick opening and closing rates and this is one of the advantages of using a roller cam. However, very quick opening flat tappet cams are available. Try to make your comparisons within the product line offered by one manufacturer. Because of the different measuring standards used, comparing one cam made by a particular company to another cam made by another company is sometimes misleading.

% CYLINDER VOLUME VS CRANKSHAFT ANGLE

See page 112, column 2 for details

DEG	%VOL	DEG	%VOL	DEG	%VOL	DEG	%VOL	DEG	%VOL	DEG	%VOL	DEG	%VOL	DEG	%VOL	DEG	%VOL
0	0	20	4	40	15.1	60	31.2	80	49.4	80	66.8	60	81.2	40	91.7	20	97.9
1	0	21	4.4	41	15.8	61	32.1	81	50.3	79	67.6	59	81.9	39	92.1	19	98.1
2	0	22	4.8	42	16.5	62	33	82	51.3	78	68.4	58	82.5	38	92.5	18	98.3
3	0.1	23	5.2	43	17.3	63	33.9	83	52.2	77	69.2	57	83.1	37	92.9	17	98.5
4	0.2	24	5.7	44	18	64	34.8	84	53.1	76	70	56	83.7	36	93.3	16	98.7
5	0.3	25	6.1	45	18.8	65	35.7	85	54	75	70.7	55	84.2	35	93.7	15	98.8
6	0.4	26	6.6	46	19.5	66	36.6	86	54.8	74	71.5	54	84.8	34	94	14	99
7	0.5	27	7.1	47	20.3	67	37.5	87	55.7	73	72.3	53	85.4	33	94.4	13	99.1
8	0.6	28	7.7	48	21.1	68	38.4	88	56.6	72	73	52	85.9	32	94.7	12	99.3
9	0.8	29	8.2	49	21.9	69	39.4	89	57.5	71	73.7	51	86.5	31	95	11	99.4
10	1	30	8.7	50	22.7	70	40.3	90	58.4	70	74.5	50	87	30	95.4	10	99.5
11	1.2	31	9.3	51	23.5	71	41.2	89	59.2	69	75.2	49	87.5	29	95.7	9	99.6
12	1.4	32	9.9	52	24.4	72	42.1	88	60.1	68	75.9	48	88	28	96	8	99.7
13	1.7	33	10.5	53	25.2	73	43	87	61	67	76.6	47	88.5	27	96.2	7	99.7
14	2	34	11.1	54	26	74	43.9	86	61.8	66	77.3	46	89	26	96.5	6	99.8
15	2.3	35	11.7	55	26.9	75	44.9	85	62.7	65	78	45	89.5	25	96.8	5	99.9
16	2.6	36	12.4	56	27.7	76	45.8	84	63.5	64	78.7	44	90	24	97	4	99.9
17	2.9	37	13	57	28.6	77	46.7	83	64.3	63	79.3	43	90.4	23	97.3	3	100
18	3.2	38	13.7	58	29.5	78	47.6	82	65.2	62	80	42	90.9	22	97.5	2	100
19	3.6	39	14.4	59	30.4	79	48.5	81	66	61	80.6	41	91.3	21	97.7	1	100
																0	100

STROKE = 4.00 **ROD LENGTH = 6.135** **RL/S RATIO = 1.53**

mance cams in the first place? The main reasons are parts longevity and production cost. A more conservative approach is taken when you want a street cam that will need to last 100,000 miles. A street driven vehicle needs to have a decent idle, and work well with an automatic transmission and vacuum operated accessories. Hydraulic cams and lifters don't usually require lash adjustment and so maintenance is low. If you are looking for greater performance in a race engine, you must be prepared to periodically check valve springs, valve lash and to replace worn parts. You are asking the race parts to take more abuse and to make more power. The trade off is that for that increased power, more vigilance is required on your part. With higher lifts and more aggressive profiles, stiffer valve springs are required to control the lifters. These increased pressures will reduce the life of the camshaft and lifters, especially flat tappet varieties. The roller cams give you more life, because the roller lifter can handle the increased spring pressures up to a point. Some of the 600-800 psi spring pressures required on the most aggressive drag race style roller cams will require that you constantly check for broken valve springs, rocker arms, lifters, and pushrod deflection. Its part of the price you pay to keep a race motor running at extreme levels.

CONFUSED? Unsure of what you need in a cam? I remind you to call the cam makers direct. They have made these types of choices thousands of times and can help you get the right cam, the first time.

STATIC COMPRESSION VS. DYNAMIC COMPRESSION AND CAM SELECTIONS

There are two types of compression measurements. "Static" compression ratio is a measurement that takes into account the bore and stroke of the motor, combustion chamber volume, head gasket volume and thickness, the piston deck to block deck height and the piston dome or dish volume. In other words, static compression is a measurement that includes the total volume inside a particular cylinder and cylinder head chamber combination. This is a very useful measurement. You need to know that if you change any of the above items the static compression ratio will be changed. For example, if you changed from a head gasket that was .022" thick to a gasket that was .038" thick, the total volume would be increased and the static compression would decrease. If you changed from a domed piston to a flat top piston the static compression would decrease, because the total volume above the piston increased. These are all measurements of total cylinder volume made while the engine is "static" and not running. Okay so far? Good, now let's go on.

"Dynamic" compression is the measurement of compression while the motor and it's particular combination of parts is running. What you have to understand is that the parts you choose for your engine have a direct effect on the "dynamic" compression of the running motor. One of the most important part selections is the cam timing. Remember that the cam does not "instantly" open or "instantly" close the valve when the piston is at the top or bottom of the stroke. It can't physically be done. Most camshafts start to open the intake valve before the piston actually reaches top dead center (TDC) on the exhaust stroke. The piston then swings over and starts down the cylinder on the intake stroke and since the intake valve is open, air and fuel are pulled into the cylinder. When the piston reaches bottom dead center (BDC) and actually starts to move up on the compression stroke, the intake valve is starting to close, but it is still open a little. On the compression stroke the piston will get up the cylinder wall a little, before the intake valve actually closes and then the piston starts to compress whatever volume is in the cylinder. This means that at lower rpms, while the intake valve is open on the upward compression stroke, part of the air/fuel mixture and residual exhaust gases may get pushed back out of the intake valve towards the intake manifold. At higher rpms there is not enough time for this to happen. After the intake valve closes, the piston keeps going up on the compression stroke and the spark occurs a little before the piston reaches TDC. The cylinder fires and the piston heads back down the cylinder on the power stroke.

Here's an example of what I'm getting at. You might have noticed that when a high lift, long duration camshaft is put into an otherwise stock low compression engine, that the engine runs differently. The motor may not want to idle at a low speed, say 750 rpms. It coughs and spits or just dies when you try to set the idle. You "blip" open the throttle and as the rpms go up, the motor seems to smooth out and then really jumps right up on the tach. You find out that you have to set the idle up around 1500 rpms before the motor will sit there and idle but it still may idle roughly with a distinctive lope and "rump" "rump" sound from the exhaust. You heard that putting a bigger, longer duration cam into a motor would make it run better but, now that the change has been made, it seems that the performance isn't what was expected. It's hard to pull away from a traffic light, the motor wants to stall when you put it in gear and you have to coax it along until the rpms get up and then the motor pulls smoothly, until the next traffic light and the fight to keep the motor going starts all over again. You're starting to think there has been a mistake. There has been a mistake. The parts combination isn't working well.

Now let's go back to that intake valve closing point. You figured out your static compression ratio, but a cam with a very late intake valve closing point was installed. The static compression measurements told you that you had a certain amount of volume in the cylinder, but because the intake valve was open on part of the compression stroke, the actual volume trapped in the running motor is less than what you calculated when you did your static compression ratio formulas. The piston had already moved up part of the cylinder before the intake valve closed. This pushed some of the incoming air/fuel mixture back thru the intake valve and back up the intake passages towards the carb. The air/fuel mix in the intake passages doesn't know whether it's coming or going. More to the point, the actual volume that the piston compresses is less than what you thought it would be because dynamic compression is not the same as static compression. You may have a motor with a static compression of 8.5-1, but if you chose incorrect parts the dynamic compression of the motor may be as little as 3.0-1. That's why the motor pulls so badly. There's hardly any dynamic compression or cylinder pressure on the example motor when it was running at lower rpms. The motor picked up at higher rpms because there was not enough time for much of the mixture to be pushed out of the intake valve. The cylinder pressure, developed when the piston reached TDC on the compression stroke and the mixture was fired, was lower because it had less total volume to compress. Cylinder pressure is also a way of thinking about torque. The more cylinder pressure, the more torque the engine will have. The less cylinder pressure, the less torque the motor puts out. Increases in cam duration will necessitate increases in the static compression of the engine in order to keep the cylinder pressures where they should be. The way to do this would be to increase the piston dome and/or use a smaller combustion chamber volume in the head.

When you choose a camshaft take a look at the information supplied by the cam maker. When does the intake valve close in relation to BDC? The cam makers show this measurement in degrees of crank rotation. When the piston is at bottom dead center (BDC), it's at 180 degrees, when the piston is at 0 degrees, it's at top dead center (TDC). For example, a cam card may show that the intake valve closes at 82 degrees after bottom dead center (ABDC). Now think about this for a moment, 82 degrees after bottom dead center. That means that the piston is on it's way up on the compression stroke and is almost half the way up the cylinder before the intake valve is totally closed. Halfway up the cylinder would be 90 degrees

BIG BLOCK Quick Facts & References

CAM DURATION

If you know when cam timing events occur, you can determine the amount of duration that particular cam has. Just add the two intake events together and add 180 degrees to that figure to determine duration. The same is true for the exhaust. Just add the two exhaust events together and add 180 degrees to determine exhaust duration. If the figures for the timing events were taken at .050", then you will find duration at .050". If the timing was calculated for seat to seat timing the duration you find will be "advertised duration".

Here's an example. Let's say you have a cam with the following timing at .050":

Intake opens at 29 degrees Before Top Dead Center (BTDC)
Intake closes at 59 degrees After Bottom Dead Center (ABDC)
Exhaust opens at 67 degrees Before Bottom Dead Center (BBDC)
Exhaust closes at 25 degrees After Top Dead Center (ATDC)

Add the two intake events together (29 + 59 = 88), now add 180 degrees (88 + 180 = 268). So the intake duration at .050" on this cam is 268 degrees.

On the exhaust side the two events are (67 + 25 = 92) then add 180, (92 + 180 = 272). The exhaust duration at .050" is 272 degrees.

If you have a cam with unknown timing specs, you can use a degree wheel and a dial indicator to figure out the timing events and then calculate duration.

CAM TIMING EFFECTS

In other parts of this book there has been discussion of the effects of cam timing, duration and static compression ratios. Here are six computer generated dyno tests made using the "Desktop Dyno" Version 2.0G developed by Motion Software, Inc. of Brea, CA. This program, which is distributed by Mr. Gasket Corporation, is very helpful because it is easy to use and produces pretty accurate results without turning a wrench or paying dyno rental fees. It is most helpful when you are trying to make comparisons or spot trends between engines with different components. That's what we are going to do here. I've built a computer run 427 big block and bored it out .030 over for a total of 433 cubic inches. Stock big block heads with the small 2.06/1.72 valves are used along with a dual plane intake with a 700 CFM carb. The motor will also have small tube exhaust headers and mufflers. There are only two things that will change during these six tests, the hydraulic lifter camshafts and the compression on the last two tests. The first four tests will have static compressions of 9.5-1, suitable for use with pump gasoline.

TEST 1

This cam is a very short duration street cam, advertised duration is 248 and it has a valve lift of .453"/.484". Cam timing measured "seat to seat" is IO-15, IC-53, EO-61, EC-19, LSA-110, and the cam is ground 4 degrees advanced with an intake centerline of 105. As you can see the intake closes very quickly and traps a high percentage of cylinder volume so cylinder pressures are very high at low rpms. At 2000 rpms this motor is making 502 ft./lbs. of torque, but the torque begins to fall as rpms go up. Horsepower peaks at 333 at 4000 rpms and hangs on to almost the same amount at 4500. After 5000 rpms the show is over in terms of horsepower. If you were building a street driven car with an automatic transmission and a 3:08 gear ratio or if you were building a truck that was going to be pulling some weight, this cam would give excellent service. The cam timing and heads limit this application from off idle to 5000 rpms and it will run on 90-92 octane pump fuel.

TEST 2

This cam has an increased advertised duration of 260, a little more valve lift of .484"/.515" and the cam timing looks like this IO- 19, IC-61, EO-69, EC-23, LSA is 112 and the intake lobe centerline is 107 degrees.Since the intake valve closes a little later than in Test 1, we would expect that the cylinder pressure and torque will be lower because less cylinder volume will be trapped. At 2000 rpm this engine has 477 ft./lbs. and at 2500 the peak torque is found at 482. We have lost 20 ft./lbs. of peak torque compared to Test 1, but on the horsepower side we gained 17 horsepower at the same rpm, but now horsepower peak is at 4500 rpm with a total HP gain of 29. The extended cam timing is helping the motor breathe better at increased rpms. At 5500 rpm HP is going down but the engine is making 52 more HP that the Test 1 motor at 5500. The torque curve on this motor is almost flat from 2000 to 4000 rpm. Still it is a very streetable cam on pump gasoline.

TEST 3

We replace the cam and use a 272 advertised duration with more lift, .515"/.510". Cam timing is IO-25, IC-67, EO-75, EC-29, LSA is 112, intake lobe centerline is 107. As you can see torque at 2000 is lower still and peaks at 467 ft./lbs. at 3500, but is fairly flat from 2000 to 4400. HP peaks at 5000 at 385 and hangs on decently until 5500 and represents a 52 hp peak gain over Test 1, but a peak torque loss of 25 ft./lbs.

TEST 4

This our last cam change. This one has an advertised duration of 278 and a lift of .529"/.525". IO-26, IC-72, EO-80, EC-30, LSA- 114, intake centerline is 109. The intake valve closes later still and lift opens more but there is only 4 more horsepower compared to Test 3 and torque is less than that developed by Test 3. This engine is starting to suffer from the lack of enough static compression. Not enough cylinder pressure is being built, power is falling off.

TEST 5

In this test we use the same 278 cam that was in TEST 4, but we drop the static compression down to the stock compression ratios (8.25-1) found on many engines made after 1972.The 9.5-1 static compression ratio found in Test 4 wasn't enough and if this cam was put into a totally stock motor Test 5 shows that things get even worse. Horsepower and torque fall off and are less. Peak HP is 362 and torque peak only reaches 431. This is what I was talking about when there was a discussion of dynamic compression and cam selection. The static compression is not matched to the dynamic compression requirements of this cam.

TEST 6

Now let's optimize the static compression for the cam used in Test 4 and 5. The intake valve closes at 72 degrees after Bottom Dead Center. This leaves 72.5% of the cylinder volume in this particular motor. When we do our calculations to optimize and match this cam we find that this cam requires a static compression ratio of 12.0-1 and higher octane race fuel. Remember all we did was increase the static compression and that forces us also to change fuel types. The following combination will not run well on lower octane pump gasoline.Horsepower climbed to 430 at 5000, but look at the torque curve. Peak torque at 4000 is 495 and from 2000 to 5000 rpms the torque is over 450 ft./lbs. When we compare these results to Test 4, you can see the difference when the static compression matches the cam timing requirements. There was a 41 HP increase at the same 5000 rpm level and peak torque increased 39 ft./lbs., and torque was up at all other rpm levels up to 6000. We got back almost all of the torque found in Test 1, but at 4000 rpm instead of 2000 and there is almost 100 more horsepower on tap compared to Test 1.

Test 1

RPM	HP	Torque
2000	191	502
2500	236	496
3000	277	485
3500	312	468
4000	333	437
4500	331	386
5000	309	325
5500	266	254
6000	211	184
6500	147	118

Test 2

RPM	HP	Torque
2000	182	477
2500	230	482
3000	274	479
3500	318	476
4000	350	460
4500	362	423
5000	352	370
5500	318	303
6000	266	233
6500	207	167

Test 3

RPM	HP	Torque
2000	169	444
2500	217	457
3000	263	460
3500	311	467
4000	352	462
4500	380	444
5000	385	404
5500	363	346
6000	322	281
6500	268	217
7000	207	155

Test 4

RPM	HP	Torque
2000	161	424
2500	210	442
3000	254	445
3500	303	455
4000	348	456
4500	381	444
5000	389	409
5500	372	355
6000	336	294
6500	286	231
7000	224	168

Test 5

RPM	HP	Torque
2000	153	401
2500	199	418
3000	240	420
3500	286	430
4000	328	431
4500	358	417
5000	362	380
5500	340	325
6000	301	264
6500	249	201
7000	185	139

Test 6

RPM	HP	Torque
2000	175	459
2500	227	478
3000	276	483
3500	328	492
4000	377	495
4500	415	485
5000	430	452
5500	420	401
6000	389	340
6500	342	276
7000	282	211
7500	226	158

ROLLER CAMS

If you are bucks up and you can get away with spending additional money on your favorite race engine without incurring the murderous wrath of your "life partner", use a roller camshaft. The money will buy you the roller cam, roller lifters, new pushrods, new valve springs to work with the cam, 10 degree valve retainers and locks, needle bearing roller rocker arms, a cam thrust button and a bronze distributor drive gear. Depending how extreme you get, you may also need a "rev kit", rocker arm stud girdle and "tall" valve covers. You can tell your "partner" that it will make the engine go faster and besides it really is only about $600.00 more than what you would have spent for a flat tappet cam, lifters, springs, retainers, etc. If you survive telling that story without having a kitchen knife stuck in your chest or receiving a certified letter from one of the prominent divorce lawyers in your area, here are some of the other benefits of using a properly selected roller cam.

Since one of the key limiting factors in building a high performance pushrod motor is valve opening and closing speeds, a roller cam is a big improvement over a flat tappet style cam. The roller lifter is better able to follow the more radical lobe shape on a roller cam. The rate at which a valve can be opened or closed is much quicker. At all rpms more air can get in and out of the cylinder. In addition, since the roller lifter isn't sliding over the cam lobe face like a flat tappet, but is instead rolling over it, friction is reduced compared to a flat tappet cam. The roller lifter needs higher spring pressures to keep it in contact with and riding directly on the cam lobe. If a roller lifter is put into "valve float" where the valve spring can not keep up with the rpm demands placed on it, the lifter quickly self-destructs. Now the spring pressures on a "street" hydraulic roller cam are not too great, but neither are the anticipated max rpm levels. However, on some of the race roller cams, with huge amounts of lift and where max power rpm levels are high, the spring pressures can get enormous (700-850 psi valve open pressures) compared to the spring pressures used with a flat tappet cam. At these maximum effort race levels, the amount of force generated on the cam lobe surfaces are extreme and any frictional savings made with the roller lifter are welcome. The roller cam, when compared to a similar flat tappet cam, will always rotate with less frictional losses. These high spring pressures are why roller cams are made from steel billets instead of the cast iron materials used on most flat tappet cams. The roller cam lobe shape, roller lifters, a "rev kit" and stud girdle make it possible to use very high lifts and very quick valve opening rates. At these high race roller spring pressures pushrod deflection can become a problem, rocker arm studs also want to bend (which is why you may need the stud girdle) and a "rev kit" will help keep the lifters on the cam lobe. Small base circle cams with these super high race valve spring pressures can actually twist a little along the cam center axis from front to back, throwing off valve timing from one cylinder to another.

Now many of the less radical roller cams offered by cam makers don't experience the problems mentioned above. It depends on how much cam you want. All of the roller cams will give you the advantages talked about earlier. So how much additional power can you get for your additional $500-600 dollar investment? Well, it depends. If you compare a flat tappet cam and a roller cam of equal duration, lift and valve timing, the roller cam will get the valve open and closed faster and put more area "under the lift curve". That means more cylinder pressure and more power. Let's look at some examples.

TEST 1

In Test 1 I used a flat tappet solid lifter cam with the cam timing as indicated in Test 2. Using the "Desktop Dyno" I was able to substitute the same cam timing, lift, duration and lobe centerlines on to a roller cam and run it again in Test 2. This cam has .595/.621 lift with 258/268 degrees duration at .050. Seat to seat timing is 48/78 and 93/43 degrees with 110 degree LSA and the intake centerline at 105°.

TEST 2

With everything the same, but a switch to a roller style cam lobe and lifters with the same cam timing, this particular motor picked up 35 peak horsepower and 25 ft./lbs. peak torque which is about a 5.75% increase in HP on this engine. If you look at any rpm point from 2000 to 7500, both HP and torque are higher than that found with the flat tappet cam. Also notice that because the roller cammed motor is breathing better, it tends to hang on longer at higher rpms than the flat tappet cam. More HP and torque are being made by the roller cam because of quicker valve action and a reduction in friction.

FLAT TAPPET (TEST 1 above)

RPM	HP	Torque
2000	143	375
2500	198	416
3000	248	434
3500	318	477
4000	394	518
4500	470	548
5000	533	559*
5500	578	552
6000	601	526
6500	609*	492
7000	592	444
7500	570	399
8000	522	343

ROLLER TAPPET (TEST 2 above)

RPM	HP	Torque
2000	148	388
2500	205	430
3000	256	448
3500	329	493
4000	409	537
4500	489	571
5000	556	584*
5500	606	578
6000	632	553
6500	644*	520
7000	627	471
7500	608	425
8000	561	368

Now does this mean that flat tappet cams are not worth messing around with. Definitely not! It means that sometimes additional power on any motor costs more money. I'm not a rich man, so I try to think of what will give me the most power for the cost. A properly selected flat tappet cam can produce a lot of power and generally costs less than the roller cam combination. However, if you are racing in a class where roller cams are the norm and not the exception, you may be forced to spend the money in order to be competitive. On some motors, the horsepower increase can be considerably more and if you have a poorly selected flat tappet cam and move to a better roller cam selection, the increase will also be larger. Likewise on motors making a large amount of power, the differences between a flat tappet and a roller cam can be much more than that shown in this example. It is extremely important that you match the cam to your engine, the intended engine use and operating power band. Also remember to check the open valve to piston clearance, especially if you are using a very high lift or long duration cam.

On late model big blocks using a production hydraulic roller cam, the spring pressures used can limit the rpm of the motor. The relatively heavy stock hydraulic roller lifter can cause early valve "float". If you install higher pressure valve springs to increase the rpm before valve float occurs, the plunger in the production hydraulic roller lifter can be affected. The rpm range of the motor and the high rpm power can be increased by using a "Hydra-Rev kit" from Air Flow Research that exerts pressure on the outside of the lifter body without affecting the plunger of the production hydraulic roller lifter.

CAM SELECTION AND COMPRESSION RATIOS

David Vizard is an engineer, race engine builder and a prolific automotive author with wide experience. I always find his books and articles to be thoughtful, enlightening and very informative. One of the things which I found thought provoking were his calculations for finding the optimum static compression ratio for any given camshaft. Vizard attributes the initial observations about this to Dema Elgin, of Elgin Cams. The problem is to match the timing specifications of a camshaft to a static compression ratio which produces the best dynamic compression and thus engine performance. If you have read the section on Dynamic Compression, one of the issues is the intake valve closing point and its effect on dynamic compression in the motor. We want to trap as much of the intake charge as possible to maximize cylinder pressure and torque. We have to balance the static compression ratio with the intake closing point in order to do this.

Elgin's observation was this, most carbureted motors seem to run best with a dynamic compression ratio of 8.5-1 when used with 90- 92 octane pump gas. If higher octane gasoline race fuels are used a dynamic compression of 9.0-1 can be used. Remember these are dynamic compression ratios, not static compression ratios. There are also some assumptions. First these formulas are for normally aspirated motors that are not operated at abnormally high coolant temperatures. Higher operating temperatures will raise the possibility of detonation. If the operating coolant temperatures are high you can reduce the fuel constants in the formula (8.5 for pump fuel or 9.0 for race fuel) slightly to compensate and thus reduce cylinder pressures. In order to use the formulas Elgin and Vizard give us, we have to know a few things first.

What is the swept volume of the individual cylinder in cubic centimeters?

When does the intake valve close on a particular camshaft?

What percentage of the cylinder volume is in the cylinder when the intake valve closes?

What type of fuel will be used, pump gas or high octane race fuel?

The individual cylinder swept volumes on a 454 bored +.030, for example, are easy.

460 divided by 8 = 57.5 cubic inches per cylinder

Convert that to cubic centimeters by multiplying by 16.387

57.5 cubic inches x 16.387 = 942.25 cc's total swept volume per cylinder

The cam spec card you get with a camshaft will give you the "seat to seat" valve timing and will show in crank degrees when the intake valve closes. We want the "seat to seat" timing here not the "@-.050" valve timing. A call to the cam maker will also get you the information or you can sometimes look this figure up in their catalogs. You can also determine this if you don't have a cam card, by using a degree wheel. You can mock up the engine, using the crank, one rod and a piston, the camshaft and a solid lifter. Then measure how many degrees (usually after bottom dead center) occur until the intake valve closes. Let's say that we find that the cam card tells us that the intake closes at 48 degrees After Bottom Dead Center, it's usually shown like this "Intake closes - 48 ABDC" or "IC - 48 degrees ABDC".

Okay so far? Now if the intake valve closes at 48 degrees after bottom dead center what percentage of the total cylinder volume is left when the intake closes? See crank angle chart on page 108.

This chart shows the percentages of volume for 180 crank shaft degrees and piston travel from the top of the cylinder to the bottom. This particular chart is for a 4" stroke using a 6.135" connecting rod. If the stroke or rod length are changed the percentages also change. This one will work for our example 460 cubic inch motor. The chart reads from the upper left hand corner to show that at Top Dead Center (TDC) the crank has moved 0 degrees and there has been 0.0 change in cylinder volume. This represents the piston at the top of the cylinder. All the way over in the lower right hand corner, the piston is at the bottom of the stroke, Bottom Dead Center (BDC) and the cylinder is 100% filled.

We know that the intake valve on this particular cam closes at 48 degrees after bottom dead center, so starting in the lower right hand corner of the chart, read up each column until you get to 48 degrees, then look at the percentage next to it. We find that the cylinder has 88.0% of its volume when the intake valve closes, using this cam and this stroke and rod length. You can put this into inches if you like by multiplying the stroke length by the percentage.

4" stroke x 0.88 = 3.52"

Then subtract that sum from the total stroke length and we find that the piston has moved 0.48" up the cylinder wall before the intake valve closes, just under 1/2 an inch of travel. (The 0.48" of piston travel and 48 degrees are just a coincidence in this particular example.)

So now what do we have? We have a cylinder swept volume of 942.25cc and an intake valve closing point that leaves 88% of the cylinder volume. In this instance we are going to use pump gasoline so we will use the "8.5" gas type factor. We have to now make two calculations to find the optimum static compression ratio with this particular cam and intake valve closing point.

$$\text{TOTAL CHAMBER VOLUME} = \frac{\text{TOTAL SWEPT VOLUME x \% OF CYLINDER VOLUME}}{\text{GAS TYPE - 1}}$$

$$\text{TOTAL CHAMBER VOLUME} = \frac{942.25 \times 0.88}{8.5 - 1 \text{ for pump gas}}$$

$$\text{TOTAL CHAMBER VOLUME} = \frac{829.18}{7.5}$$

TOTAL CHAMBER VOLUME = 110.55733cc

Next we do the following in order to find the best static compression ratio for this particular cam:

$$\text{STATIC COMPRESSION RATIO NEEDED} = \frac{\text{TOTAL SWEPT VOLUME + TOTAL CHAMBER VOLUME}}{\text{TOTAL CHAMBER VOLUME}}$$

$$\text{STATIC COMPRESSION RATIO NEEDED} = \frac{942.25 + 110.55733}{110.55733}$$

$$\text{STATIC COMPRESSION RATIO NEEDED} = \frac{1052.8073}{110.55733}$$

STATIC COMPRESSION RATIO NEEDED = 9.5227272 or rrounded off to 9.5-1

In other words, a parts combination that gives a static compression ratio of 9.5-1 will work best when this cam is used along with pump gasoline. You will have to juggle the parts combination to get to the required static compression. The ENGINE MATH chapter has the calculation for determining static compression ratios.

If we had a longer duration race cam with an intake closing point of 78 degrees After Bottom Dead Center, what static compression would work best? Looking at the chart again and starting in the lower right hand corner, we find that if the intake valve closes at 78 degrees there is 68.4% of the total swept volume left in the cylinder. We can also change the fuel type factor to 9.0 for higher octane race fuel.

TOTAL CHAMBER VOLUME = $\dfrac{942.25 \times 0.684}{9.0 - 1 \text{ for race fuel}}$

TOTAL CHAMBER VOLUME = $\dfrac{644.499}{8}$

TOTAL CHAMBER VOLUME = 80.56cc

Now solve for the required static compression ratio.

STATIC COMPRESSION RATIO NEEDED = $\dfrac{942.25 + 80.56}{80.56}$

STATIC COMPRESSION RATIO NEEDED = $\dfrac{1022.81}{80.56}$

STATIC COMPRESSION RATIO NEEDED= 12.696251 Round that off to 12.7-1

Using these formulas will optimize the cylinder pressures and torque for any given parts combination. When you do the calculations for a street driven combination with a static compression of 9.5-1 or less, you will find that the cam timing is very short. Advertised durations of 245 to 250 or shorter with lower valve lifts will work best. These street cams will have a duration "@ .050" in the area of 195-205 degrees. The engines will have a flat torque curve and make large amounts of torque from idle to around 3500-4000 rpm and will make peak horsepower around 4000- 4500 rpms. This is exactly what you need in a street driven vehicle with low octane pump gasoline, "street" rear gears and an automatic transmission. The large amounts of torque will accelerate a heavy vehicle and also work with heavy towing loads in trucks or RV's. Keep in mind the octane requirements. At static compressions above 9.5-1 you may need to use an octane booster or race fuels.

You may have a street driven vehicle with a static compression level at or below 9.5-1 and after doing these calculations may not want to use a very short duration cam. If you decide to use a longer duration cam remember that the bottom torque will decrease, while torque peak will occur at a higher rpm. Peak torque and cylinder pressures will also occur at a higher rpm. Manual transmissions will be more forgiving of cam duration increases than an automatic in terms of driveability. Also if you are using a longer duration cam it will probably work better with a lower ratio rear gear. The lower gear will help the motor get to the best power rpm range quickly.

Generally speaking, as cam timing increases, (longer durations and later intake valve closing points), the low and mid rpm range torque amounts will be relatively less, the torque peak will move higher in the rpm range and peak horsepower will occur at a higher rpm. Also the horsepower and torque curves will get "peakier". In other words, with a longer duration cam, the best power operating rpm will be over a higher but more narrow rpm range.

After working with these formulas, I found that any cam which has an intake closing that leaves 88.0% of the cylinder volume when it closes will operate best at a static compression of 9.5-1 on pump gasoline. If more than 88% is still in the cylinder when the intake closes, the static compression needed can be less than 9.5-1 using pump gasoline of 90-92 octane. 94.0% cylinder volume will work with 9.0-1 static compression. If there is less, for example if 84.0% is in the cylinder, the engine will require a 10.5-1 static compression ratio and high octane fuel. Cams which leave 80.0% in the cylinder will require a static compression ratio of 11.0-1. Cams that leave 70.0% will need 12.4-1 static compression and high octane race fuel.

Now when you have determined which is the best static compression ratio for a particular cam and then try to make a parts selection to attain that static compression, you may find that you can not match it exactly. You should try to come as close to the ratio as you can. If for instance, you do the calculations and you need 9.5-1 but the best parts combination you can come up with is 9.45-1 static compression, that is okay and will work. Try to match the recommendation if you can, but don't go over the needed static compression ratio thinking that more is better. You can juggle head gasket thickness, piston dome or dish cc's, head combustion chamber size, or piston deck to cylinder block deck height.

If you have used these formulas you already know what the "Total Chamber Volume" has to be. Any combination of head combustion chamber volume, head gasket volume, deck height volume and piston dome or dish volume that provides an amount equal to the "Total Chamber Volume", calculated above, will work. For instance in the above bracket race cam example, the formula called for a "Total Chamber Volume" of 80.56cc. So if you add a combustion chamber volume of 110cc, a head gasket volume of 8.9606801cc's and a deck height volume above the piston deck of 5.895183cc's, you get 124.85586cc. Now subtract from this amount the 80.62cc "Total Chamber Volume" figure to get the piston dome amount needed and the result is a piston dome volume of 44.295cc. This will give you the 12.7-1 static compression ratio indicated by our prior calculations. Other combinations can give you 12.7-1 CR. It depends on the type of cylinder combustion chamber size, type of piston, etc. If you plan to use a head with a smaller combustion chamber volume the size of the piston dome can be reduced. Depending on the parts combination you may be able to use a flatop or dished piston and still achieve the static compression ratio you need for your cam.

Keep in mind that you need to have a minimum clearance of .035- .040" between the piston deck and the cylinder head deck. The deck height clearance between the piston deck and the block deck should be 0 at TDC. Remember to check open valve to piston deck clearance, which for a high rpm race motor should be a minimum of .080" for the intake and .100" for the exhaust valve when using steel rods.

I have to inject some uncertainty and compromise into this discussion. One of the main problems in talking about camshafts is the fact that the various measurements regarding cam timing, durations, and lift are made differently by cam companies. Solid lifter cams are measured differently from hydraulic cams or roller cams. Different cam companies measure differently and have their own measurement criteria. This causes a cam buyer some difficulty in making comparisons between cams and even more difficulty when you compare one cam made by one company against another cam made by a different company. In order to reduce the uncertainty and the problem of possibly of comparing "apples to lemons", choose one company's products to review. There are many fine and reputable cam companies. Make your comparisons within their product line and cam types. At least then all of the choices you make within a particular type of cam, will be made knowing that the measurement criteria is the same.

after BDC. In this example, the volume in the cylinder bore that will actually be compressed by the piston is almost 1/2 of what was in the cylinder when the piston was all the way down (BDC). If you had a different cam that had an intake valve closing 35 degrees after bottom dead center (ABDC), the piston would compress more cylinder volume because the piston would barely have moved upward when the intake valve closed. Dynamic compression and cylinder pressure would go up and the motor would make more power. Starting to make sense?

The idea is to get the best combination possible. (See the sidebar on CAM SELECTION AND COMPRESSION RATIOS.) Generally speaking, if you build a stock, carbureted, street motor with 8.25-1 to 9.0-1 static compression, you need to use a cam with an intake closing point that is compatible with this static compression. That means a short duration cam with an intake closing point that occurs quickly after the piston reaches bottom dead center (BDC). This will trap more volume in the cylinder to be compressed. Cylinder pressures will go up where they should be. Now you should understand why sticking a long duration cam into an otherwise stock, low static compression motor doesn't usually work out. The stock static compression of the motor may be 8.25-1, but with a long duration cam, the intake valve closes much later, not enough volume gets trapped in the cylinder and the dynamic compression of the motor decreases so much that the motor no longer runs decently. Cylinder pressure and dynamic compression are so low that the motor can't get out of it's own way until the rpms of the motor reach a point where there isn't enough time to bleed off as much cylinder pressure.

The faster you spin a motor, the less time there is for everything to work properly. At higher rpms, say above 4000, the valves are open for less time, air and fuel has less time to move down the intake passages and the piston is moving up and down in the cylinder much more quickly. There is less time for the exhaust to be pushed out of the engine. In order for the motor to make more power, you have to figure out a way to get more air/fuel into and out of the motor at these high rpm speeds. You also want to develop as much cylinder pressure as your parts combination will allow while avoiding detonation. So now you can keep the valves open longer by increasing the cam duration, this is the amount of crank rotation that the valves are open. You can also increase how far the valve opens, so you use a cam with a higher lift. You can increase the speed of the valve opening and closing by using a roller cam.

Now there is another thing going on. In a performance motor with a good intake and a properly setup induction, the air/fuel mix has a certain velocity, mass and momentum. On the compression stroke, at higher rpms, the mixture can continue to pack into the cylinder while the intake valve is open, even though the piston has started to move upwards. This is one of those small things that happen and will increase the power of an engine parts combination set up to take advantage of it.

Now you realize why the dynamic compression of the motor can be too low if the intake valve closes later. You have to increase the static compression on the motor in order to make up for this. You add domed pistons and/or you decrease chamber volume by putting cylinder heads on that have smaller combustion chambers. There is less total volume now in the cylinder, but the cylinder volume that is still in there is now squeezed into an even smaller space. Now the static compression is up and the dynamic compression and cylinder pressure are also higher. You add tube exhaust headers so that the exhaust can flow out quicker. You can add better heads and a better design intake manifold. Even though there is less time to get air into the cylinder, the better heads and better intake ports on the manifold allow you to get more air and fuel in the motor quicker at higher rpms by taking into account the inertia of a fast moving column of intake charge. Now the combination starts to work and produces more power than the stock combination, but at higher rpms.

You still have to consider the static compression that your motor will develop. If you plan on using unleaded pump gasoline the static compression should be 9.5-1 or less. Race fuels with higher octane ratings (105 and above) resist detonation at higher compression and cylinder pressure levels.

Think about an engine with a supercharger or a turbocharger. When you check the static compression on many of these engines, it might be as low as 7.0-1. These two types of "blowers" push more air/fuel into the motor so the cylinder volume is artificially and mechanically increased. Dynamic compression is increased when the blower or turbo motor is running. Cylinder pressures go up, more power and torque are made. Nitrous-oxide injected motors do the same thing, but instead of mechanically increasing cylinder volume, nitrous motors do it with chemistry, by making the air/fuel mixture cooler and denser and providing additional oxygen molecules to burn additional fuel. Along these lines, blower or nitrous motors need different cam timing. Because of the artificially increased cylinder volume created, more exhaust cam timing and a bigger exhaust valve may be beneficial along with a larger exhaust system. Cam lobe separation angles on a blower motor should also be wider.

Now you see why cam makers ask so many questions when you speak to them about cam selection. They want to know what kind of vehicle you have, how big is the engine, how much the car weighs, what kind of transmission you are going to use, and what rear end ratio you have? What is the static compression of your motor, what kind of heads, intake, carb, exhaust system and most importantly, what do you want to do with the motor? Drive it on the street and use pump gasoline? Pull a trailer? Get better mileage?

Production "piddle valve" solid lifters are for all solid lifter cams running stock valve gear where an appropriate amount of upper engine oiling is necessary.

Go drag racing, stock car racing or road racing and use higher octane racing fuels? Do you want a smooth idle? Do you want to make power at low rpms? Or do you want a motor that will make power at high rpms?

There is much more to learn about camshafts and how they work, much more than I can cover in this book. If you want to learn more there are a number of fine books that discuss this area. Buy one or go to your library. If you have a computer, Crane Cams and Competition Cams each have a cam selection program on floppy disc that you can buy. These two discs also have some very helpful utility sections that can help you figure static compression ratios, rear gear selections, and total swept volumes and much more. In the mean time, call your cam maker. They can help get you thru the selection process.

LIFTERS

There are a number of flat tappet lifter types made for the big block. (PN-5232720) is a production "hydraulic" lifter for use with all "hydraulic" flat tappet cams. (PN-17102353) is a hydraulic lifter used on the '91-'95 Gen.V 454 HO and 502 HO motors. 1996 "Vortech 7400" 454's use a roller cam with hydraulic roller lifters and the Gen.VI 454 and 502HO also have them.

For solid lifter flat tappet cams, (PN-5232695) is a "solid" lifter which has a "piddle valve" to meter the amount of oil sent up the pushrod to the rocker arms. It should be used when a flat tappet solid cam is used with stock style stamped rocker arms. A second type of "solid" lifter is also called an "edge orifice" lifter (PN- 5231585). This solid lifter will reduce the amount of oil sent to the rocker arms by 10-20%. Because it reduces the amount of oil, you should only use this lifter with needle bearing roller rocker arms. If you use the "edge orifice" lifter with the stock stamped rocker arms and rocker balls there will not be enough lubrication.

If you are rebuilding an engine and intend to use the old cam and lifters again, there are some things to keep in mind. First measure the cam lobes to make sure they are not worn out. If it passes inspection and is okay to reuse, then check the lifters for wear. If each lifter still has a crown on the foot of the lifter, they can be reused. You must take the lifters out of the motor and sort them according to which lifter bore they came out of and then they must be replaced in the same lifter bore and on the same cam lobe that they were previously running on. Otherwise you run the risk of the cam lobe and/or lifter quickly self-destructing and then spreading small metal particles all around your engine. It is much easier and safer to replace the cam and lifters when you rebuild. Which reminds me, putting used lifters on a new cam is definitely not something you should contemplate or do.

Aftermarket roller lifters are linked in pairs. These links keep the roller lifter from rotating in the lifter bores and in their correct position in relation to the roller cam lobes. If you are using an aftermarket roller cam, it is possible to buy roller lifter pairs that can be separated at these links. If one of the roller lifters is damaged or worn, you only need to get one roller lifter. Other roller lifter pairs with links that can't be separated, will require you to purchase the pair.

PUSHRODS

Three pushrod diameters have been used in the big blocks. Low performance engines received either a 5/16" or 3/8" diameter, while some very high performance engines received 7/16" pushrods and each pushrod diameter size has a matching diameter guideplate.

Mark IV big block, short deck, intake pushrods are 8.285" long (3/8" PN-10134307) and exhaust pushrods are 9.256" long (3/8" PN- 10134308). Pushrods used on tall deck truck motors are 0.4" longer to match the 0.4" taller block deck heights. Gen.V 454 and 502 intake pushrods are 8.171" long (3/8" PN-14097068) and the exhaust pushrods are 9.151" (3/8" PN-14097070) and fit their non-adjustable valvetrain. If you have a Gen.V style non-adjustable valvetrain and you switch to a Mark IV style adjustable valvetrain, remember to switch from the Gen.V pushrods to the Mark IV pushrods which are slightly longer.

Pushrods for use with "Bowtie" head casting #-861 or 862 heads are 8.285" for intakes and 9.256" for exhausts. Symmetrical "Bowtie" head casting #-128 and 129 require even longer pushrods of 9.100" intake and 10.0" exhaust.

A motor designed and built for everyday stock street use, with stock low performance cams and low pressure single springs, can use the 5/16" pushrod and matching guide plates. If you are going to use a camshaft with more than .475" lift at the valve, I suggest you use the 3/8" pushrod and matching guide plate (PN-3860038). At cam lifts greater than .600", where you want the most strength possible and the least pushrod deflection, you can move up to the 7/16" 1010 steel (intake PN-10134304, exhaust PN-10134303) or 7/16" 4130 chrome moly steel pushrod (intake PN-10134306, exhaust PN- 10134305) and corresponding 7/16" guideplate (PN-3879620). All of the GM pushrods listed above are a one piece design and most have wall thicknesses of .080". The one exception is the Gen.V pushrods for use with non-adjustable valvetrains, their wall thickness is .058" for the 3/8" diameter variety.

I like the 3/8" pushrods because on most motors, I can get increased strength without the increased weight of the 7/16" one. Of course, if you are using a race cam with lifts over .600" at the valve and have high valve spring open pressures, you should be willing to add more weight in return for the maximum strength and minimum deflection found in the 7/16" pushrods. Tall deck blocks and some "raised port" heads will require a long pushrod and in these cases a 7/16" pushrod will help prevent deflection with high valve spring pressures.

Some pushrods are made from two pieces and some have a hardened section on the pushrod. Make sure that when you install such pushrods so that the hardened end goes on the rocker arm end and passes thru the guideplate. Some pushrods are marked "up" or have arrows indicating "up". If you put the hardened end down against the lifter, the metal guideplate will wear out the non-hardened surface of the pushrod.

One note of caution, the stock big block pushrod guideplates don't fit some Bowtie cylinder heads. The raised intake ports and the reinforcement ribs on the rocker arm stud bosses, found on some aluminum Bowtie heads, interfere with the guideplates. You can grind the guideplates or the reinforcement ribs so that they fit, but another solution is to use flat, hardened washers under the rocker arm studs and guideplates. These washers will act as shims and raise the guideplates so that they fit. Aftermarket guideplate makers may offer a solution to this problem.

ROCKER ARM ASSEMBLY

Screw-in rocker arm studs are used on all big block heads. Mark IV studs are 7/16" in diameter, however, Gen.V and Gen.VI heads have 3/8" rocker bolts. You should also note that stock, production Gen.V and Gen.VI heads for street motors have a non-adjustable valvetrain. The pushrods, lifters, rocker arms, rocker arm bolts, and rocker arm balls for this non-adjustable valvetrain are completely different than the Mark IV adjustable valvetrain components. The Gen.V and Gen. VI non- adjustable valvetrain can be altered to a Mark IV adjustable valvetrain by tapping the Gen.V head rocker stud bosses to accept the larger and stronger Mark IV 7/16"-14 NC rocker arm studs (PN- 3921912), and then using a kit with Mark IV long-slot rocker arms, balls and nuts (PN-10112680). You must also change to the slightly longer Mark IV pushrods and use the matching diameter pushrod guideplates. You can also use PN-10198929 adjustable stud and PN-10198930 nut to convert a Gen.V head to an adjustable valvetrain.

Aftermarket suppliers are making rocker arms studs for Gen.V and Gen.VI heads that have a 3/8" thread on the portion of the stud that goes into the Gen.V head, but is 7/16" on the portion that an adjustable rocker arm can ride on. With this aftermarket rocker arm stud there is no need to tap and rethread a Gen.V or Gen.VI head rocker arm stud boss in order to use an adjustable Mark IV style rocker arm assembly. However, Chevy only recommends using this 3/8" stud if your open valve spring pressures are less than 450 lbs.

Symmetrical port "Bowtie" heads (#-128) use a rocker arm ratio of 1.8-1.

Chevrolet also produces Mark IV rocker arms (PN-3959182) that have a longer than stock slot to enable you to use up to .600" lift camshafts without the rocker arm hitting the rocker arm stud. These Mark IV long slot rocker arms are marked with an "H" or "L". Gen.V rocker arms

Roller tip rocker arms, like these fromCompetition Cams, help decrease side loads on the valve guides and stem, reduce some friction and they have accurate rocker arm ratios. They can be used with open spring pressures of up to 300-350 pounds maximum.

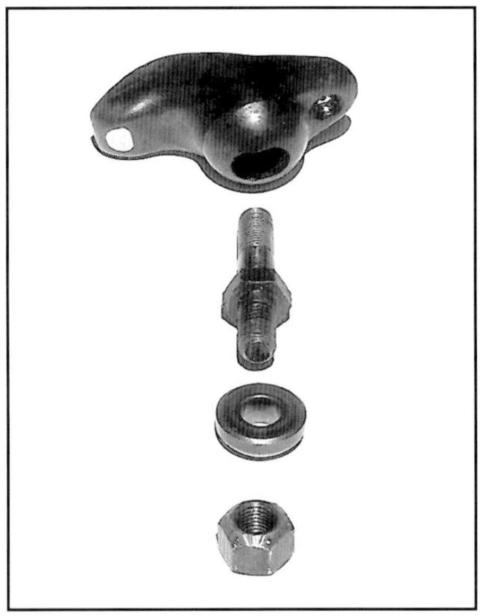

This is an adjustable 7/16" Mark IV rocker arm assembly. The Gen.V and Gen. VI "non-adjustable" production rocker arm assemblies use a rocker arm bolt instead of a stud, use a different rocker arm and use different rocker arm ball.

have the same 1.72:1 ratio, but are made to work with the non-adjustable valvetrain of the Gen.V engines and are not interchangeable with the Mark IV rocker arms. A long-slot Gen.V rocker arm is available for non-adjustable valvetrains as (PN- 12508879) and is used on the Gen.V and Gen.VI 454 HO and 502 HO crate motors.

Chevy now also offers a Crane Cams made needle bearing roller rocker arm set for big block heads with 7/16" adjustable valvetrain rocker arm studs. (PN-12361323) includes 16 CNC machined, aluminum roller rocker arms with an accurate 1.72-1 rocker arm ratio. Individual "Bowtie" roller rocker arms are found under (PN-12361330). The CNC machined aluminum roller rocker has a very accurate rocker arm ratio.

A rocker arm that has a roller tip will

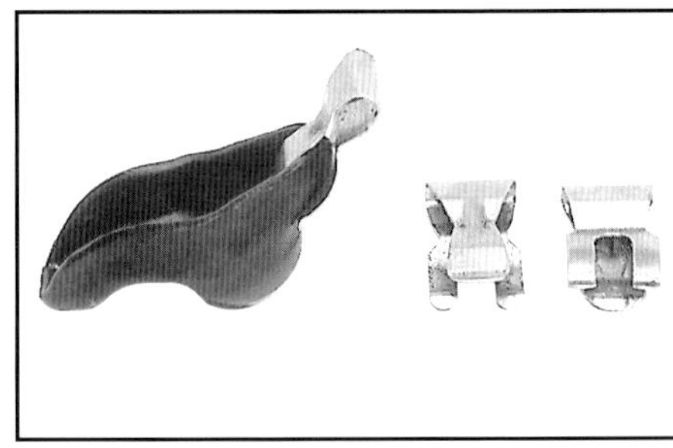

These little clips will keep hot oil from squirting all over when you set the valve lash on a running motor.

help diminish side loading of the valve guides especially when using a high lift camshaft. A roller rocker arm whether it is made from aluminum or steel, will reduce friction and can live for a while in a reduced lubrication environment. Most quality aftermarket roller rocker arms have accurate ratios and increased strength to stand up under very high valve spring pressures with little or no deflection.

If you are going to use "poly locks" on an adjustable rocker arm, make sure that the top of the rocker arm stud is flat. Often the ends of the studs have irregular surfaces left from when they were made. When the poly lock screw bears down on an irregular surface, there is not full surface contact and the locking screw can loosen and then back off. Take a good look at the ends of your rocker arm studs and make sure they are flat. You can file or grind any surface irregularities off.

When using a performance camshaft with more than .500" lift at the valve, you should consider using a rocker arm that has a long slot, roller tip and/or needle bearings. If the open spring pressures exceed 350lbs. move to a rocker that has a roller tip and a roller trunion. Many types are available to suit your engine's requirements. Different rocker arm types and rocker arm ratios are also available from aftermarket suppliers.

What ever rocker arm you use make sure that when the rocker arm is at full lift, that there is still some clearance between the rocker arm slot and rocker arm stud. Also check for clearance between the rocker arm and the spring retainer. Make sure that when the rocker arm is at full lift that there is still some clearance between the spring coils. Finally check that the bottom of the spring retainer does not hit the top of the oil seal or the top of the valve guide when the valve is at full lift.

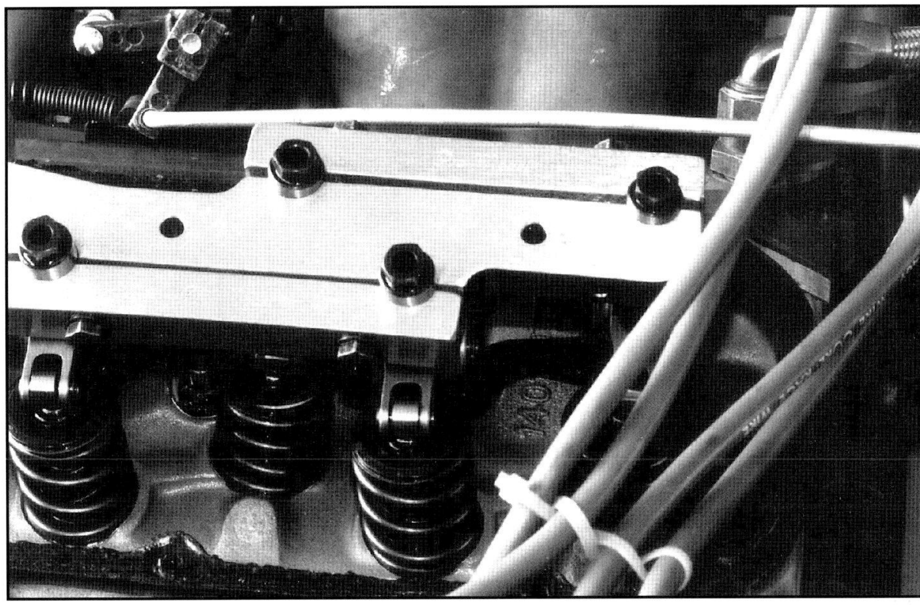

Stud girdle, steel needle bearing roller rockers and triple valve springs are needed to keep a high lift, long duration drag race roller cam under control at high rpms with high valve spring pressures.

VALVE SPRINGS, RETAINERS, LOCKS, SEALS

Correct valve springs are extremely important to the proper, trouble-free operation of an engine. The spring must match the operational requirements of the camshaft that is installed in the motor. Installed height, open and closed spring pressures, spring outside diameter, open height, coil-bind height and operating rpm range must be considered in spring choice.

Make sure that whatever valve spring you use that it will fit in the stock diameter of the valve spring pocket cast into the head.

If it is wider in diameter, as some aftermarket springs are, you may need to have the diameter of the valve spring pocket in the cylinder head enlarged or use a spring which looks like an inverted cone with a smaller spring coil diameter on the head side and a larger spring outside diameter on the retainer end.

Chevy used three types of production valve springs, but had a number of variations of these springs to match Chevy cam requirements. One spring (PN-3859911), for low performance camshafts, has a single round spring with a flat wound damper spring inside it. The outside diameter is 1.487" and it is the stock, low performance, production spring used with low lift factory cams.

(PN-3970627) dual spring was used in most Chevy performance motors, including the LS-6 and has two round springs. It has an outside diameter of 1.487". Use steel retainer (PN-3964264). This dual spring has an installed height of 1.88" and 105 lbs. of pressure at this height with the valve closed. Coil bind is at 1.28".

A third spring has two round springs and a flat wound damper spring and was used in heavy duty high performance motors offered by Chevrolet. This triple spring (PN-3916164) was used on L-88, LS- 7 and ZL-1 motors. Use steel retainer (PN-3879613). This triple spring has an outside diameter of 1.487" and an installed height of 1.88". Closed valve spring pressure is 116 lbs. with an open valve pressure of 310 lbs. at .560" valve lift. Coil bind occurs at 1.20".

Another "second design", heavy duty valve spring (PN-3989354) uses steel retainer (PN-3989353) and was used on L-88 and ZL-1 motors. It has two round springs and a flat wound damper spring. The outside diameter is 1.538". It uses an installed height of 1.90" and gives 128 lbs. of closed valve pressure and 330 lbs. of open valve pressure. Coil bind occurs at 1.26".

Some later model Mark IV stock production low performance cylinder heads came with short valve spring installed heights (1.78"- 1.820"). These work fine when you use the factory installed stock cam which only has a valve lift of .398". However, if you put in a higher lift cam and larger valves and then change the valve springs to match, you may find that you do not have enough installed height room between the bottom of the spring retainer and the valve spring pocket on the head. You may find that your new valve springs call for an installed height of 1.88", but you only have, say, 1.78". There are a number of solutions. You can have the valve spring pockets cut

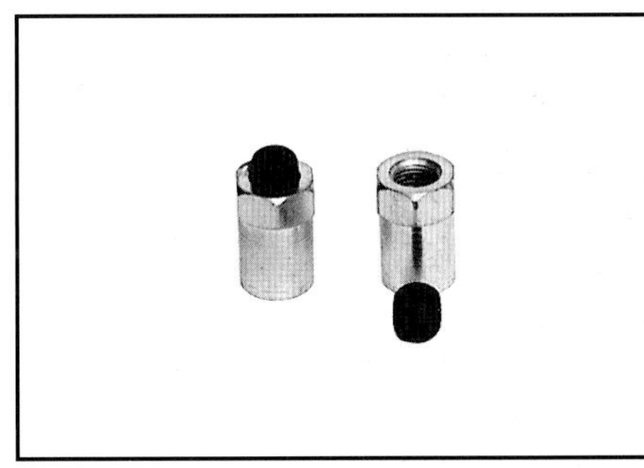

Aftermarket rocker arm nuts have an Allen screw that helps maintain lash adjustment. Make sure the rocker arm stud that this Allen screw butts up against is perfectly flat. Often the top surface of the rocker arm stud is irregular from manufacture and will cause the Allen screw to loosen and then the adjusting nut can back off. The top surface of the stud should be spot faced flat and perpendicular to the axis of the stud.

deeper down, up to a point. You don't want to cut into a water jacket. Or you can try a different aftermarket retainer which may give you more room. There are also aftermarket retainer locks which allow a slightly different installed height. You can get a valve with a longer stem or you can use a different spring with a shorter installed height. Usually one of these or a combination of these will get you the installed height your new springs need.

Remember to check the valve train geometry when you are done. You may need to change the pushrod length in order to keep the geometry correct. Aftermarket cam and tool makers can supply a number of different adjustable

BIG BLOCK Quick Facts & References

CHECKING INSTALLED SPRING HEIGHT

Here's a tip for checking installed height on valve springs. You can use a dial indicator caliper or a spring height micrometer specially made for this task or you can use a piece of very stiff wire or rod to make the measurement. Determine what the correct installed height of the spring should be. Most Chevy big block springs install at 1.88" or 1.90", (follow the manufacturer's suggested installed spring heights).

The installed spring height is the distance from the surface of the spring pocket on the head to the bottom of the spring retainer. So put a valve in the head and put on the retainer and locks that you are going to use.

Let's assume that the installed spring height you need is 1.88". Cut the wire or rod to that length plus a little extra and double check it with a dial caliper. Make sure that the ends of the wire or rod are flat. If you have to grind or file the wire or rod, make sure the length is still correct. Keep removing material until the length is exactly 1.88".

Now use the rod or wire to check the distance between the head and the retainer. If the retainer has a step on the bottom surface, measure the retainer surface farthest from the head, near the side edge of the retainer. How is it? If there is additional space, you can use spring shims to reduce the distance. Shims come in various thicknesses, .015", .030" and .060". If the distance is shorter or less than the 1.88" required, make sure that the retainer is pulled up tight against the locks. If you are still short there are other types of aftermarket retainers which will sit slightly higher so that you can get the necessary installed spring height. On some combinations you may have to use a valve stem that is longer than stock in order to get the installed height you need. Also if you have a truck head and you have removed the exhaust valve rotators from beneath the exhaust valve springs, the rotators should be replaced with a Federal Mogul rotator eliminator spacer (VS1360S) or Competition Cams rotator eliminator kit (Competition Cams PN–4779–8).

Remember to check the valvetrain geometry, no matter what combination of parts you use.

pushrod length checkers. Manley offers a big block Chevy pushrod length checker (PN-42133) that is inexpensive and slips on the rocker arm stud. It will tell you whether you need a different pushrod length or a different length valve stem. If you want to very accurately determine correct geometry, contact Ferry's Aluminum Heads and Block Repair in Dallas. They make a valvetrain geometry tool that can accurately measure the distance a roller tip rocker moves across the valve stem. Knowing this you can determine whether pushrod length or valve stem length changes need to be made. (See the sidebar on Valvetrain Geometry.)

The Gen.V 454/502 H.O. motors use a dual spring (PN-14097002) with a outside diameter of 1.487". Use retainer (PN-14096274).

It has an installed height of 1.88" and has 110 lbs. of pressure at this height. At .500" of open valve lift it gives 400 lbs.of pressure. Coil bind occurs at 1.290".

A second design retainer and seal kit for Gen.V and Gen. VI 454/502 is (PN-12360874) and this gives better oil control than the previous design. Use with spring (PN-14097002).

Valve spring shims are sold by Chevy under (PN-3875916, .015" thick) and (PN-3891521, .065" thick).

IT IS EXTREMELY IMPORTANT THAT YOU MATCH THE CAMSHAFT AND SPRINGS AND USE THE CORRECT RETAINERS, LOCKS AND OIL SEALS. THE SPRINGS MUST THEN BE INSTALLED AT THE CORRECT HEIGHT AND MEASUREMENTS MUST BE MADE TO INSURE THAT NO COIL BIND EXISTS WHEN THE VALVE LIFT IS AT MAXIMUM. ROCKER ARM GEOMETRY MUST BE MAINTAINED CORRECTLY. IF YOU ARE USING A HIGH LIFT CAM, MEASUREMENTS MUST BE MADE TO INSURE THAT PUSHROD LENGTH AND VALVE STEM LENGTH ARE CORRECT. DO NOT USE MORE VALVE SPRING PRESSURE THAN YOUR CAMSHAFT REQUIRES. CHECKS NEED TO BE MADE THAT INSURE THAT THERE IS NO INTERFERENCE BETWEEN ROCKER ARM, ROCKER ARM SLOTS, ROCKER STUD, RETAINERS, SPRING COILS, GUIDES, OIL SEALS AND VALVE STEM TIPS.

Aftermarket supplies can supply chrome silicon, H-11 tool steel, high tech coated steel valve springs and other steel alloys that can withstand or reduce heat fatigue and increase spring life. Some Pro Stock race springs are being made of titanium to reduce valve spring weight and provide increased open valve spring pressures, but the cost of these coils are extremely high. For most people steel valve springs will do the job and be far less expensive.

Aftermarket retainers are made with steel, aluminum or titanium depending on requirements and match the spring that they are to be installed on. Be careful with aluminum retainers. Valve springs with high lift and high spring pressures tend to cut into aluminum retainers and wear away the spring side of the retainer surface.

If you have a cylinder head that came from a truck, it may have valve rotators underneath the original stock exhaust springs. These rotators are meant to rotate the valve on its seat and continually clean deposits from the seats and valve margins. If you switch to a high performance cam and matching high pressure springs, the open spring pressures will be too much for the rotators. Take them out and use an aftermarket spacer available from Federal-Mogul (PN-VS1360S) or Competition Cams (PN 4779-8). These "rotator eliminators" or spring cups, fit perfectly in the stock valve spring pocket, are the same thickness as the stock rotators and will keep the installed height of the spring near the correct height. You can then use shims, if necessary, to get the correct installed height for your spring.

Each spring type requires its own retainer, locks and oil seal types. Various types of valve stem oil seals are made. Chevrolet sells an "umbrella" oil seal for the big block as (PN-460527). However, most dual and triple springs styles do not have enough room to install an umbrella seal inside the inner spring coils. Then you have to use a seal with a smaller diameter. Positive seals mounted on the valve guides can all be installed with some types being better than others at keeping oil out of the combustion chamber. Remember that if you are using an oil pan vacuum extraction system on a race motor sometimes the vacuum system works so well that not enough oil reaches the valve guides. Some engine builders will leave the oil seals off of the exhaust valves completely and use a seal on the intake guides that allow for oil control.

Locks or "split keys" need to match the retainers and "hardened" locks (PN-3947880) are available for extra strength. Stock 7 degree retainers and locks or 10 degree locks and retainers for racers needing more holding power, are also made by aftermarket suppliers.

VALVE COVERS

Factory installed Mark IV stamped steel short valve covers were either painted or chrome. Some of the early valve covers had oil dripper rails installed on the interior roof in order to bring splashing oil directly to the rocker arms. Some covers had spark plug wire looms built into the exterior surface of the cover. All of the factory installed Mark IV covers are short covers and are interchangeable. A painted short style Mark IV valve cover is (PN-14095840). Valve cover bolts are (PN-94339930)..

BIG BLOCK Quick Facts & References

ROCKER ARM RATIOS

Big block production Mark IV rocker arms (PN–10112680) have a "theoretical" rocker arm ratio of 1.72–1. The term "theoretical" or "nominal" is used because while the blueprints call for a 1.72–1 ratio, the actual stamped factory rocker arms may fail to achieve this ratio. It is very difficult to mass produce a stamped rocker that produces the ratio in every part made. When actually measured the rocker arm may only operate at a ratio of 1.69–1 or another may measure out at 1.70–1. The CNC machined aluminum roller rocker (PN–12361330) has a very accurate rocker ratio.

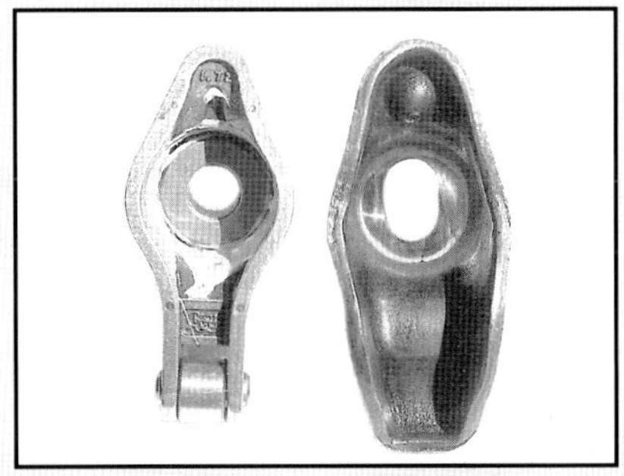

Now, on a daily driven, stock motor, who cares? Nobody. The fact that some of the rocker arm ratios are off on individual rocker arms is of no consequence and will not affect the way in which the motor runs or its longevity. However, if you have purchased a performance camshaft or plan to race an engine, keep the rocker arm ratio in mind. If the measured ratio is less than 1.72–1, the amount of lift to the valve will be less than you expected. Some valves may open more or less than others. Some aftermarket suppliers can provide rocker arms that provide a true, measured 1.72–1 ratio on their stamped steel big block rocker arms. They can take the manufacturing time to make sure the ratios are accurate.

BIG BLOCK Quick Facts & References

VALVETRAIN PARTS

	MARK IV Adjustable	GEN.V Non-adjustable
Int. pushrod 3/8"	10134307	14097068
Exh. pushrod 3/8"	10134308	14097070
Int. pushrod 7/16"	10134304	
Exh. pushrod 7/16"	10134303	
Int. pushrod 7/16" chrome moly	10134306	
Exh. pushrod 7/16" chrome moly	10134305	
Guideplate 3/8"	3860038	14097005
Guideplate 7/16"	3879620	
Rocker arm	10112680	12523976 w/ball
Rocker arm long slot	3959182	12508879
Roller Rocker Arm, Mark IV	12361330	
Rocker arm stud 7/16", Mark IV	3921912	
Rocker arm bolt 3/8", Gen.V		10114123
Rocker arm ball	5232762	10114122
Rocker arm nut	3896648	
Hydraulic lifter	5232720	17102353 Gen.V 454HO/502HO

You can use this parts list when converting a Gen.V non-adjustable valvetrain to a Mark IV adjustable valvetrain. Pushrods for a Mark IV short deck motor are 8.282" and 9.256" long, while the pushrods for the non-adjustable Gen.V and are 8.171" and 9.151" long. Pushrods for tall deck blocks are 0.4" longer than stock. Pushrods for the Symmetrical Port Bowtie heads are longer still.

BIG BLOCK Quick Facts & References

VALVETRAIN GEOMETRY

Valvetrain geometry often gets overlooked when making parts changes to improve engine performance. When you move to a higher lift camshaft, deck the block or heads, change to a different head casting, use a different head gasket with a different compressed thickness, change valve stem length, use a different pushrod length, change installed valve spring height, or change to a different rocker arm ratio, the relationship of the rocker arm tip to the tip of the valve stem can be altered. Remember that the rocker arm tip rotates in an arc as it pushes the valve stem down. The rocker arm tip motion is not straight up and down. It looks like it is, but its not. It rocks on the stud and the tip of the rocker sweeps across the face of the valve stem. If you look very closely at the motion and the actual contact point between the surface of the rocker tip and the valve stem tip, you will see that as the rocker starts to open the valve, the contact point is at a point slightly outboard of the center of the valve stem. As the valve is opened further, the contact point moves toward the center of the valve stem tip. Opening the valve still further towards full lift, the contact point now continues to move slightly inboard of the center of the valve tip.

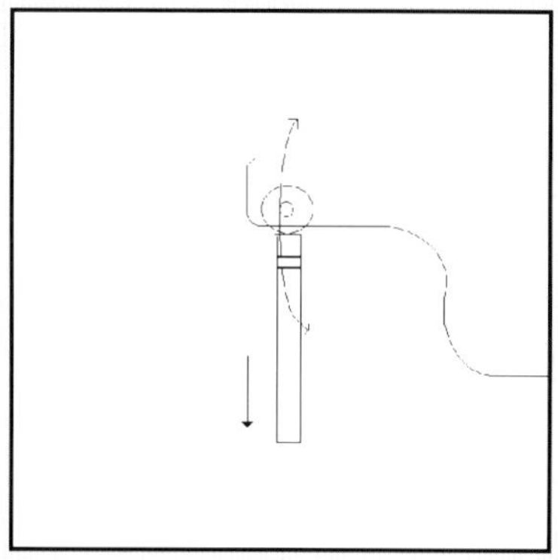

If the geometry is okay, the rocker arm will be at 90 degrees to the valve stem center when the valve has been opened one-third of the total lift. The other two-thirds of the valve lift will be accomplished by the remaining motion in the rocker arc as the contact point moves inboard. Now needless to say this is very difficult to see with your eyes. So after you look at it and can't really see anything, you start thinking that it doesn't matter and besides the geometry is probably okay. The problem is that if the geometry is not correct a number of problems can occur. Certainly accelerated valve guide and valve stem wear is one. Side loading of the guide will be worsened by improper geometry. More importantly, the rate of acceleration given to the valve opening will not be optimum. This is where power can be lost by improper geometry. Keep in mind that the cam lobe has a certain amount of lift and the rate of lift and acceleration is transmitted to the cam lifters, then the pushrod, then to the rocker arm which also multiplies the lift and acceleration motion by its rocker arm ratio. The rocker arm ratio actually changes as the rocker arm goes thru its opening motion. With some high pressure spring combinations, deflection (bending) of the pushrod, rocker arm stud, rocker arm or even the valve stem, will change this relationship.

So what do we do next? You can use an adjustable length pushrod (available from cam makers) to determine which length you need. Or you can get an inexpensive length checker from Manley (PN-42133) that slips on the rocker arm stud in place of the rocker arm and will help determine whether you need a different length valve stem or different length pushrod.
Ferry's Aluminum Heads and Block Repair has a great tool that uses a little jig and a dial indicator that actually measures the movement of a roller tipped rocker across the tip of the valve stem. This enables you to accurately determine movement and the position of contact relative to the center of the valve stem. You can then determine the optimum pushrod length.

Generally, increasing the valve lift or increasing the valve stem length to get more spring installed height, will require that you use a longer pushrod. However, as indicated in the first paragraph other factors are involved. Custom length pushrods are available in off the shelf lengths or just about any particular length you need.

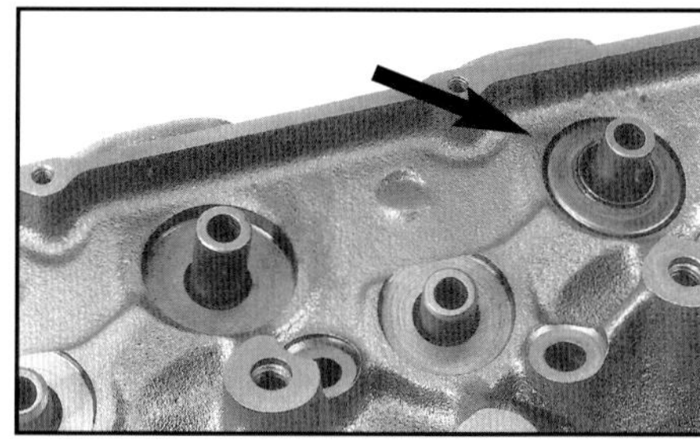

Stock exhaust valve rotator must be replaced with a "rotator eliminator" when you use higher than stock valve spring pressures. Competition Cams or Federal-Mogul make the eliminator kits.

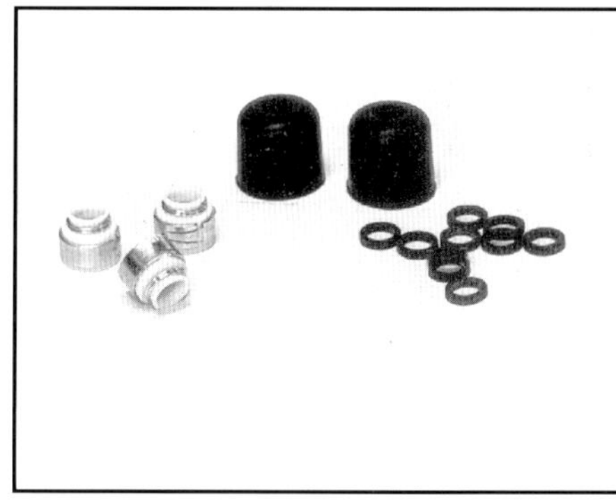

Three styles of valve stem oil seals. If you use the positive style (left) which slip over the valve guides, the top of the guides will have to be cut to fit them. If you use a dual valve spring, dual valve spring with a flat damper spring or a triple spring, this type of seal will fit inside of the spring coils. The umbrella seals (middle) will fit inside single springs with a flat damper and some dual springs. The O-ring (right) type seals are not generally used on big blocks.

Stud girdles help prevent rocker arm stud flex when using very high spring pressures. You will need a tall valve cover to clear these.

Chevrolet makes a chrome "Bowtie" valve cover in both a short version (PN-12342093) and a tall style (PN-12342099). Valve cover stud (PN-14082320) and nut (PN-14051876) will help hold the valve cover gasket in place. Valve cover gasket (PN-14085759) is a steel reinforced gasket that won't squeeze out and crack like some of the other gasket styles.

Keep in mind that if you install a very high lift cam, roller rockers or a stud girdle, you may need to use a "tall" valve cover for valvetrain clearance room. The tall covers may not fit if you have production A/C brackets or other engine room clearance problems with brake booster vacuum canisters, large A/C or heater core boxes.

Tall valve cover from Street & Performance will clear high lift cams, roller rockers and stud girdles. (Photo courtesy of Street & Performance)

Gen.V short style valve covers (PN-10183745 LH & 10183746 RH) were redesigned to match the machined gasket rails on Gen.V heads. These cast aluminum covers, matching gaskets and the Gen.V machined rail heads do a better job of sealing than the Mark IV system, unfortunately, they cannot be used on a Mark IV head without machining off parts of the valve cover gasket sealing surface.

GM does have an "over the counter" cast aluminum short valve cover, introduced in 1996, which will fit either Mark IV, Gen.V or Gen.VI heads under (PN-12363958). It uses a standard gasket and has oil drippers to get additional oil directly to the rocker arms and valve springs. (PN-12342099) is a "tall" chrome cover that will work with any '65 to '97 head.

If you use a very high lift cam, aftermarket rocker arms or use a stud girdle you may need to use a "tall" valve cover in order to provide enough clearance for the rocker arm assembly. However, keep in mind that "tall" valve covers may not fit if you intend to use the stock A/C brackets and compressor. There may also be a clearance problem with some power brake vacuum boosters, firewalls or air-conditioning/heater core boxes on the passenger-side firewall.

VALVE ADJUSTMENT

Valve adjustments are quick and easy if you use the exhaust opening/intake closing method. Simply rotate the engine while watching the rocker arms for each cylinder. When the exhaust valve just starts to open; set the intake valve lash. Then continue rotating until the intake valve has opened fully and is just starting to close. Now set the exhaust valve. This method assures that each lifter is on the base circle of the cam when you make the adjustment.

INITIAL FIREUP

The first few minutes that a new engine is run are extremely important. Nothing can match the frustration and disappointment of having something go wrong on a new motor. One of the things that can go wrong is having a cam lobe go flat. This is not uncommon and you should do everything you can to prevent it from happening. It is especially a concern with camshafts with high lifts and valve spring pressure, since a higher load is being exerted on the cam lobe and lifter surfaces. Make sure that before the cam is inserted into the motor that the surfaces of the cam are coated with a moly based lube or the cam lube supplied by the cam maker. The foot of the lifters should also be coated with lube. Chevrolet makes an Engine Oil Supplement (EOS) (PN-992869) which has excellent high pressure lubrication qualities. Throw a pint into the engine before initial fire up. Use some lube or EOS on the rocker arms and rocker arm balls during assembly. If a high lift cam is being used, you can temporarily replace the stiff valve springs with weaker stock type springs. These will enable you to break in the motor and lower the chance of lobe failure. Crane Cams also makes a special low ratio rocker arm that can be used during initial break in. The rocker arm opens the valve enough for the motor to run at relatively low rpms but keeps the valve spring pressures from getting too high. These rocker arms are easier to switch after break in than having to switch all of the valve springs.

Before you start the motor, prime the oil system. Use a single grade 30W oil instead of a multigrade oil or a synthetic oil for the initial break-in. Run the oil primer tool until oil is coming out of all of the rocker arm bleed holes. You may have to rotate the crankshaft by hand in order to expose all of the oil passages in the engine. Make sure that the distributor is placed correctly, so that the engine will fire up immediately and not crank over for a long time without starting up. Get the initial valve lash correct before starting. Have enough fuel so that the motor can run for twenty minutes without running out of gas. Fire it up and immediately bring the rpms to about 2500 rpms and keep it there for about twenty minutes. While its idling at 2500 you can check for leaks, make sure the oil pressure is okay and check for overheating.

When you are done with the break in, drain and change the oil and put a new filter on. Its okay to use a multigrade or synthetic oil after the cam and lifters have been run in. Recheck the valve lash and adjust it and the ignition timing accordingly.

Induction

Photo courtesy of Racing Head Service

All intake manifolds interchange between Mark IV, Gen.V and Gen.VI short-deck engines. Remember that intakes that came on tall-deck engines are wider than the intakes for short-deck motors. GM cast iron manifolds have a casting date showing when they were made, located on one of the rear intake runners behind the carb plenum. Casting numbers, identifying the manifold type are in the same area.

Casting dates for factory aluminum intakes are located on the top or bottom of the manifold, under the plenum area or may be hidden under an oil splash/heat shield riveted to the bottom of the manifold. In addition, factory aluminum intake manifolds and factory aluminum heads, were cast at an outside manufacturer, the Winters Foundry. The Winters Foundry used a logo of a "snowflake" and a "W" cast on all of the aluminum intakes (and aluminum cylinder heads) it made for Chevrolet. Many people who notice the "W" on factory intakes think that the "W" is an abbreviation for "Weiand" a company that has been in business for many years making excellent aftermarket intake manifolds and blowers. The "W" is the logo for Winters Foundry.

Stock production, four barrel, intakes are made from cast iron or aluminum. Chevy did make a single, two-barrel, cast iron, oval port intake for full-size, passenger cars in 1969 only and they are seldom seen these days. Big block Corvettes from 1967-1969 had optional 427 engines that

An old trick with factory split plenum manifold still works. If you want to increase the higher rpm operating range of a split plenum intake, you can remove one-inch from the plenum divider wall to increase the amount of air and fuel available to the motor at higher rpms.

used three two-barrel Holley carbs. These aluminum "tri-power" intakes came with either oval ports or rectangular ports depending on the engine horsepower rating.

Most four-barrel cast iron manifolds had bolt patterns and holes for Rochester Quadrajet carburetors. Some early cast iron intakes had carb mounting holes for Holley four-barrels. Cast iron, oval port intakes were made in hi-rise and low-rise forms and were used on low performance and medium performance engines. All cast iron intakes have small or large oval intake ports. No cast iron factory intake manifolds were made with rectangular intake ports.

Most Chevy big block, aluminum four-barrel intakes have rectangular intake ports, with Holley square pattern carb mounting bolt holes. Aluminum four-barrel, rectangular port manifolds were made in hi- rise (PN-3933163) or low-rise (PN-3977609) castings. These two manifolds have split plenums. One exception is the aluminum, low- rise, large oval port, four barrel manifolds used on 1968 and 1969 Corvette 427/390HP engines, (Casting #'s 3947801, 3919849 and 3937793). These aluminum, large oval port factory manifolds used Quadrajet carbs and are relatively rare. If you are restoring a 68'-69' Corvette then try finding one if you need it, but otherwise don't bother. There are plenty of fine aftermarket aluminum intakes for oval port heads.

Chevy never used a true open plenum manifold on their big block production engines. Even on the limited production 427 L-88 and ZL-1 engines, all Chevy did was cut down the center divider of their square port aluminum split plenum manifold, in order to increase plenum volume at higher rpms. The factory aluminum manifold requires that an 850 cfm carburetor be "stagger-jetted" because of the unequal lengths of the stock manifold intake runners and the unequal length of the intake ports in production big block heads. Some cylinders receive more of the air-fuel mixture than others with this factory manifold. Holley still makes a carburetor that is set up for the factory big block aluminum high-rise intakes. It is Holley list number 4296. It is an 850 cfm, double pumper, with different mixture tabs in the throttle bores and comes "stagger-jetted" from Holley.

The last rectangular-port aluminum intake manifold that was factory installed on production vehicles was in 1971 on 454 LS-6 motors. Since then all factory installed four-barrel intake manifolds have had oval ports and were made of cast iron.

If you use the only currently offered factory aluminum high-rise rectangular port manifold (PN-3933163) on "Bowtie" cylinder heads, you will have to match the intake runners of the manifold to the "raised" intake ports found on these heads.

Late model production big blocks, 1987-95, utilized electronic fuel injection (EFI) which is an on-board computer controlled, "Throttle Body Injection" (TBI) system induction. These use unique intakes designed for the TBI system. These TBI intakes have small, oval ports. In 1996, factory installed “Vortech 7400” 454's used a sequential port fuel injection system with a two-part aluminum and composite intake manifold.

A number of aftermarket suppliers are producing kits that enable you to increase the performance of late model factory fuel injected big blocks. These kits include throttle body injection (TBI) or tuned port injection (TBI) systems with larger port manifolds, larger intake runners, larger fuel injection nozzles, different PROMS for your on-board computer and come with whatever wiring harnesses that may be required. Aftermarket kits are also available to put a late model fuel injected big block into early model musclecars or other vehicles. In late 1995, Chevy introduced a 502 Gen.VI crate motor (PN-12360893) that can be used as an emissions legal, direct replacement motor for '88-'94 454 truck conversions.

Various intake manifold choke appliance bosses, holes and EGR valve bosses and holes were used over the years. If your carb has a particular style choke or you need to hook up an EGR valve, make sure you get a manifold which has the features you need.

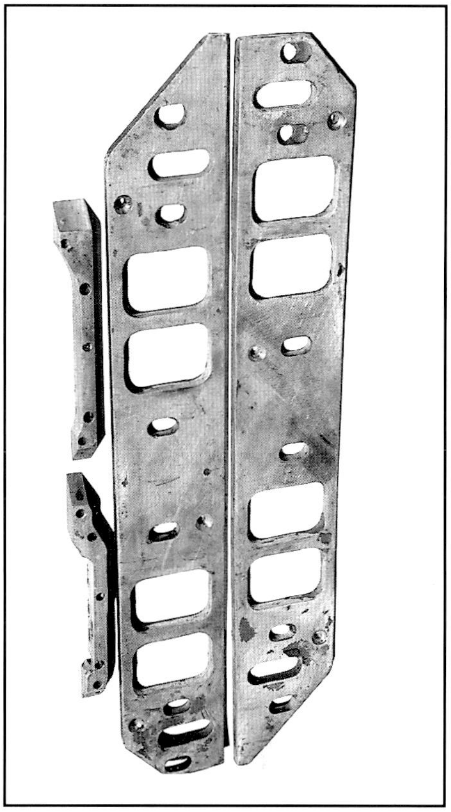

A manifold spacer kit will help fit a short deck intake to a tall deck block. A number of aftermarket performance big block intakes are made wider so that a spacer kit is unnecessary when using a tall deck block.

Aftermarket intake makers can supply many different styles to meet your requirements, including spacers to mount short deck intakes on tall deck blocks or wide, high performance manifolds to fit tall deck blocks, or lowered distributor holes on wide manifolds so that you can use short deck distributors on tall deck blocks. On some factory race heads like the symmetrical port Bowties or those from Pontiac, you will need a custom intake. Hogan is one company that can provide sheet metal intakes of various designs.

Remember to match the intake port style of the intake manifold to the port style of the heads you use. A small oval port intake should be used with small oval "truck" port heads. Large oval port intakes go with large oval port heads and rectangular port intakes with rectangular port heads. Dart has a large oval port race manifold that they recommend using with rectangular port heads in a full bodied heavier vehicle. This is an attempt to keep velocities high while using a square port head.

PLENUMS

Generally four-barrel intakes have two types of plenums. One is called a "split", dual-plane or divided plenum. The divider splits the plenum in half and this type mani - fold works well on the street because it helps create low and medium rpm power. Carb size on these manifolds can cover a larger cfm range because each cylinder "uses" only 1/2 of the total plenum volume.

The second type of manifold is an "open" plenum. These manifolds have no center divider and consequently, each cylinder "sees" a much larger plenum volume to draw upon. These open plenum manifolds therefore enable the cylinders to receive more volume at high rpms, but lose some of their low rpm power and response. Open plenum manifolds are more sensitive to carb cfm size and tend to work better within a narrow range of carb cfm sizes. Extending the port divider walls in an open plenum manifold and bringing them closer to the center of the plenum seems to add power. Extending the divider walls use to be a hand fabricated affair, but a number of race intakes are now manufactured with this detail. Brodix has some. Brodix offers aluminum, high rise, open plenum race manifolds with a various feature to fit large oval or square port heads, short or tall deck blocks and standard square flange or Dominator style carbs. Brodix also offers an intake with a "Dittmer" style round plenum/carb opening that uses a Holley Dominator carb and also has a "turtle" installed in the plenum floor. This "turtle" helps direct fuel to the individual port runners.

So a split plenum manifold allows the use of a carb cfm which is substantially larger than what might be considered minimum. A split plenum generally keeps engine vacuum higher at low and middle rpms, so carb booster signal and throttle response are very crisp at low and middle rpms. However, as rpm's get very high the split plenum manifold's smaller plenum volume gets to be too small to feed the high rpm airflow demand of the engine. Power falls off. A larger than normal carb may help the engine in this situation by providing additional volume and less restriction. Notching of the plenum divider will extend the rpm range of these manifolds.

The open plenum style manifold generally has opposite characteristics. The engine has a large plenum volume to pull from so at low engine rpms the vacuum signal to the carb is not as strong as a split plenum, but as engine rpms increase the open plenum starts to show some advantages. It's better able meet the demands of a high winding or large cubic motor because the open plenum provides more plenum volume for the cylinders to draw from.

Modern intake manifold design has eliminated some of bad characteristics found in early designs. Split plenum, dual plane manifolds have been created that work at higher rpms than the old designs by improving intake runner and plenum design.

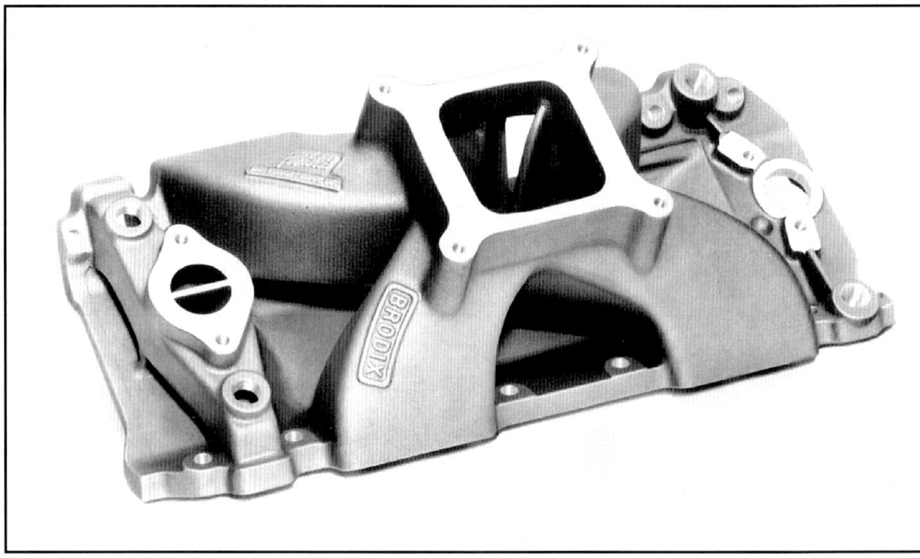

Brodix also makes a line of aluminum intake manifolds for big block Chevys. This "HV-2001" has curved runners, open plenum and extended wall dividers in the plenum. Brodix makes aluminum intakes that will fit short or tall decks, oval or square port heads and with carb flanges to fit Holley standard square flange or Dominator carbs. (Photo courtesy of Brodix)

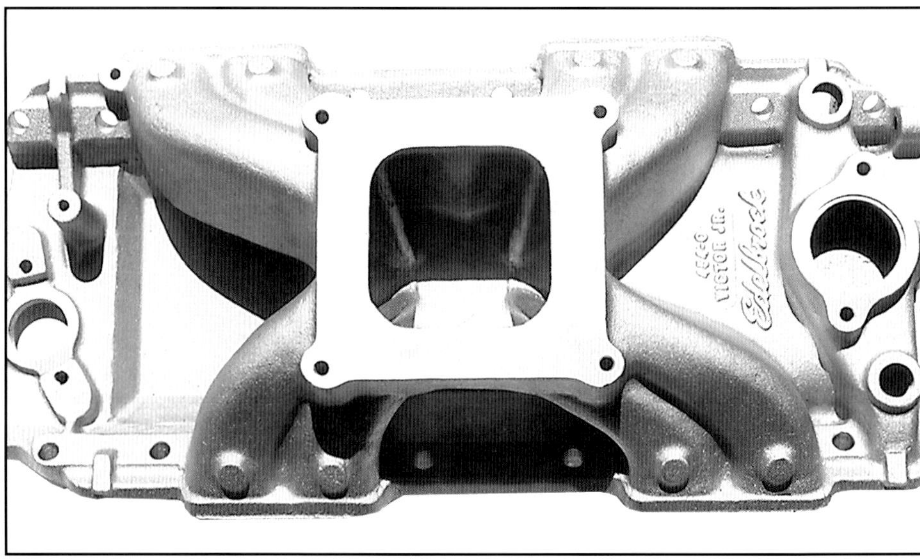

A modern design open plenum, aluminum intake makes an excellent choice for bracket racing. This Edelbrock "Victor Jr. 454-O" intake has an air-gap under the plenum for cooling, nitrous- oxide nozzle bosses on the port runners and curved runners which meet the ports on the heads at a good angle. Oval port, square port, short deck, tall deck, Dominator flange and square carb flange models are available.

MANIFOLD POWER RANGE COMPARISONS

The accompanying charts show different types of intake manifolds to illustrate the predicted effects the manifolds have on power and the rpm range where they work best. Nothing was changed on the motor except the intake manifold style and in the case of the tunnel ram the total carb CFM was increased from the 750 CFM used on the single carb manifolds, to 900 CFM to reflect the use of two 450 CFM carbs commonly found on "street" tunnel ram combinations.

The first motor (top chart) used was basically stock, with oval port heads and 2.06"/1.72" valves, 8.5-1 CR and small tube headers and mufflers. The hydraulic cam had an advertised duration of 272 degrees and lift figures of .515"/.510". As you can see in this comparison the split plenum manifold made substantially more power at lower rpms than the other manifolds. At 3500 rpms the other manifolds catch up.

The 460 bracket race motor in the lower chart had pocket ported heads, 850 cfm carb, 2.19"/1.88" valves, 12.5-1 CR and a solid cam with a duration of 256/266 at .050", lift was .595"/.621". The first tunnel ram had two 450 cfm carbs while the second used two 600 cfm four barrels. Notice that since the heads and cams used in both of these charts were not changed, the torque peak and peak horsepower occurred at the same rpms no matter which intake manifold was used. The split plenum manifold held its own and made more power until about 3500-4000 rpm when the other manifolds caught up. At 4000 the open plenum and tunnel rams start to come alive compared with the other manifold. Now these examples were done to isolate the effect of manifold style. If the motors had been optimized for their respective component parts the figures would be different, but would still show the individual operating ranges and power trends of these induction systems.

If you have an alcohol (methanol) fueled motor, a tunnel ram with alcohol carbs or a mechanical fuel injection unit will help provide the necessary fuel volume needed with this race fuel.

Whatever intake manifold you use, make sure that your street vehicle's hood will close. A hi-rise intake, with carb and air filter may sit too high in some vehicles. You may be forced to cut a holes in the hood and then add a hood scoop to make things fit. Take your time and determine what will work before you make any purchases or make any changes.

	DUAL PLANE		OPEN PLENUM		TUNNEL RAM	
RPM	HP	TORQUE	HP	TORQUE	HP	TORQUE
2000	173	454	159	418	160	420
2500	222	465	209	439	210	441
3000	265	465*	258	451	259	453
3500	310	466*	308	462*	310	465*
4000	347	455	350	459*	353	463*
4500	365*	426	374*	436	378*	441
5000	356	374	368	387	373	392
5500	323	308	340	325	346	330
6000	269	235	291	254	296	259
6500	206	166	232	187	238	192

	DUAL PLANE		OPEN PLENUM		TUNNEL RAM		TUNNEL RAM	
RPM	HP	TORQUE	HP	TORQUE	HP	TORQUE	HP	TORQUE
2000	155	407	143	375	143	376	144	377
2500	212	445	198	416	198	416	199	417
3000	258	451	248	434	248	434	249	436
3500	322	482	318	477	318	477	320	480
4000	390	512	394	518	395	518	398	522
4500	456	532*	470	548	470	548	476	555
5000	508	533*	533	559*	532	559*	542	569*
5500	543	518	578	552	577	551	591	564
6000	556*	486	601	526	599	524	617	540
6500	555*	448	609*	492	607*	491	629*	508
7000	541	406	592	444	598	449	622	467
7500	523	366	570	399	587	411	614	430

INTAKE MANIFOLD MILLING

If your big block cylinder heads have been milled to insure flatness or to reduce combustion chamber volume, or if you have decked the block to get it flat or to reduce the piston deck to block deck height to "zero", you may have to mill the intake manifold surfaces. This will keep the intake ports lined up properly and also the manifold bolt holes will line up with their respective holes in the head. This chart does not apply to head deck surfaces that have been "angle milled".

Total milled from blockand/or head surfaces	Intake Manifold Sides	Intake Manifold Bottom
.005"	.004"	.010"
.010"	.007"	.019"
.015"	.011"	.029"
.020"	.014"	.038"
.025"	.018"	.048"
.030"	.021"	.057"
.035"	.025"	.067"
.040"	.028"	.076"

The same can be said for open plenum styles. Better design has made them more tractable at low and middle rpm ranges. In older design manifolds the angle at which the intake manifold runners met the intake ports on the head was too sharp. More modern single four barrel designs have intake manifold runners that curve more and meet the head port at a better angle and improve flow. Many of the newer style manifolds are cast with a floor that isolates the hot oil found in the lifter valley from the intake manifold. This air gap passage above this manifold base floor helps cool the top surface of the floor and keeps the temperature of the manifold port runners cooler.

Maximum effort drag race engines that operate in the 6500 to 10,000 rpm ranges are another story. Motors with race prepared heads, efficient large tube race exhaust headers and high rpms can use large Holley Dominator carbs, and open plenum race manifolds, tunnel rams, or mechanical fuel injection.

Whatever you do, don't put a tunnel ram manifold on top of a street-driven motor with stock heads unless you want to be driving a dog.

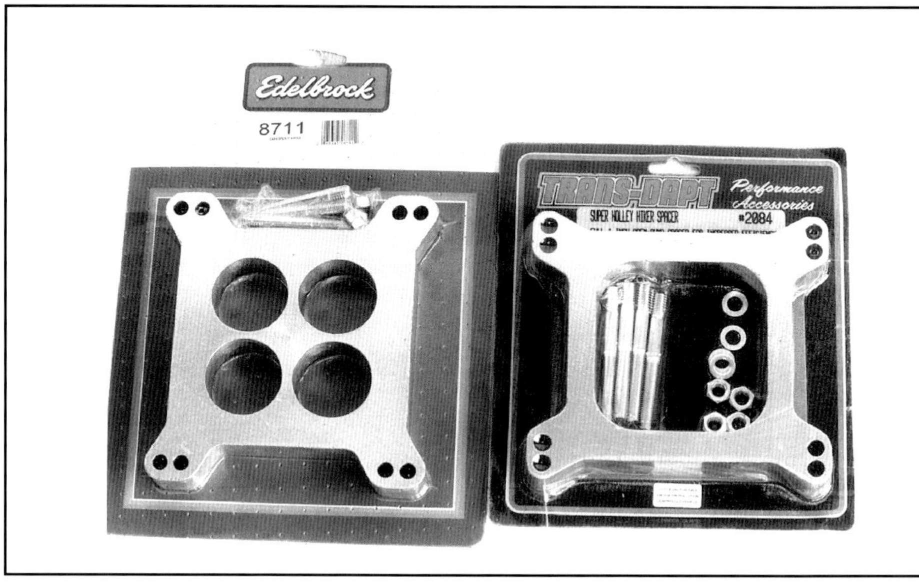

Various carb spacers are available from many companies. Four hole, single hole, wood, plastic and metal spacers are made. Not all engines gain from the use of one. You'll have to experiment to see if your motor wants or needs one.

These types of manifolds are strictly for high rpm race motors which can effectively use the volume of air made available by tunnel rams. It's a bad combination in a street motor. I know that there are a number of "street tunnel ram" setups using small cfm carbs, but they still don't work well on the street. A number of street rod builders use them to enhance the looks of their engine compartments, but this combination kills bottom end torque, which is just what you need on a street driven vehicle. Stick with a single four barrel split plenum manifold on the street and you will be much happier with the results.

CARB SPACERS

Carb spacers do add volume to a plenum and may help air flow thru the plenum when it makes the turn into individual port runners. However, you need to experiment with these spacers. Some engine/intake manifold combinations like them, some do not. You can try different size spacers or no spacer at all and then see what happens. Also since you are changing the total plenum space when you use a spacer, the metering signal to the carb may be altered. You may need to change jet sizes and accelerator pump cams and shooter sizes to get things "dialed in".

Spacers are made from aluminum, laminated wood or plastic and have one, two or four holes. The wood or plastic models seem to conduct less heat to the carb. They are available from aftermarket suppliers.

CARB SIZES

Carb size selection is important. Its easy to overcarburate any motor by thinking that the bigger the carb, the faster the car will be. It simply is not true. The carb must match the air/fuel requirements of your particular motor and how you plan to use your vehicle.

366 to 427 street engines should limit carb size to 600-750 cfm. 454 to 502 street motors can use 650-850 cfm carbs. Street engines should use a carb with vacuum operated secondaries and a split plenum intake manifold. Mileage will be decent and throttle response will also be excellent.

Race engines, that are never driven on the street, can use from 600 to 1450 cfm carbs and open plenum intakes. Double-pumper, mechanical secondary style Holley carbs are strictly for racing, and if used on a street motor, will not get decent gas mileage and throttle response may not be sharp and crisp. Those that plan on using tunnel ram manifolds on a race motor should know that these manifolds, using two carbs, only start to work at around 4500 rpm and then can give some benefit to 8500 rpm or more. Low speed driveability is poor. The dual carb setup does reduce intake restric-

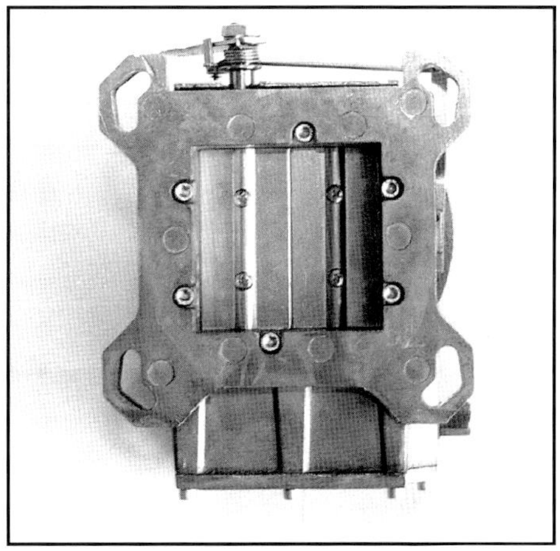
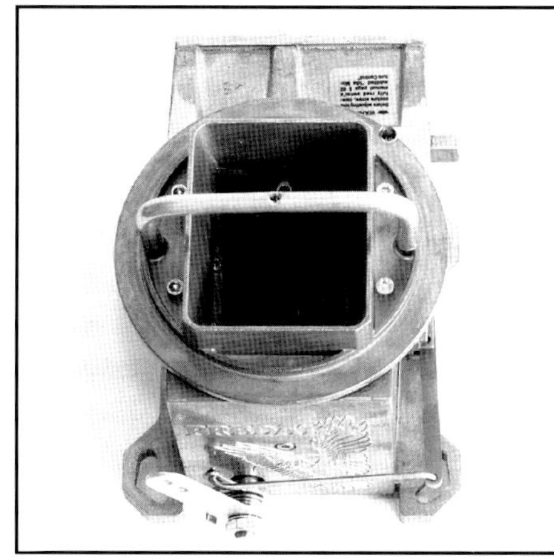

Predator carbs offers simple unique design and can be used on a wide variety of engine sizes because it will only allow as much air into the motor as it it needs by using a variable venturi. It can flow up to 930 cfm.

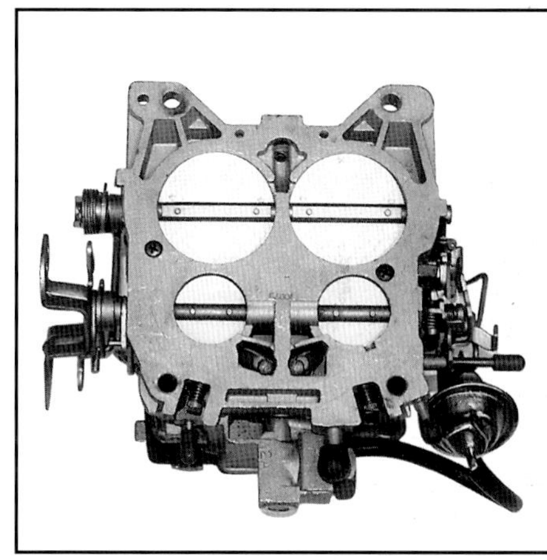

Edelbrock offers their own "Performer and Performer RPM" versions of the Quadrajet four barrel carb in sizes up to 850 CFM.

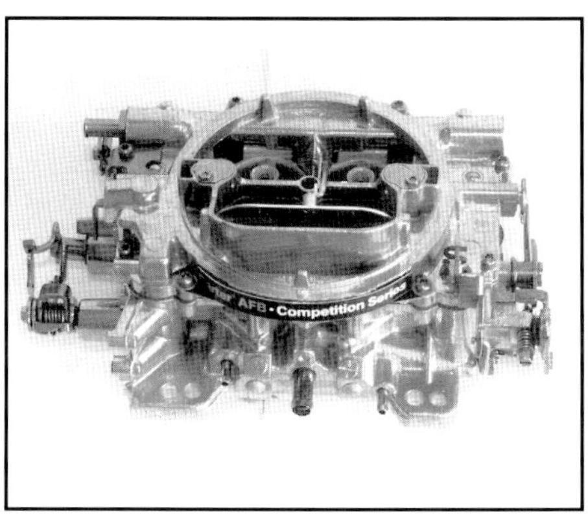

Carter Competition Series AFB is another four barrel carb with different size flows up to 750 cfm.

tions in a very high rpm motor, but you need an engine parts combination that is suited for this type of intake and high rpms to gain any advantage. A properly set up high rpm race motor with a tunnel ram will make power, but it is at the expense of low speed (below 4500 rpm) torque and power and driveability. There isn't enough signal at the carbs to get the main circuits working properly at low rpms. This loss of low engine speed power and manners is one of the reasons some of these tunnel-rammed race cars get pushed to the starting line and towed back to the pits after a run.

Check the ENGINE MATH chapter for the formula for figuring out minimum carb CFM size with a particular volumetric efficiency.

All aftermarket carburetor makers also offer numerous parts to dial in your carburetor for your particular combination. You should know that most carburetors are very close to what you will need right out of the box. If you find that you need to make a wide range of changes to the calibration of the carb to get it right for your engine, chances are you have the wrong carburetor. For instance, I have seen people who have put a carb on that is too large for their motor, falling into the trap of thinking that "bigger is better". The motor now runs too rich. They try putting in smaller jets in order to reduce the amount of fuel. Sometimes reducing by 10 jet sizes! It's a mistake. Try to think in terms of which carb size is best for your parts combination and usage.

What rpm range is your motor going to run in? What type of intake manifold are you planning to use? What are the pros and cons of that particular setup? If you are using a Holley four barrel, make sure that you get the proper power valves in the carb.

CARB TUNING

If you have picked the correct carb size for your engine size and volumetric efficiency, there isn't much that needs to be done. For a Holley carb, you need to know that Holley calibrates its carbs for sea level operation at 59 degrees Fahrenheit. This means that if you are at a different altitude or air temperature, the air/fuel calibration of the carb can be changed to your conditions for maximum performance. Since there is less air density as you increase altitude, you can decrease the main jet sizes one jet size for every 2000 feet in altitude above sea level. This will keep the

POPULAR HOLLEY CARBS

Here is a list of some of the more popular Holley four barrel carbs that can be used with big block Chevy engines. Holley carbs have a "List-#" stamped into the front surface of the choke horn. They also have a date code stamped in the same location. On Holley carbs that were originally installed by Chevrolet, there is a Chevrolet part number for that particular model carb also stamped into the choke horn area. On Model 4500 Dominators, which have no choke horn, the "List-#" is on the flat portion of the airhorn. Matching "List-#'s" are also sometimes stamped into the top surface of metering blocks. As improvements or changes are made to existing model "List-#'s" Holley adds an additional hyphenated number to the "List-#". It will appear like this "List-3310-2".

Many of the competition style carbs have various features for racers, these include carbs with annular or short boosters, progressive linkage, 4-corner idle mixture settings, two accelerator pumps, choke plates removed or the entire airhorn removed and the main body top contoured, Dominator style fuel bowls, high flow needle and seats and much more. A careful look at what is available from Holley will help you make the right choices.

In 1992 Holley started adding power valve blowout protection to some of its competition carbs. If you have a carb that does not have this feature, Centec makes a power valve shield that is simple, inexpensive and easy to install.

This list will only concern itself with the Holley "List #" and the most popular Chevy carbs available on the aftermarket. Holley made and continues to make, hundreds of different "List-#'s" that have many different features and will fit various types of engines. Holley has been around for many years and continues to lend active support to car restorers, racers and customizers. Contact them if you need help.

LIST #	CFM	VAC. SEC.	DOUBLE PUMPER	SQUARE FLANGE	SPREAD BORE
1850	600	Yes		Yes	
3310	780	Yes		Yes	
3310-2	750	Yes		Yes	
3418	855	Yes		Yes	
4118	725	Yes		Yes	
4776	600		Yes	Yes	
4777	650		Yes	Yes	
4778	700		Yes	Yes	
4779	750		Yes	Yes	
4780	800		Yes	Yes	
4781	850		Yes	Yes	
4788	830		Yes	Yes	
6210	650		Yes		Yes
6211	800		Yes		Yes
6212	800		Yes		Yes
6853	650		Yes		Yes
7002	650		Yes		Yes
7010	780		Yes		Yes
7320	1150		Yes	D	
8082	1050		Yes	D	
8156	750		Yes	Yes	
8162	850		Yes	Yes	
8896	1050		Yes	D	
9380	850		Yes	Yes	
9381	830		Yes	Yes	
9375^	1050		Yes	D	
9377^	1150		Yes	D	
9379^	750		Yes	Yes	
9380^	850		Yes	Yes	
9381^	830			Yes	
9645*	750		Yes	Yes	
9646*	850		Yes	Yes	
80186^	750		Yes	D	
80496	950		Yes	Yes	
80497	950	Yes		Yes	
80498*	950		Yes	Yes	
80776#	600		Yes	Yes	
80777#	650		Yes	Yes	
80778#	700		Yes	Yes	
80779#	750		Yes	Yes	
80780#	800		Yes	Yes	
80781#	850		Yes	Yes	
84010	600	Yes		Yes	
84011	750	Yes		Yes	
84012	600		Yes	Yes	
84013	750		Yes	Yes	
84014	650	Yes			Yes
84015	800	Yes			Yes
84016	650		Yes		Yes
84017	800		Yes		Yes
84021	650	Yes			Yes

HOLLEY CARB CASTING DATES

Holley carbs have a "List #", such as "List # 4781" stamped into the front of the choke horn which tells you what type, model and cfm size that carburetor is. If there have been changes made to that model over the years, the particular version will have a number added to the "list #". It will look like this "4777-2". There is also a date code which tells you when the carb was made. The date code usually is three digits long. It might look like this "743". The first number is the year of decade (7). The second digit or letter is the month of the year (4). Holley uses a month sequence of 1-2-3-4-5-6-7-8-9-0-A-B for January thru December. The last digit stands for the week of the month (1 thru 5). So our example of "743" translates to a carb made in the third week of April, in the seventh year of the decade. You might have to look at the other features of the carb to give you clues as to which decade it was made. Holley has also used a four digit code, where the first three digits are the day in the year (from 1 through 365) and the fourth digit being the year of the decade.

Finally, on factory installed Holley carbs, between 1964 and 1972, there is a GM part number stamped in the same choke horn area. These part numbers were assigned depending on what type of vehicle, engine/carb size and transmission combination was used.

If the Holley carb was bought on the aftermarket there will not be a GM part number stamped into the choke airhorn.

*=Manufactured and calibrated for Alcohol fuels, #=Power valve blow out protection, ^=Annular discharge boosters D=Model 4500 Dominator race carbs which require a unique aftermarket intake manifold and carb bolt pattern for the 4500 Dominator series. Dominator carb/manifold bolt hole spacing square pattern is 5.380" on each side.

QUADRAJET CASTING DATES

Quadrajet four barrel carbs were made by Rochester, but some early Quadrajets were built for GM by the Carter Carburetor Company. The ones built by Carter, for GM, are marked Carter. Both makers put part numbers and manufacture dates, either on small aluminum tags attached to a bowl screw or they were stamped into the driver's side of the carb. The stamped date codes are usually four digits with the first three digits the sequential day of the year and the last digit the year of the decade. It might look like this "2349", which means the "234th" day of the year, in 1969 or 1979.

air/fuel ratio where it should be. Likewise, for each 35 degree temperature increase above 59 degrees you can decrease one jet size. Power valve changes can also compensate for altitude increases. For every 3000 foot increase in altitude, the power valve opening point can be reduced 1.5-2.0 sizes. The best way to choose a power valve is to put a vacuum gauge on the carb or intake manifold at idle rpm and then choose a power valve with a rating 1/2 to 1 step below the reading you get on the vacuum gauge. For instance, if the vacuum gauge shows a reading at idle of 6.5 then use a power valve with a 5.5 rating. Don't take the power valve out and then plug it. If you do this it throws the fuel metering way off and you will have to increase the main jet size 6 to 10 sizes to compensate. Part throttle mileage will decrease if you remove the power valve and plug it and at idle or part throttle the air/fuel ratio will be too rich. Spark plug fouling can become a problem at low rpms.

For drag racers you can also experiment with the accelerator pump cams and shooter size to improve the engine's performance off of the line. If you are using a vacuum operated secondary style carb, you can use different springs in the secondary vacuum canister to change the way the secondaries are actuated. Make sure however, that you begin any experimentation with the stock calibrations and that you note what they are, so that you can go back to the original setting

BIG BLOCK Quick Facts & References

SELECTING HOLLEY CARBURETORS

Holley four-barrels are one of the most versatile carburetors available for high performance work. They are available in a broad range of CFM ratings and Holley offers special configurations for street, high performance and racing applications. A Holley four barrel on top of your engine is still a universal sign that you mean business. This 990 CFM version was prepared by the Carburetor Shop in Rancho Cucamonga, California.

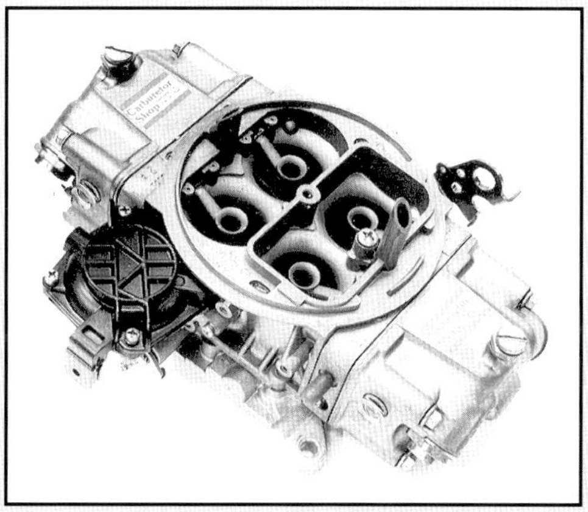

Holley's List 3310-vacuum secondary four-barrel is by far the most desirable Holley carburetor for street performance, RV applications and some circle-track racing. The 3310 has vacuum secondaries that allow a crisp transition from a well tuned primary circuit. It can be tailored to suit the unique requirements of nearly any application. If you want good drivability and throttle response in a street carburetor, the 750 CFM 3310- is your best choice.

For racing applications and "outer limits" street machines, the Holley Double Pumper is the leading contender. Available in sizes ranging from 600 CFM to 950 CFM, Double Pumpers feature an accelerator pump on both the primary and the secondary metering circuits. They can be configured with adjustable 4-corner idle circuits for precise tuning, and they are easily modified to suit any gasoline or alcohol racing application.

Holley Dominators are pure racing carburetors designed to flow massive amounts of air. Most of them feature an intermediate idle circuit in addition to the two standard metering circuits and they are quite tunable. The Carburetor Shop builds them in a range of sizes all the way up to 1425 CFM. The 750 CFM Street Dominators can be modified for strong street performance or racing.

if necessary. If your vehicle leaves the line very quickly, you may need main jet tube extensions on the secondary main jets. These extensions keep the main jets covered by fuel under hard acceleration. Using secondary jet extension tubes may require that you use a different float assembly in the secondary fuel bowl so that the float will clear the main jet extensions.

If you are going to use a "used" carb that you don't know the history of, I suggest you take it apart and check to see what main jet sizes are in it, what power valve, gaskets, the adjustment on the accelerator pump lever, etc. Often the previous owner(s) of the carb experimented or used inferior parts and a "used" carb can be far away from its original calibrations and settings. Holley publishes a listing of all of its carbs, their "List #'s", CFM sizes and their respective calibrations, stock jet and power valve sizes, which can be very helpful in establishing the "baseline" calibrations of your carb. I have seen many carbs which were described by their owners as "junk" or dismissed as "not the right one" for their motor. Many times these carbs had been previously "rebuilt" by someone who had used the wrong jet sizes, incorrect power valves or power valve plugs and a bunch of other race "tricks". With the correct parts these carbs worked just fine.

For those of you that are using a very "lumpy" camshaft, it often causes trouble getting the carb's idle mixture circuit to work properly. The fix was to drill small holes in the throttle plates to let in more air at idle and still have the throttle plates properly positioned near the idle transfer slots in the throttle bores. Holley has a number of carbs, in various cfm sizes, that have an adjustable secondary idle circuit built in. These "four corner idle" carbs make it easy to adjust the idle mixture in both the primary and secondary bores and usually solves the idling problem when using a race style, long duration cam. Some of these Holley carbs have holes drilled in the throttle plates in addition to the "four corner idle" system.

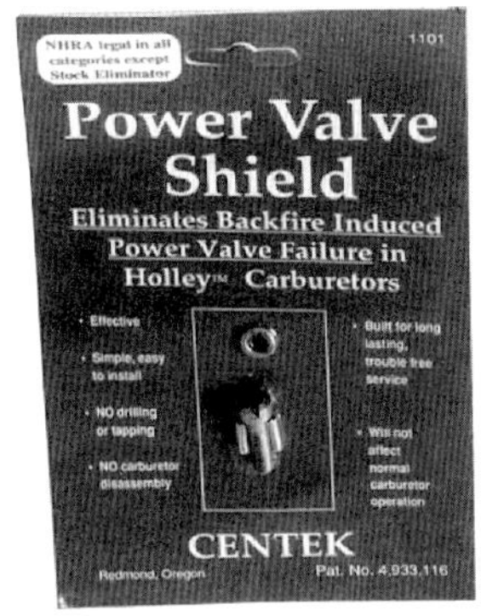

Centek makes a simple, easy to install power valve back fire protector for older Holley carbs. Many Holleys manufactured after 1992 are made with power valve back fire protectors built in.

One of the problems that Holley carb users noted was the possible failure of the power valve when there was a engine backfire thru the carb. When the power valve failed the carb would run rich and the power valve would have to be replaced. However, since 1992, Holley has incorporated backfire protection into the power valve circuit on a number of their performance carbs and has eliminated this problem. In addition, a number of aftermarket kits are available that allow you to quite simply and cheaply provide backfire power valve protection to older carbs that do not have this provision.

Speaking of making things easier, Brad Urban's Carburetor Shop, is making a replaceable metering block for Holley carbs called a "Speed Block". It bolts right in place of a stock metering block but has relocated the main jets on extensions that enable you to change main jets quickly and easily, without taking the carb apart, even with the motor running at idle. These "Speed Blocks" use regular Holley jets, gaskets and power valves and come in a wide assortment to fit your particular gasoline or alcohol carb.

A lot of the information here concerns Holley carbs and most of my experience is with these carbs. However, many other brands of carburetors are available. Edelbrock is manufacturing Quadrajet style carbs up to 850 cfm along with a series of carbs similar to the Carter AFB. Federal-Mogul, thru its Carter

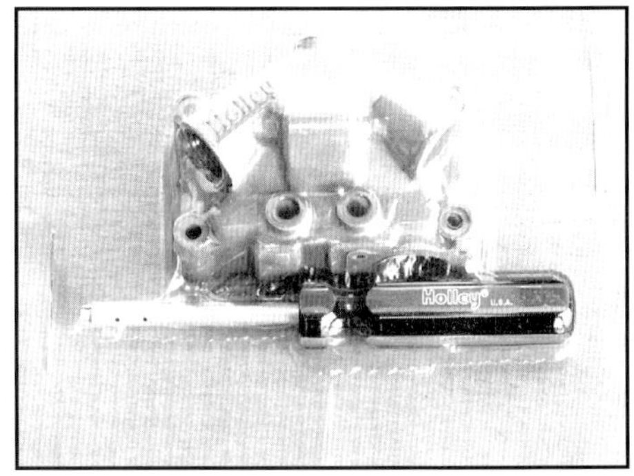

Holley offers a fuel bowl kit that allows you to change main jets without removing the fuel bowls. Kit includes a special jet tool.

Carburetor line, also makes a series of AFB-style four barrels in various sizes. Individual parts are available to "tweak" these fuel mixers to suit your needs. Predator carbs are also interesting because of their unique simplicity and their ability to be used in a wide range of engine sizes.

FUEL PUMPS

GM offers a number of mechanical and electric fuel pumps for the big block Chevrolet. These high performance mechanical pumps listed here will provide enough fuel volume to feed engine sizes from 396 to 454 and will fit most chassis applications. The mechanical pumps listed here all have rotatable lower housings which can be very helpful in making them fit your application. Keep in mind that Gen. V production engine blocks do not have a provision for mounting a mechanical fuel pump, so an electric pump must be used. Gen.VI blocks do have a boss for a mechanical fuel pump.

(PN-6415748) is a high capacity mechanical fuel pump and will meet the requirements of a street driven 396-454 Mark IV big block. It does not require the use of a separate fuel pressure regulator.

(PN-12355618) is a street use chrome mechanical fuel pump with a free flow volume of 70 gph and a pressure of 8.5 psi. It will work with engine sizes of 396 to 454.

(PN-12355614) is a high capacity mechanical pump. Pressure is set a 7 psi and it will flow 100 gph. It has 3/8" inlet and outlet ports. It will provide enough volume for engine sizes 396 thru 454. It does not require an additional fuel pressure regulator.

(PN-12355615) is another high capacity mechanical fuel pump. It has a free flow rate of 100 gph, but has a fuel pressure of 9 psi. It requires the use of a separate fuel pressure regulator. It also has 1/2" inlet and outlet ports and is considered, by GM, to be a pump for competition.

(PN-12355619) is marine competition mechanical fuel pump. It has a free flow rate of 130 gph at 9 psi. It uses 1/2" inlet and outlet ports. It also has a tube on the pump which will direct any fuel leaked to a flame arrestor or carburetor port to limit the possibility of fire onboard a boat. This pump also comes with specially hardened parts and a corrosion resistant finish to help it live in a marine environment.

The fuel pump pushrod used in Chevy big block engines with a mechanical fuel pump is (PN-3704817). The same fuel pump pushrod is used in the smallblock Chevy engine and is interchangeable.

When using a fuel pressure regulator with a carburetor, the pressure should be set so that a maximum of 6 or 7 psi is at the carburetor inlet ports and that the pressure does not fall below 4 psi at wide open throttle. Any pressure regulator should be mounted as close as possible to the carburetor and after any fuel filters which can clog and restrict fuel flow.

Chevrolet offers two styles of electric fuel pumps. (PN-25115899) is a 12-volt pump that free flows 72 gph at 6-8 psi. This pump should be mounted in the fuel line near the gas tank so as to "push" the fuel to the engine. (PN-25116279) is a late-model "in tank" electric fuel pump. It provides 14 psi and should be used with a fuel pressure regulator if you are using a carbureted engine.

Holley makes a number of mechanical and electric fuel pumps which give excellent service for all types of applications whether they be used on the street or for racing.

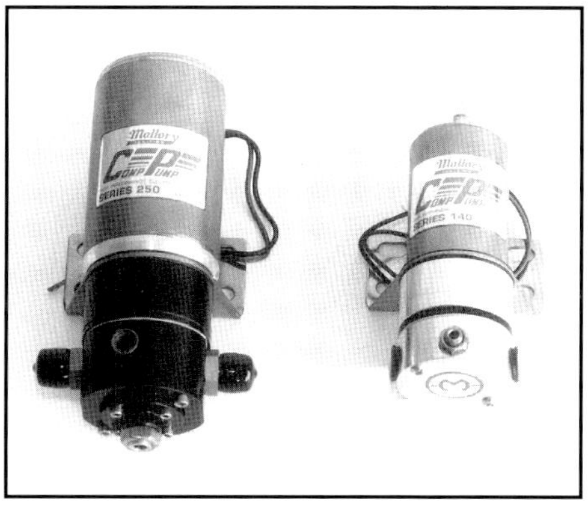

Mallory has a line of fuel pumps for any engine need from 110 gallons per hour flow capacity to 250 gph.

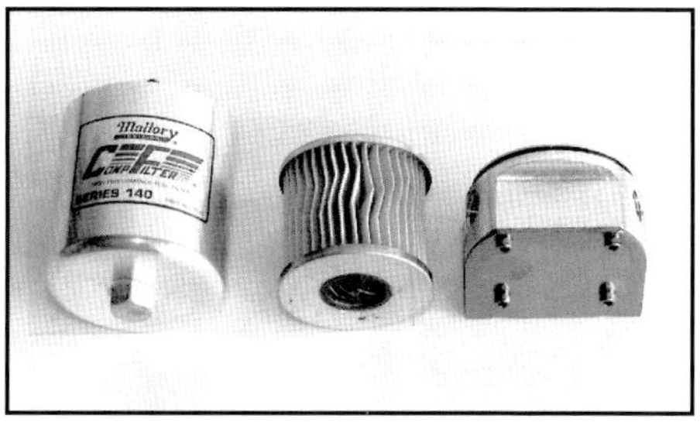

Make sure that you use a high flow capacity fuel filter on any high performance vehicle.

These pumps are rated to provide various amounts of gallons per hour and at different amounts of fuel pressure depending on the intended uses and requirements. (See the section on Fuel Volume in the ENGINE MATH chapter.) The "red" pump (Holley PN-12-801) delivers 97 gph, while the "blue" pump (Holley PN-12-802) gives a free flow volume of 110 gph and must be used with a pressure regulator. Two other Holley "Volumax" "blue" electric pumps give 160 gph (Holley PN-12-705) and 250 gph (Holley PN-12-706). Get a pump that provides the amount of fuel you need at the pressures required and use a fuel pressure regulator (Holley PN-12-803, 12-704 or 12-707 for dual carbs). Use a fuel filter and fuel line sizes that will handle the volume of fuel you plan to use.

Fuel lines in any application should be steel tubing and/or braided steel hose. You should try to use the minimum amount of neoprene rubber fuel hose when making connections. (The NHRA rules, as an example, only allow a total maximum of 1 foot of neoprene fuel line to be used on the entire fuel system from tank to carb.) I know, you are saying to yourself that braided steel hose is expensive, especially the fittings. But the fact is that an engine fire is much more expensive. Besides the braided hose looks great. Never use copper tubing to make fuel connections, they don't last. Make sure that fuel lines stay as far as possible from exhaust pipes and manifolds, fan blades and other rotating parts.

FUELS

Which type of fuel to use depends on what type of engine you are building and the uses it is intended for. Fuel octane ratings give you an idea of the fuel's ability to resist detonation. When you increase cylinder pressures and cylinder temperatures, fuel has tendency to "self ignite" or detonate. Fuels can be chemically formulated to resist pressure related self-detonation and this resistance is given as its octane rating. As static compression ratios go up you will need a higher octane rated fuel. A 8.0-1 to 9.2-1 engine can use pump gasolines which have ratings from 86 to 92 octane. Ratios of 9.5-1 are going to need premium unleaded fuels of 90-92 octane. Above 9.5-1 pump fuels may not provide protection from detonation and that leaves you with octane boosting additives or race fuels. The additives can provide some increased octane levels to the fuel (6-8 additional octane points), but they may not be enough, depending on the static compression ratio of your motor. Motors with high compression ratios above 10.5-1 are going to need octane levels only currently available with race fuels. These fuels are expensive, two or three times more expensive than pump gas per gallon. Make sure that you consider this when you are planning what type of engine to build. If the vehicle will be solely used to race, then race fuel costs are just one of the additional costs to go racing. However, if the vehicle will be used mainly on the street with weekend passes at the racetrack, then you are better off building a motor with a compression ratio that can use unleaded pump gasolines and generally this means a motor with a static compression ratio of 9.5-1 or less. In a related matter, if you plan to use unleaded pump fuels make sure that the cylinder heads you use have hardened valve seats that will survive the use of unleaded fuels. Early big block heads made from '65 to about '74 and designed for leaded gasoline, did not originally come with hard seats. Hardened seat inserts can be installed in your heads by a competent machine shop.

Full time race engines with compression ratios that are at the extreme levels, say 13.0-1 and higher, may get some benefits from using alcohol based race fuels (methanol). There are pros and cons to using methanol as a race fuel. On the "pro" side a well set up methanol motor can make upwards of 10% more mid range torque than a motor run on gasoline fuel. The motor will run cooler and provide a more dense air/fuel mixture and show less tendency towards detonation, even at high cylinder pressures. Because of this cooling effect, the motors seem to be more consistent in terms of various outside temperatures and atmospheric conditions at a racetrack. Since the motors run cooler, "cool downs" between races are reduced. If you are using a "filled" block or "solid" heads, methanol can help keep engine temperatures under control. There are a number of large CFM Holley carbs that have been manufactured and calibrated for use with methanol. A tunnel ram intake with two Holley alcohol carbs or a mechanical fuel injection setup may be an easier solution to the increased fuel volume requirement.

Methanol fuels are cheaper per gallon than other race fuels, however, you need much more fuel volume than when you use gasoline. Some of the other "cons" relate to fuel delivery costs. Since a methanol fueled engine needs a lot of fuel, you need a fuel delivery and induction system that can deliver that kind of volume from the tank thru the carb. These costs can run from $1,500- $3,000 to set up. Everything has to have a larger fuel delivery capacity, fuel pumps, fuel lines, filters, and carbs or fuel injection. In addition, because methanol attracts water, methanol use can be corrosive to certain materials normally used in gasoline based fuel systems, so you have to use materials which will not be eroded when you use methanol. Maintenance of a methanol fuel system is required on a regular basis. The fuel system needs to be flushed and the system drained of remaining methanol fuel if the engine will not be run for a while. Careful attention must be given to using fresh methanol fuel, since methanol is very susceptible to water contamination. Water is absorbed by methanol from air, so keep fuel cans or barrels well sealed. Also the use of a "top lube" mixed with the methanol can help lubricate fuel system parts.

Alcohol motors generally will respond well to increased ignition advance because alcohol based fuels are harder to ignite and burn slower. In a related matter, try to keep the piston dome on a methanol motor to a minimum or use a flat top piston. A piston dome will slow down the flame front in the combustion chamber and the slower burn time with methanol only makes this situation worse. You can reduce the chamber volume on the head to increase compression, instead of using a domed piston. A longer connecting rod will increase the amount of time the piston stays in the area of Top Dead Center and give the flame front a little more time to burn the fuel and build cylinder pressure. Also keep in mind that since an alcohol motor uses so much more fuel volume, the fuel actually displaces the amount of air that can be carried in the port runners. Moving to a larger port volume will help increase the volume of the air/methanol mix and increase power.

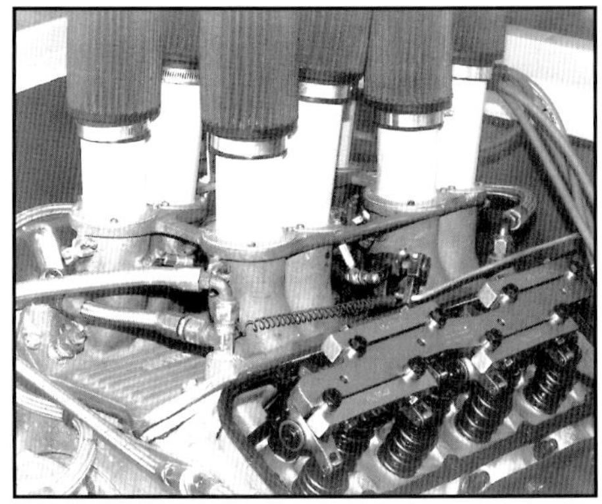

Hilborn mechanical fuel injection unit (above) delivers alcohol to this big block. Throttle response is crisp to say the least.

FUEL INJECTION

In the mid 1980's carburetors disappeared from factory installed big block induction systems and were replaced by electronic Throttle Body or Sequential Port injection systems. This was done to improve throttle response, lower emissions, cold starting, gas mileage and provide better low and mid range engine torque for street driving. On board computer systems control fuel delivery and ignition. These factory fuel injection systems can be adapted to earlier carbureted engines. A number of aftermarket companies are making wiring harnesses, fuel supply systems and computer controls that make the swap possible. They are relatively expensive compared to a standard carb and intake, but they do provide the benefits associated with modern fuel injection systems.

Holley makes a "Pro-Jection 4D" throttle body system in various CFM sizes up to 900 cfm (which come with 85 pound per hour injectors), that will bolt to many standard carburetor flange type intake manifolds, (including tunnel rams), and they come with the appropriate wiring harness, sensors, adjustable fuel pressure regulators and an electronic control unit that gives you a range of tuning options for the fuel system. The "Pro-Jection 4Di" (interactive) system can be fuel curve map programmed with a laptop computer and Holley provides a fuel map calibration program disk and computer harness that you install in your laptop.

Accel Performance Products is making a 1200 CFM TBI mounted on a standard square flange hi-rise Accel aluminum intake. The intake manifold has electronically controlled port fuel injection nozzles and bosses incorporated. Each piece can be purchased separately or you can get the entire package with throttle body, intake manifold, programmable computer control and wiring harness.

Street & Performance has complete Tuned Port Injection (TPI) systems for big block motors. These TPI systems offers large increases in torque, crisp throttle response and excellent mileage. Many have programmable electronic controls that enable you to dial in the engine operating parameters that you want. Custom "chips" can be produced that will allow the use of the pre-1994 GM "black box" computers. Street & Performance can also custom program '94- '95 style computers.

Most electronic fuel injection systems require high fuel pressure and so the appropriate fuel pumps must be used. Don't forget to add a fuel pressure bypass line back to the fuel tank. Clean, large flow capacity fuel filters are also a must.

Autotronic Controls Corp., the makers of the MSD ignition line, also has an MSD Fuel Management division. The division offers a wide range of high performance and race EFI parts including competition injectors, high pressure fuel systems, and engine management sensors. MSD also makes the hardware, electrical connectors and parts to fabricate and mount injectors and fuel rails to your manifold. There is also an MSD line of distributors to be used with computer controlled EFI systems. These distributors provide an ignition signal and an injection synchronization signal to the ECU computer without having to use a crank or cam trigger.

Race mechanical fuel injection systems are available from Hilborn, Kinsler, Ron's Racing Products and Enderle and these companies have been in business for many years. They can provide entire gasoline or alcohol mechanical injection systems for Chevy big blocks and know what will work and what won't.

Another interesting thing happening with mechanical fuel injection units is the marriage of a mechanical unit with modern EFI units. The old style port fuel injection nozzles are replaced with electronically controlled nozzles and then computer controlled. A laptop computer can provide the necessary fuel curve programing to provide a very specific fuel amount and emphasize those factors you feel important whether it be power or fuel mileage. Similar things are being done by Barry Grant Fuel Systems with carburetors by replacing the fuel bowls and metering blocks on Holleys carbs with electronic fuel injectors. These "Venturi Fuel Injection" hybrids will bolt to conventional carburetor intakes and can be programmed to supply the fuel curve you wish.

NITROUS-OXIDE, SUPERCHARGERS, TURBOCHARGERS

There are a number of nitrous and blower systems that are designed for street use with a stock production motor. The amount of nitrous used in these systems is smaller and the resulting horsepower increases are in the range of 100-150 horsepower. Likewise some of the street-use blower and turbo systems offer increases in the same range and will work well with stock parts.

Problems occur when the end user gets comfortable with the increased power and decides he or she wants more. The amount of nitrous used goes up or the boost on the turbo or blower system gets turned up in an effort to make more power and go quicker. At some point, a stock motor has enough and decides to take a vacation. This is the type of situation you can avoid when you are planning on building a motor that will use these systems. If you use stronger parts, better suited to stress and abuse, the chances of engine damage will be lowered. You will be building in an extra measure of durability that will help the motor survive your temptations to "see Elvis".

All of these systems do at least one thing, they artificially increase the amount of volume in the cylinder. In carbureted or injected systems, normal atmospheric air pressures fill the cylinders. With nitrous or blowers, the cylinders are more densely packed, cylinder pressures are increased and power goes up in the motor. If you are contemplating such induction systems, make sure you use the strongest engine parts available. You should use a four-bolt block, forged crank, forged pistons and heavy duty connecting rods with quality rod bolts. Replace the stock vibration dampers with a high quality unit like those made by Fluidampr which are sold by Fluidampr and GM. High quality engine studs, instead of bolts, in conjunction with high quality head gaskets, may be required because of their superior clamping power and resistance to high cylinder pressures. Cylinder pressures can be very high with these types of induction and stock production, street stock parts will not last if the boost is turned way up or you use large doses of nitrous.

NITROUS OXIDE

Nitrous oxide is an interesting fuel additive because of its chemical properties. It has two nitrogen atoms and one oxygen atom which form the nitrous-oxide molecule. Inside a running engine temperatures are high enough to breakdown the molecule and free the atoms and this adds additional oxygen to burn additional fuel. The percentage of oxygen in nitrous is larger than the percentage of oxygen in normal air available to a normally aspirated motor. Also when nitrous is stored in the bottle it is in a liquid state. When the nitrous is injected into the engine it immediately begins to turn from a liquid to a gaseous state. When it does, the temperature of the nitrous gas drops tremendously. This helps deliver a much denser and cooler charge to the cylinders. So you get more oxygen, more volume and a cooler charge. The key though is to provide the additional volume of fuel so that the correct fuel ratios are maintained. In fact it doesn't hurt to run a nitrous motor a little on the rich side when the nitrous is on in order to further cool cylinder temperatures and keep them under control to avoid detonation. Not providing enough additional fuel will result in a lean mixture and big trouble. Don't forget to retard the total ignition timing when using the nitrous system. The flame front travels quicker across the piston surface in the presence of nitrous and the motor needs less total ignition timing.

There are a number of different nitrous delivery systems on the market. One type uses a plate which fits between the carb and the intake manifold and has spray bars which direct addition fuel and nitrous into the plenum. Another system plumbs nitrous and supplementary fuel lines to nozzles mounted directly in individual intake manifold port runners. Electric solenoids control the flow of nitrous and additional fuel.

With nitrous oxide the travel speed of the flame front is quicker than with gasoline alone. This means that the total ignition timing does not need to be as advanced as when the motor is run on gasoline alone. As nitrous is mixed with the gas, total ignition advance timing should be reduced. If you have multiple stages of nitrous the timing advance should be reduced as each stage comes in. MSD makes an add-on ignition module that will retard the total timing on the motor as the nitrous is switched on. The amount of ignition retard is controlled thru the use of plug-in chips that operate at different engine rpms.

As you increase the amount of nitrous, you must also introduce additional fuel. This will require upgrades to the entire fuel delivery system. Keep in mind that most of the nitrous delivery systems provide a constant amount of additional fuel and nitrous, no matter what rpm the motor is running. On any engine, running at low rpms, the amount of time that an intake valve is open is longer than when the motor is at high rpms. With a nitrous assisted motor this means that if you were to activate the system at low rpms more nitrous and fuel would be ingested by each cylinder than is ingested at high rpms. This can cause a real problem with low rpm speed detonation because too much nitrous is present in the cylinder. Most manufacturer's advise that you not activate the nitrous until the engine rpms are in the mid range.

On a performance motor the increases in power can be very large and for the money, properly setup nitrous systems seem to give you the most power per dollar of almost anything currently out there. Cylinder pressures are raised substantially and this results in more engine torque. But this must be tempered with thoughts about the increased costs of building a stronger motor and drivetrain if the intended power increase is large. There is a limit to how much power any particular part can take and before you set up a system that will deliver large amounts of nitrous to your engine, make sure that your parts combination can withstand the increased power you are looking for. You can get the nitrous, air and additional fuel ratios absolutely correct and then start bending or breaking connecting rods or blowing the deck off of the block because the parts

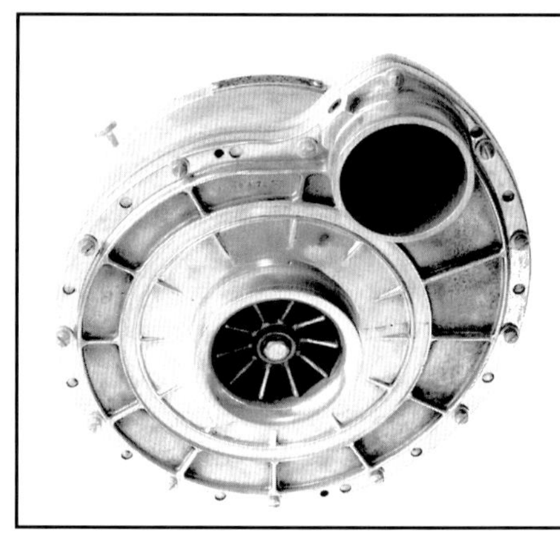

Paxton centrifugal supercharger can be used to provide additional power and will fit under the stock hood of many vehicles.

You can really step things up with a 502, a roots style blower and mechanical fuel injection.

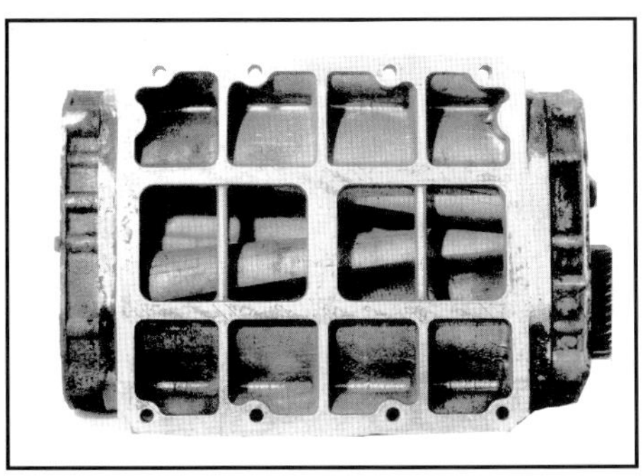

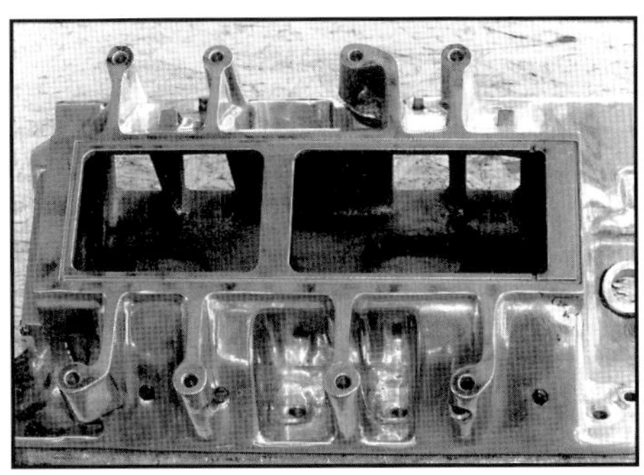

6-71 Blower rotors can be seen from this top view. Large roller ball bearings control the ends of the rotors in the case. Bottom view of 6-71 case and rotors with adjacent view of corresponding manifold top.

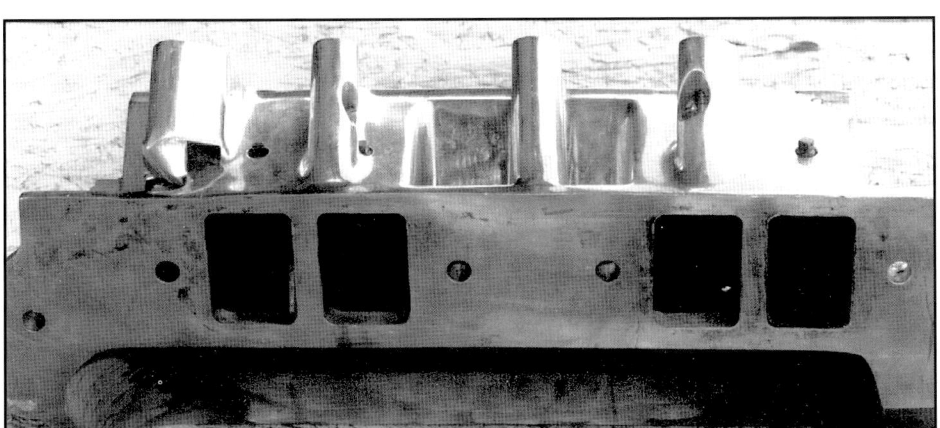

Blower manifold does not offer much in the way of restriction.

you chose couldn't take the large increases in nitrous and resulting cylinder pressures.

SUPERCHARGERS

Superchargers increase power on a motor by forcing additional volume into each cylinder at pressures that are above the normal atmospheric pressure. The intake charge is denser and cylinder pressures go up along with increases in torque. However, as the blower vanes compress the air, the temperature of the compressed air and fuel goes up. In addition, the faster the blower is turned, the more power it takes to turn it and the hotter the incoming air is made. At some point the additional power the supercharger makes is offset by the amount of power it takes to turn the blower over. There is a point with each blower system where the amount of boost given generates the most power. At boost levels above that power declines. The idea is to strike a balance where the blower motor runs best consistent with what your intentions are as far as engine use and power band. Keeping boost to 5 lbs. or less will enable you to use fairly stock parts if you wish to. At boost levels above this you will need premium rotating parts capable of withstanding higher engine output.

The are a couple of different supercharger types. A centrifugal blower compresses air only by using an impeller spun at very high rpm. The compressed air is then sent thru a carburetor or port fuel injection system where fuel is introduced to the system. Generally these types of blowers are compact and can be mounted in different positions as long as they can be driven by a belt. This makes them very useful in a vehicle that retains a stock hood. They can be hooked up to various types of intercoolers in order to drop the temperature of air being introduced to the impellers which allows for a denser charge, cooler impellers and more power. ATI,

STATIC COMPRESSION AND BLOWER BOOST

Here is a chart that will help you determine changes in static compression when using a Roots type supercharger. Total engine cubic inch displacement, cam timing, valve sizes and head port dimensions will also have an effect on boost pressure. Cam timing will also have an effect on the "dynamic" compression of any combination. Determine your actual static compression ratio. Don't guess. Generally speaking a maximum of 4-5 lbs. boost can be used with stock parts. If you go above this amount you need to use stronger rotating assembly parts. Ignition timing should be limited to a maximum of 32 degrees total timing to avoid detonation. You can also use an ignition retard device to further retard total timing as boost pressures go up.

Static C/R	2	4 lbs.	6 lbs.	8 lbs.	10 lbs.	12 lbs.	14 lbs.	16 lbs.	lbs.
7.0-1		8.0	8.9	9.9	10.8	11.8	12.7	13.6	14.5
7.5-1		8.5	9.5	10.6	11.6	12.6	13.6	14.6	15.7
8.0-1		9.1	10.2	11.3	12.4	13.4	14.5	15.6	16.7
8.5-1		9.7	10.8	12.0	13.1	14.3	15.4	16.6	17.8
9.0-1		10.2	11.4	12.7	13.9	15.1	16.3	17.6	18.8
9.5-1		10.8	12.1	13.4	14.7	16.0	17.3	18.5	19.8

Larger capacity blowers will put out more boost at a lower blower speed. A smaller capacity supercharger can be made to put out similar amounts of boost pressure, but at a higher blower rpm. Remember also that the faster the blower rpm, the more engine power it takes to drive the blower. Superchargers also heat up incoming air because the air is compressed and pressurized. Anything you can do to lower incoming air temperatures will add power to the motor. Air scoops or cool running fuels like alcohol or nitrous, will help keep the intake charge cooler and denser.

Paxton and Vortech make centrifugal systems along with all of the plumbing pieces you will need.

Roots type blowers use a pair of two or three lobe rotors which compress air and fuel between the spinning lobes. Carburetors or fuel injection, mounted on top of these blowers, provide fuel and air to the rotors. These are considered "wet" systems and the fuel has a cooling effect on this type of supercharger. However, this makes for a tall assembly and holes thru the hood are often required. There are some systems that can use side mounted carbs and this lowers the overall induction system height. Roots type blowers have many different cubic inch blower displacement sizes.

A Screw type supercharger has a blower case similar to the Roots type but the rotors are screw-shaped and push the compressed mixture from front to back in the case.

Different drive systems for blowers are available along with various pulleys sizes. Changing the pulley diameters will alter the amount of boost and the drive ratios that are required to get that boost. To find the pulley ratio divide the number of teeth in the bottom pulley by the number of teeth in the top pulley. Keep in mind that the size of the blower displacement will determine how fast it must be driven to give the boost you require. For instance, a 4-71 or 6-71 style can give "X" amount of boost volume at a certain rpm, but an 8-71 with a larger volume capacity will give the same amount of boost at a lower blower rpm and with a reduction in blower generated heat because of its increased volume. But since the engine gods have decreed that there is "no free lunch", at higher rpms, the larger the blower capacity, the more power it takes to turn it over. So you are back to doing your balancing act again. Careful attention needs to be given to how fast the motor and blower will spin, the cylinder volume and the volume of the blower, the amount of boost and the point where the law of diminishing returns come into play with regard to the amount of power generated vs. the amount of power to drive the blower. How much heat is being added to the air/fuel mix, how much boost and cylinder pressure can your parts take before they start thinking about giving up? Alcohol based fuels can help with mixture temperatures by helping to cool things down. So would nitrous-oxide. When you work these things out, the engine torque can be ground shaking, instantaneous and off idle with a supercharger.

Don't forget the exhaust side of motors with these types of induction. Because more air and fuel is sent to the cylinders you can get additional power benefits from larger exhaust valves in the heads and more exhaust cam timing so that this increased exhaust volume can also get out of the motor. This also requires a larger exhaust system capable of handling more volume, whether it be larger diameter header tubes or larger diameter exhaust pipes and mufflers. Pistons will need lower static compression ratios and camshafts are designed differently for an engine that will use these systems. Cams with wide LSA's of 112-114 degrees will help keep the blower from pushing the intake charge right on past the open exhaust valve during the valve overlap period.

RAT MOTOR Performance Tips & Tricks

RING END GAPS FOR NITROUS AND BLOWER MOTORS

The increased heat and cylinder pressures generated by the use of nitrous-oxide or when using a blower can cause the top and second piston rings to expand more. You can allow for this increased heat when you gap the rings for these motors. Speed-Pro makes the following recommendations for end gaps with iron rings.

TOP RING END GAP (ductile iron)

Nitrous Oxide - Street	.006" per inch of bore diameter
Nitrous Oxide - Drag Race	.008" per inch of bore diameter
Supercharged	.006" per inch of bore diameter

SECOND RING END GAP (cast iron)

Nitrous Oxide - Street	.005" per inch of bore diameter
Nitrous Oxide - Drag Race	.0065" per inch of bore diameter
Supercharged	.005" per inch of bore diameter

These are recommendations only! Check with your piston and ring maker for his recommendations when using these types of fuels and inductions. When a motor is freshened or pulled down, always check to see if the ring ends have been butting together. If they have, you will have to open the gaps further. If there is no evidence of ring end butting, then you can think of reducing the gap. This requires that you keep accurate records of what the end gaps were at the time the motor was built. Be careful. There is some power to be found by getting the ring gaps just right for your particular combination. However, the motor can be damaged if the gaps are too tight.

Most engine damage with these systems, seems to be caused by incorrect engine parts combinations, inadequate fuel delivery systems and air/fuel ratio problems that result in very lean conditions. Detonation is the enemy to avoid. Total timing advance needs to be kept to a maximum of 32 degrees with a blower. MSD makes a number of ignition boxes and ignition retard accessories that will control ignition timing and help avoid detonation.

Systems that provide boost above 5 psi may require forged blower pistons with thicker decks, forged cranks and high quality rods. Make sure that your fuel delivery system has increased flow capacity to match the increase in volume made possible by these systems.

Also check that there is room to install water lines and the distributor and cap you intend to use. With some blower intake manifolds you may have to use a different water neck or distributor to get the clearance you need to install them.

TURBOCHARGERS

Turbochargers in a way are similar to centrifugal blowers except that the power to run them is generated by the pressure of the outgoing engine exhaust flow. Pressure from the exhaust is routed to a finned turbine which in turn spins another turbine impeller. A wastegate in the system controls the amount of exhaust pressure driving the impeller and controls the amount of boost on the system. Compressed air can be blown thru a carburetor or introduced after the carburetor or used with a fuel injection system. Turbo installations can be done without altering the hood and intercooler systems will lower incoming air temperature and help produce more power.

Be Careful! Talk to your aftermarket supplier and follow their recommendations carefully and completely. Weiand, Paxton, B&M and Blower Drive Service (BDS) are good places to shop for supercharger systems. Nitrous Oxide Systems (NOS), Nitrous Works, and Compucar can provide nitrous oxide systems of many different sizes and power levels. Gale Banks Engineering provides turbo charged systems for many different applications.

AIR FILTERS

GM used metal canisters, with various inlet snorkels, to hold the air filter. Some high performance

PERFORMANCE AND WEATHER

When you are at the races and you start to get some idea of what your vehicle is capable of and how to be consistent in everything you do as a driver and tuner, Nature, Fate, God, Luck, Karma and all the other unseen forces that can have an effect on us human beings can pull up along side of you. While you can't deal with Fate, God, Luck or Karma, you can cope with Nature. Short of a rain or snow storm shutting down the track, it is possible to make "dial in" allowances for the moment to moment changes in weather. If you have been paying attention, you know that your motor will run better when the air is cool and denser. More oxygen molecules can get into the cylinders. The question is how to take advantage of the weather conditions and optimize performance. The problem is especially tough for folks who have to "dial in" their expected ET or run on the class "index" because running the wrong ET can lead to a loss. If you spend the warm daylight hours in time trials and then the elimination rounds begin after the light is gone and air temperatures decrease, predicting what the elapsed time will be is difficult.

The first thing you need is information and experience. What does your particular car do? You have to keep some kind of record. When you make a run, what is the air temperature? What time of day or night? What kind of condition was the track surface in? If you can, find out what is the barometric pressure? What is the humidity? What is the density altitude? You can spend the money to buy a "weather station" and computer. These will tell you what the weather conditions are and form a basis on which to make comparisons and predictions. Changes in air temperature, barometric pressure, humidity, and density altitude are determined and any correction factors computed. When compared to other conditions accurate changes can be made to run on your "dial in". There are a number of companies who make these weather stations. If you don't have one or you wish to spend your hard earned money on something else, I have suggestion. Park your race car in the pits near someone who does have a weather station and ask them what the conditions are. If they are friendly they will tell you. Sometimes it helps if they are running in another class and are not your competition. In any case, keep a record. After a while you will know what to expect for any given change and can alter your tuneup or "dial in" appropriately. If your future plans include using a "throttle stop", your records will help you make the correct adjustments to run the elapsed time you want to run.

Now you may say to yourself, "that's too much to do. I just want to run my car. I don't like all that record keeping and that electronic, computer 'mumbo-jumbo". That's okay. It's still possible to win without knowing this stuff or taking it into consideration. Luck, Fate, God and Karma are still out there and they can still have an effect on all human endeavors including racing. The only problem is when you need them, Luck, Fate, God and Karma are usually too busy helping someone else that needs more help than you do.

BIG BLOCK Quick Facts & References

ELECTRONIC FUEL INJECTION

Aftermarket performance fuel injection systems have seen huge changes, innovation and consumer demand in recent years. One company that does it all is STREET & PERFORMANCE of Mena, Arkansas. This SEMA award winning company has a large inventory of electronic fuel injection systems and offers complete packages for late model vehicles and conversion setups for rodders. They fabricate their own parts and have the CNC milling equipment and chrome and polishing facilities to turn out the parts you need. Custom computer chips, TPI, TBI and EFI induction systems for oval port or square port heads, wiring harnesses, V-belt and serpentine accessory bracket systems, pulleys, thermal coated tube headers, Chevy crate engines and blower setups are just some of the things available.

engines used a 14" open element style with a 3" tall air filter element along with a chrome metal top and bottom plate (PN-12341859). You can also get a 4" tall air filter element as (PN-8997189) (AC#-A697C). These open element filter assemblies are less restrictive than the canister type air filter assemblies and will fit over Holley or Quadrajet four barrel carburetors. Some of the best air filters on the aftermarket are sold by K&N. They are free flowing and can be easily cleaned and reused. Make sure that your hood will still close with the taller filter elements. Some vehicles with combinations of hi-rise intake manifold, carb and air filter won't fit. You may have to cut a hole in the hood and add a scoop or use a different intake, carb, air filter combination in order to use the stock hood.

"Cowl induction" cold air systems were also available on many different Chevy vehicles. L-88 427 Corvettes and ZL-1 Camaros and Corvettes received a special cold air induction system that used a screen and a foam-sealed metal base for cold air induction. Other Chevy models used similar cold air packages. Some passenger cars and trucks use a flexible plastic hose that is mounted to a hole in the radiator support to bring cold air to the air cleaner canister.

(PN-25043641) is a fully enclosed canister air cleaner assembly offered for smallblock 350 HO four barrel carbureted Camaros. It has two large air inlet snorkels which can be hooked up to large diameter flexible hose ducting to route cold outside air directly to the air filter assembly and it can be adapted to your big block combination.

Using cold outside air will make more power, but you must use some sort of filter to make sure that no dirt or foreign objects get into the engine. In fact most dyno tests have shown that an engine will make more power with a properly designed air filter assembly in place, when compared to running with no filter assembly at all. This is because the filter canisters and bases help to smooth out and channel the airflow into the motor. If there is room under the hood, you can add a taller filter element or stack two filter elements on top of another to increase the total filter surface area and lower air restrictions.

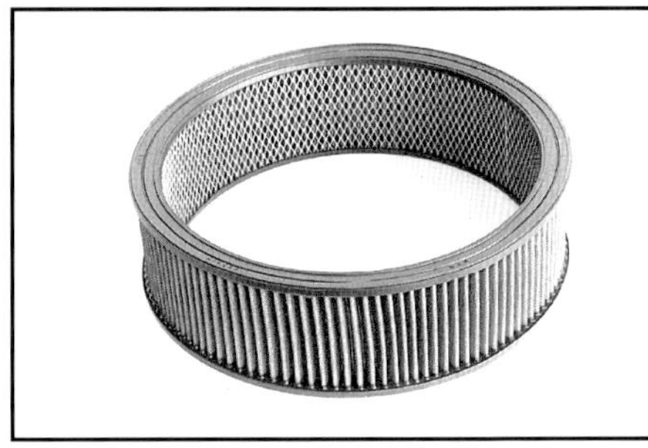

K & N makes an excellent line of air filters that are low restriction and when they get dirty, you can clean them and reuse them. They have all popular sizes and you can get added filtration and flw capacity by using the taller versions if you have the available hood clearance.

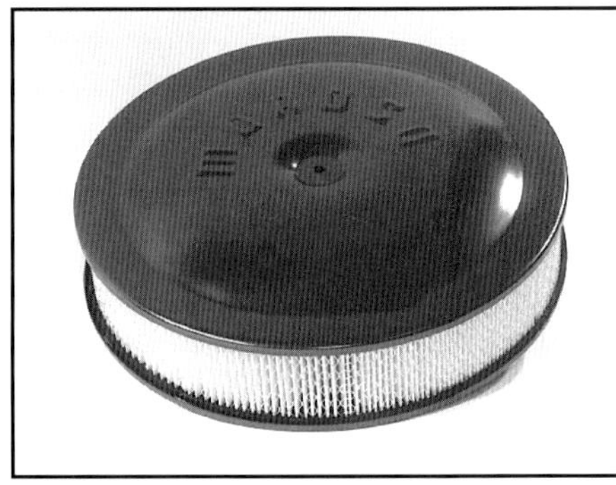

These air cleaners (two above) look good but their main function is to keep dirt out of the engine, smooth air flow into the carb and help contain flash fires if the motor spits back thru the carb. It helps if the filter is open and non- restrictive.

Inducting cooler outside air into the motor will increase power whether you use a factory setup or build you own.

Homemade cowl induction on this bracket drag car works well. Popular air gap intake manifold configuration isolates inlet charge from engine heat and the aluminum carburetor plate isolates incoming air from hot under hood temperatures. A very clean functional approach.

Original style open element 14" air cleaner is still available from Chevrolet (PN-12342079) and it is excellent addition to any engine. Many different filter element heights are available.

Racers can use a bell-mouthed velocity stack or one of K&N's Stubstacks to smooth out the airflow into the engine. Once again there is power to be gained by using some sort of device to smooth the air entry into a carburetor as opposed to not using anything at all. The Stubstack can be used inside a normal paper air filter, and different versions are made to fit two-barrel and four- barrel Holley carbs. Certainly if you are racing a mud bogger or off-road vehicle a high capacity, low restriction air filter assembly is required.

There is one other important aspect concerning air filters and air cleaner canisters, FIRE. If the motor belches back thru the carb, its very easy to light off the small amounts of fuel almost always found in the venturis and the floor of the intake plenum. Usually its no big deal and easily put out if the hood is up and you see it. Air cleaner assemblies usually contain it and the fire quickly goes out itself, either because it runs out of fuel or there is not enough air to sustain it. However without any filter device on the carb, the flames have more access to air and can continue to burn. If the hood is closed any hood insulation or paint on the underside of the hood (especially a fiberglass hood or scoop assembly) can catch fire and things can quickly go from a minor inconvenience to a major disaster.

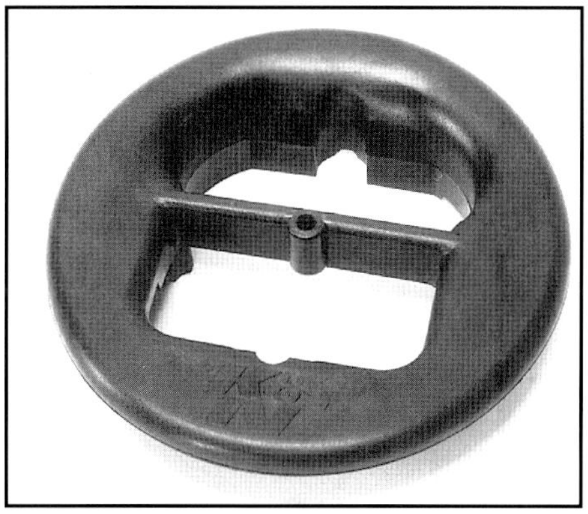

K&N's Stubstack for a four barrel Holley carb can be used alone or will fit inside an air filter canister.

BIG BLOCK Quick Facts & References

INDUCTION PREP CHECKLIST

1) **Inspect manifold for stripped holes, broken ears and any other damage.**
2) **If the manifold is aluminum, obtain proper size studs for the carburetor, water inlet housing, distributor clamp and any other items that are often removed.**
3) **Heli-coil all damaged holes.**
4) **Match ports to cylinder heads.**
5) **If your combination will require additional vacuum fittings in the manifold, drill and tap them now.**
6) **Install all plugs and fittings.**
7) **Wash manifold in clean solvent and blow dry prior to installation.**

Ignitions

DISTRIBUTORS

Chevrolet big blocks use two different distributor shaft lengths, one for the short deck big block motors and another with a longer shaft for the tall deck engines. Small block Chevy V-8 distributors are interchangeable with big block, short deck engines and that opens a lot of possibilities because so many different types of ignitions and distributors were offered over the years for the small block Chevy V-8. Single-point, dual-point, High Energy Ignition (HEI), transistor, magnetic pulse, capacitive discharge (CD) and multiple spark discharge ignitions with rev limiters were made for Chevrolet for V-8 engines. Corvette distributors thru 1974, had a mechanical cable tach drive incorporated into the distributor shaft housing. Some distributor housings were made of cast iron, but most were made with an aluminum housing.

GM/MSD

If you are building a high-performance motor for the street or to race and you will spin the motor above 5000 rpms, move up to one of the GM multiple spark discharge ignitions (PN-10037378). This is the same ignition offered by the Autotronic Controls Corporation as the "MSD 6" series box. These multiple spark discharge ignitions (MSD) actually fire many times on each compression stroke of the motor, for twenty degrees of crank rotation, while at lower rpms. As rpms go higher the ignition changes to a single spark on each compression stroke. This results in more of the air/fuel mixture being burned, more power, cleaner emissions, easier starting and better fuel mileage. These ignition boxes work with just about any type of distributor to "trigger" the ignition system. The distributor becomes a switch that tells the ignition box when to turn on and off and routes the spark to the correct plug. Use one of the electronic distributors that will reliably provide spark to the rpms levels you need. Point distributors can be modified to spin reliably above 7000 rpms, but these days its hardly worth the cost or effort. Electronic distributors with LED mechanisms or magnetic triggers work much more reliably at high rpms and there are no points to replace. Chevrolet offers a hi-performance magnetic pickup distributor as PN-1013455 or 10093387 with an adjustable height hold down collar. These two distributors do not have a vacuum advance and use a standard size Chevy cap and rotor. The cap will clear most induction systems. Small cap distributors are

MSD billet distributor (PN-8550) has a magnetic trigger, adjustable slipcollar and a locked mechanical advance. It comes with a bronze distributor gear for use with a steel roller cam. (Photo courtesy of MSD.)

available from MSD if you have an induction system that requires more distributor clearance. The Chevy coil for use with the GM/MSD ignition box is (PN-10037380) and has a maximum output voltage of 45,000 volts.

The GM/MSD style ignition amplifier boxes provide much more ignition power than the stock HEI unit and over a much larger RPM range. Spark plug gaps can be opened up from .045" to as much as .060", if you have high quality RF suppression type spark plug wires that will keep the electrical energy in the wires and prevent leakage. The GM/MSD ignition box can provide measurable increases in horsepower and torque because of it's multiple spark capability, longer duration spark, larger possible plug gaps, and better flame travel in the combustion chamber. Also because of these improved flame travel characteristics, you may be able to retard the total timing advance on the motor 2 or 3 degrees and richen the carburetor slightly.

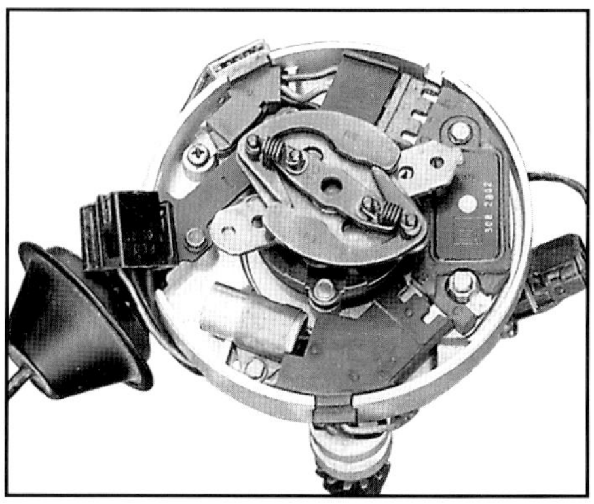

By altering the mechanical advance weights and springs, the mechanical advance curve on the distributor can be optimized for your engine and vehicle combination. Vacuum advance, if you use one, can also be altered by using an adjustable vacuum canister or by using a stock vacuum canister with a different amount of vacuum advance.

In one test, MSD built a 509 cubic inch Chevy big block. It had a 9.5-1 compression ratio and used square port Merlin heads. They used a roller cam and a Holley carb with race fuel. In the first dyno test, they used a bone stock GM/HEI. The motor was pulled to 7400 rpms and the motor made a maximum of 505 ft./lbs. of torque at 5400 rpms. Maximum horsepower came in at 585 at 6400 rpms, but at 6400 the engine was beginning to miss. The motor was left on the dyno and the GM/HEI distributor was then fitted with an MSD GM/HEI replacement ignition module and MSD coil. Torque went to 510 at 5900 rpms and horsepower reached 590 at 6400. A peak increase of 5

BIG BLOCK Quick Facts & References

IGNITION CURVE REQUIREMENTS

If you are making a significant change to your big block on which you had already established optimum mechanical and vacuum advance curves, the recommendations in this chart will help you establish a new curve to regain peak power.

Ignition Curve Requirements

Modification	Vacuum Advance	Mechanical Advance
Increase compression ratio	Retard overall curve (increase spring tension)	Less overall advance, especially at peak-torque (heavier springs and reduce advance weight movement)
Add a high-flow induction system	Remains unchanged	Less advance throughout RPM range (heavier springs and reduce advance weight movement)
Add headers	May need to either advance or retard, depending on charge temperature and exhaust contamination	If charge contamination is reduced, charge temperatures are usually lower. This often requires a slight increase in advance throughout the RPM range (Lighter springs and more advance weight movement)
Longer duration cam	Reduce spring tension in the diaphragm to bring the curve in faster	Initial advance needs to be faster and, to a lesser extent, total advance should occur at a lower RPM (both accomplished with lighter springs)
Supercharger	Needs special vacuum advance mechanism with pressure retard	Total advance should be reduced and the curve should come on slower (stiffer springs and less advance weight movement)
Turbocharger	Needs special vacuum advance mechanism with pressure retard	Reduce total advance with quicker initial part of the curve. Slow curve when turbo boost builds. (use a combination of light and stiff springs, plus reduce advance weight movement
Ported cast-iron heads	Remains unchanged	Slow high end of curve (add one heavy spring)
Aluminum heads	Reduced surface temperatures call for slightly more advance (increase advance amount and slightly reduce spring tension)	Often needs less total advance due to better cylinder filling, but faster initial advance due to lower port velocities and swirl (lighter springs and slightly reduce advance movement)
Switch to unleaded gas (same octane)*	Reduce overall advance (increase spring tension and amount of advance)	Reduce overall advance (increase spring tension and reduce advance weight movement)

*Note: Removing the lead in gasoline reduces the ignition delay time and requires reduced advance. Leaded fuels burn slower and require more advance.

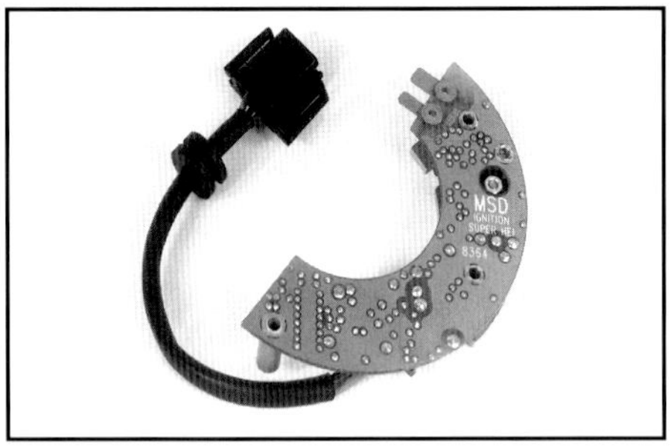

"Super HEI" ignition module and HEI coil from MSD will get your stock GM HEI distributor ready for the races by increasing its high rpm capability and ignition power. (Photos courtesy of MSD)

more foot pounds of torque and 5 more horses and importantly there was no miss on the motor up to 7400 rpm. Last the engine was fitted with a MSD 7AL-2 ignition box and put thru its paces on the dyno. Max torque hit 512 at 5600 and the engine made 605 horsepower at 6800. This ignition added 7 more ft./lbs. of torque and 20 more horsepower than the stock HEI ignition with no more engine misses at higher rpms.

The MSD boxes can also be "triggered" by magnetic pickups mounted on the front vibration damper or on the flywheel. Using these types of crank triggers helps reduce ignition timing variables or fluctuations caused by timing chain slack, camshaft twist or crank vibrations transmitted to the distributor shaft. A distributor is still used, but only to send the spark from the ignition box to the appropriate plug wire.

If you are using a distributor which has had the mechanical timing advance mechanism "locked out", MSD has two timing advance computers that will electronically provide an ignition advance curve. (MSD PN-8980) has a built in 20 degree advance and is not adjustable. (MSD PN-8981) allows a fully adjustable ignition advance curve. Both of these boxes have an adjustable high speed retard feature built in. Both units will retard the advance 20 degrees to aid in starting the engine.

The MSD retard boxes can also help in a number of different situations. In a bracket drag racer, you can set the ignition retard to come on after you get into high gear. The motor is under a heavy load in high gear

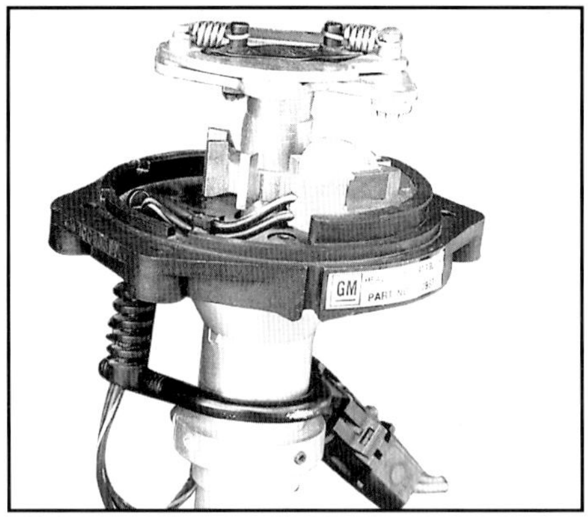

GM can provide a heavy duty, magnetic pickup distributor and coil, multiple spark discharge ignition control box and separate rev limiter. This type of ignition will out perform an old style point distributor or stock HEI.

and total ignition timing can be retarded a few degrees for more power. The retard boxes have individual plug in chips that will let you retard the timing in high gear from 1 to 4 degrees. This can be worth a couple of miles per hour. It also allows you to run slightly more ignition advance in first and second gear for more punch off the line. Another use for the retard box is with a nitrous race motor using multiple stages of nitrous injection. Total timing can be retarded a few degrees as each nitrous stage comes in. MSD has boost retard boxes to work with blower or turbo motors that will retard timing as boost pressure rises in the engine.

Finally two and three step rev limiting boxes can be added for drag racers who need or want them. You can set one rev limit for the burnout, one limit for starting line staging and a third limit for gear changes while going down the track. All of this can be easily tied in to line-locs, trans-brake buttons and other electronic gizmos you may decide to use.

Racers who have not used an MSD style ignition before, install them and then may find that after a few passes on the quarter mile, performance drops off or a engine miss develops. Sometimes the car runs great the first day of a two day weekend meet, but on the second day, problems begin. The racers hurriedly start to check all types of things trying to figure out why there is a problem with the car. Often the problem is the battery voltage and the state of its charge. Many racers who use these high-powered ignition boxes choose to remove the alternators on their cars and recharge the battery between rounds using a portable generator and battery charger. Their feeling is that they recover some lost power because there is no parasitic loss from the drive belts to the alternator and it drops a few pounds off of the front end of the vehicle. The problem is that modern race ignitions develop so much power that a fully charged battery can be taken down from 13.5 to 10-11 volts during a single pass down the track. Consistency also suffers since the battery voltage is fluctuating and falling. Electric fuel pumps are affected by the drop in battery voltage and total pump volume can decrease. Putting the alternator back on the car will improve consistency and provide full battery voltage during an entire quarter mile pass. This is especially in the late rounds, when the turn around time between elimination rounds get shorter and there is not much time to charge a battery.

An alternative is to use a 16-volt battery system on a race motor. A number of systems are available. Battery voltage will still fall without an alternator, but the voltage will stay high enough to run your ignition and fuel pumps properly. Using a 16-volt system and an alternator can help you make those quick turn around late elimination rounds when there is not much time to recharge a battery.

If you use a large (5") diameter deep groove alternator pulley, the alternator speed will be reduced, V-belts will stay on and the parasitic power loss will be probably less than 1 horsepower with a fully charged battery. Even with an alternator, during a two day meet the alternator may not be run long enough to fully recharge the battery. If

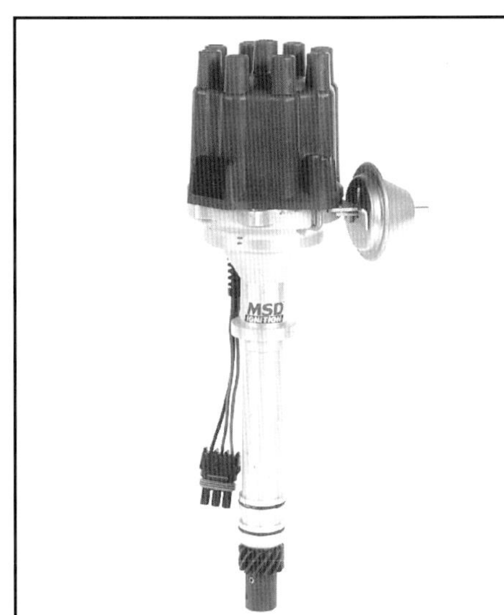

MSD Pro-Billet magnetic trigger distributor (PN-8360) has a built in ignition amplifier. Other features include ball bearings, Tuftrided, over size shaft, 10,000 rpm use, mechanical and vacuum advance and simple wire hookup. (Photo courtesy of MSD)

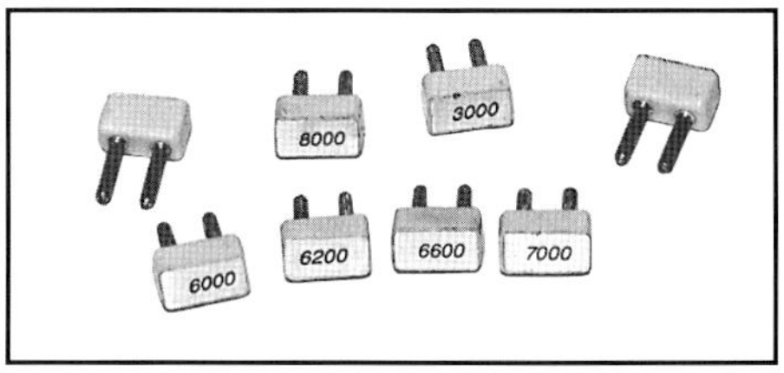

MSD rev limiters use plug-in chips to set the rpm when the limiter comes on. Similar plug in chips are used in ignition retard boxes and "two step" and "three step" rev limiters.

you have a number of other electronic gizmos (delay boxes, Two-Step rev limiter, electric fan, electric water pump, solenoid shifter, etc.) you may have to recharge the battery between rounds with a portable generator and battery charger in addition to using the highest amperage alternator you can use. As far as losing the weight, you can find something else to reduce weight. You could go on a diet and lose five pounds and accomplish the same thing or you could let your significant other drive, which would probably mean a reduction of fifty to one hundred pounds right there.

If you are using an alternator, and you have installed an MSD ignition, the engine may continue to run after you have turned off the ignition key. What is happening is that the ignition box is receiving a small amount of voltage thru the alternator charge warning light circuit. MSD provides a small

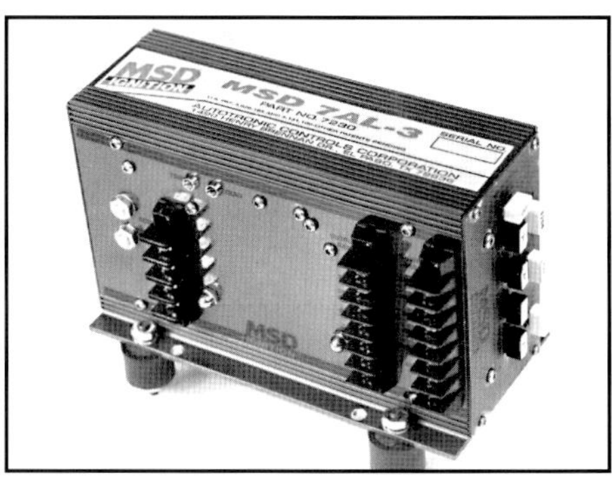

MSD 7AL-2 (PN-7220) and the 7AL-3 (PN-7230) capacitive discharge ignition boxes feature more power for drag race engines. They feature built in rev limiters and a terminal strip to hook up other accessories. Use a MSD Pro Power coil (PN- 8201) with these two boxes for faster rise time and use to 10,000+ rpms. The 7AL-3 (above), which was introduced in 1996, has 35% more power than the 7AL-2. It also has a built in Three-Step rev limiter and a multi-step retard with four possible settings. You also get a 20 degree "start" ignition retard which is great when you have a locked ignition advance or crank triggers. One other 7AL-3 feature is a rpm activated switch that can be used to activate a shift light, shift solenoid or other rpm activated feature you may have. (Photos courtesy of MSD)

diode with their installation kit that can be spliced in to this circuit and prevent this sort of engine run on.

One thing I have noticed with regard to rev limiting devices used during drag races is that some drivers use the rev limiter as if it were a "shift light". They rev the motor until the rev limiter comes on and they hear the "popping" sound that is typical and then shift the trannie. Our experience has been that allowing the rev limiter to come on even for just one "pop", consistently affected our 1/4 mile elapsed time by two-tenths of a second. Even though the limiter was only working for a small fraction of a second before the transmission went into the next gear, this momentary loss of power on one of eight cylinders was reflected in the elapsed time. We used a three speed automatic and the motor revved in first gear very quickly. The driver needed to see the last light on the "tree", leave the line, watch the tach, listen to the motor, drive the car, move his hand on the gear selector and make the shift from first to second gear. We figured he was doing all of this in less than 1.5 seconds. If he wasn't quite quick enough the motor would just begin to over rev in first gear and we would get one "pop" out of the engine and the ET would be off those two tenths. We learned our lesson. Stay out of the rev limiter once you leave the line and the car will produce a quicker ET. The rev limiter is an important safety device. You can use a shift light, or your tach needle "tattle tale" or an audible sound played thru earphones to make your shifts at the correct moment. After racing a given combination for a while, you will be able to tell just from the sound of the revving motor when the time to shift has come. If all else fails or if you think it will improve your driving consistency, you can hook up an air or electric solenoid shifting device to make the gear shifts at the proper time.

Think about the following when you consider race ignition systems. At 7000 rpms,

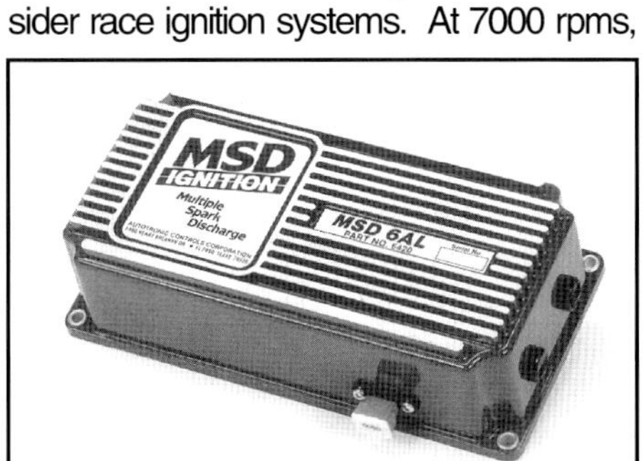

MSD "6A" ignition box will work with just about any type of big block distributor or "trigger". The MSD "6AL" version has a built in rev limiter. (Photo courtesy of MSD)

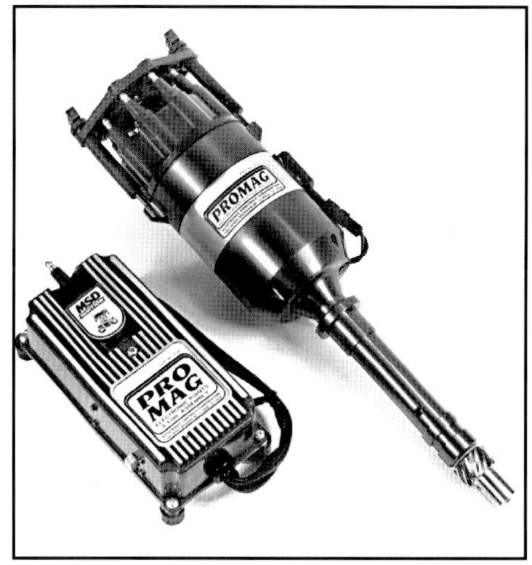

MSD "Pro-Mag" magneto and ignition box are improvements over traditional magneto ignitions. Versions for alcohol and nitro motors are available and it can be used with other MSD ignition accessories. (Photo courtesy of MSD)

during a 10 second run, each plug fires a little over 58 times a second. Multiply this times 8 cylinders and the ignition is firing 466 times a second. At 8000 rpms the ignition is firing 533 times each second. Get a good ignition, use an alternator and make sure the plug wires are in good shape.

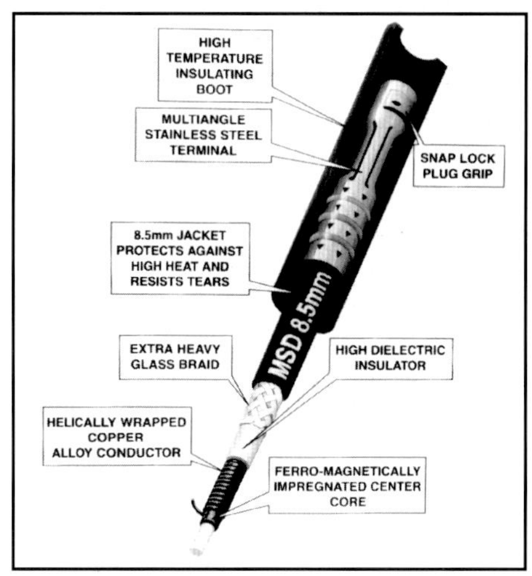

Graphic shows the construction of MSD 8.5mm Super conductor spark plug wire, which is an excellent choice for performance and race use. (Photo courtesy of MSD)

MSD "Blaster 2" coil will work with the 5,6, and 7 series MSD ignition boxes. (Photo courtesy of MSD)

BIG BLOCK CHEVY BUYERS GUIDE

A.B.S. PRODUCTS
591 W. Apollo Street
Brea, CA 92621
714-671-0798
(tools, balancing supplies)

ACCEL PERFORMANCE PRODUCTS
8700 Brookpark Rd.
Cleveland, OH 44129
216-398-8300
(ignitions, sparkplugs, plug wire, electronic fuel injection systems)

AIRBORN COATINGS
18 Emely St.
E. Hartford, CT 06108
203-528-8281
5300 SW 23rd
Oklahoma City, OK 73128
405-685-4433
(Dry film coatings, powder coating, Aluminum/Ceramic coatings)

ARIZONA SPEED AND MARINE
4834 S. 40th Street
Phoenix, AZ 85040
602-437-2510
(Electronic fuel injection systems)

ATI
6747 White Stone Rd.
Baltimore, MD 21207
410-298-4343
(Vibration Dampers)

AUDIE TECHNOLOGY
315-363-8530
(Computer programs for cam measurement, engine simulation, valvetrain geometry, flow bench)

AUTO METER PRODUCTS
413 W. Elm
Sycamore, IL 60178
815-895-8141
(Quality instrumentation)

AUTOMOTIVE RACING PRODUCTS (ARP)
250 Quail Ct.
Santa Paula, CA 93060
805-525-5152
(High quality fasteners)

AUTOTRONIC CONTROLS (MSD)
1490 Henry Brennon Dr.
El Paso, TX 79936
915-857-5200
(MSD ignitions and related controls, EFI parts, injectors, fuel systems)

BARRETT ENGINEERING
1604 N. Magnolia
El Cajon, CA 92020
619-562-3306
(Bracketless race alternator and water pump)

BILL MILLER ENGINEERING (BME)
4895 Convair Dr.
Carson City, NV 89706
702-887-1299
(Forged aluminum rods, Forged pistons)

BILL MITCHELL
35 Trade Zone Drive
Ronkonkoma, NY 11779
516-737-0372
(Big block race parts, machine work)

BORLA PERFORMANCE INDUSTRIES
5901 Edison Dr.
Oxnard, CA 93033
805-986-8600
(Performance exhaust systems and mufflers)

BRAD URBAN'S CARBURETOR SHOP
8460 Red Oak St.
Rancho Cucamonga, CA 91730
909-481-5816
(Carburetors, Carb parts, Fuel systems)

BRODIX CYLINDER HEADS
301 Maple
Mena, AR 71953
501-394-1075
(Cylinder heads, intakes, valvetrain parts)

BURT CHEVROLET/GM PERFORMANCE PARTS
5200 S. BROADWAY
ENGLEWOOD, CO 80110
800-525-5096
(MAIL ORDER GM PERF. PARTS)

CALLIES PERFORMANCE PRODUCTS
110 W. Jones Rd.
Fostoria, OH 44830
419-435-2711
(Crankshafts, Bearings)

CLOYES GEAR AND PRODUCTS, INC.
P.O. Box 287
Paris, AR 72855
501-963-2105
(Timing gears and chains)

COMPETITION CAMS
3406 Democrat
Memphis, TN 38118
901-795-2400
(Cams and valvetrain components, computer programs)

CRANE TECHNOLOGIES GROUP
CRANE CAMS
530 Fentress Blvd.
Daytona Beach, FL 32114
904-258-6174
(Cams and valvetrain components, ignitions, computer programs)

CROWER CAMS
3333 Main St.
Chula Vista, CA 91911
619-422-1191
(Cams, Valvetrain, Cranks, Rods)

DART MACHINERY
353 Oliver St.
Troy, MI 48084
810-362-1188
(Cylinder Heads, Intakes)

DONOVAN ENGINEERING
2305 Border Ave.
Torrance, CA 90501
310-320-3772
(Aluminum blocks, CNC machining)

EAGLE SPECIALTY PRODUCTS (E.S.P.)
3075 Norbrook Dr.
Memphis, TN 38116
901-345-5886
(Connecting rods)

EARL's PERFORMANCE PRODUCTS
189 W. Victoria St.
Long Beach, CA 90805
310-609-1602
(Braided hose, Fittings, Specialty hardware)

EDELBROCK CORPORATION
2700 California St.
Torrance, CA 90503
310-782-2900
(Heads, intakes, carbs, water pumps, exhaust systems)

ELGIN INDUSTRIES
1100 Jansen Farm Dr.
Elgin, IL 60123
847-742-1720
(Cams, valvetrain, cranks, rods)

ENGINE PROTOTYPE DEVELOPMENT
49 Laxalt Dr.
Carson City, NV 89706
702-246-5015
(Brodix cylinder head development "EB" series Brodix heads)

FEDERAL-MOGUL CORPORATION
26555 Northwestern Hwy.
Southfield, MI 48034
810-354-7700
(TRW and Speed-Pro pistons, rings, Federal-Mogul bearings, Carter carbs, Sterling pistons, TRW valvetrain components)

FEL-PRO CORPORATION
7450 N. McCormick Blvd.
Skokie, IL 60076
708-674-7700
(Gaskets, sealants)

FERREA RACING COMPONENTS
2600 NW 55th Court
Suite 238
Ft. Lauderdale, FL 33309
954-733-2505
(Valves, Springs)

FERRY'S ALUMINUM CYL. HEADS AND BLOCK REPAIR
710 Quietwood St.
Dallas, TX 75253
214-286-6199
(Valvetrain geometry checking tools)

GALE BANKS ENGINEERING
546 Duggan Ave.
Azusa, CA 91702
800-438-7693
(Turbo charger systems, performance packages for late model, emissions vehicles)

GM/PERFORMANCE PARTS
(See your local Chevrolet or GM/PERFORMANCE dealer)

GM/RESTORATION PARTS
(GM Catalog PN-12363125 lists all approved and licensed parts vendors for GM/RESTORATION parts)

GOODSON SHOP SUPPLIES
PO Box 847
Winona, MN 55987
800-533-8010
(specialty engine tools and shop supplies)

HOLLEY PERFORMANCE PRODUCTS
PO Box 10360
Bowling, KY 42102-7360
502-781-9741
(Carbs, carb parts, carb restoration, intakes, electronic fuel injection, fuel curve mapping computer programs, fuel pump systems)

HOOKER HEADERS
1024 W. Brooks St.
Ontario, CA 91762
909-983-5871
(Exhaust tube headers, mufflers, exhaust pipe,etc.)

HYPERTECH
1910 Thomas, Rd.
Memphis, TN 38134
901-382-8888
(Performance PROM computer chips, underdrive pulleys, ignition parts)

INTERNATIONAL HOT ROD ASSOCIATION (IHRA)
P.O. Box 3029
Bristol, TN 37625
423-764-1164

JE PISTONS
15312 Connector Lane
Huntington Beach, CA 92649
714-898-9763
(Pistons)

JESEL
908-901-1800
(Shaft rocker systems and Timing belt drives)

K&N ENGINEERING
PO Box 1329
Riverside, CA 92502
909-684-9762
800-992-3000
(Stubstack, Air & oil filters)

LUNATI
4770 Lamar Ave.
Memphis, TN 38118
901-365-0950
(Cams, valvetrain parts, Cranks, Rods, Pistons, Engine kits)

MADCAP RACING ENGINES
2201 19th St.
Denver, CO 80202
303-298-0044
(Big Block race engines, dyno service, head porting)

MALLORY
550 Mallory Way
Carson City, NV 89701
702-882-6600
(Ignitions)

MANLEY PERFORMANCE
1960 Swathmore Ave.
Lakewood, NJ 08701
908-905-3366
(Pistons, rods, valves)

MOROSO PERFORMANCE PRODUCTS
80 Carter Dr.
Guilford, CT 06437
203-453-6571
(Engine products, tools)

MOTION SOFTWARE, INC.
535 W. Lambert, Bldg.-E
Brea, CA 92621-3911
(Automotive computer simulation software)

MPG HEAD SERVICE
3881 S. Jason Dr.
Englewood. CO 80110
303-762-8196
(Exhaust port plates)

MR. GASKET COMPANY
8700 Brookpark Rd.
Cleveland, OH 44121
216-398-8300
(Gaskets, ignitions, clutches, engine computer software)

NITROUS OXIDE SYSTEMS (NOS)
5930 Lakeshore Dr.
Cypress, CA 90630
714-821-0580
(Nitrous Oxide systems and hardware)

NATIONAL HOT ROD ASSOCIATION (NHRA)
P.O. Box 5555
Glendora, CA 91740-0950
818-914-4761

OLIVER RACING PARTS
1025 Clancy Ave., NE
Grand Rapids, MI 49503
616-253-8108
(Connecting Rods)

PAINLESS WIRING
9505 Santa Paula Dr.
Ft. Worth, TX 76116
817-244-6898
(Engine and trans electrical wiring harnesses and supplies)

PERFORMANCE DISTRIBUTORS
2699 Barris Dr.
Memphis, TN 38182
901-396-5782
(Performance distributors, alternators, ignition parts)

POLYMER DYNAMICS
4116 Siegel St.
Houston, TX 77009
713-694-3296
(Engine Parts Coatings)

POWERHOUSE
3402 Democrat Rd.
Memphis, TN 88118
901-795-7600
(Specialty Engine building Tools and Measuring devices)

RACING SYSTEMS ANALYSIS
PO Box 7676
Phoenix, AZ 85011
602-241-1301
(Computer programs)

RODECK
2134 Tractor St.
Paso Robles, CA 93446
805-237-6040
(Aluminum race blocks)

ROSS RACING PISTONS
12901 S. Yukon
Hawthorne, CA 90250
800-392-7677
(Race Pistons)

SIMPSON RACING PRODUCTS
2415 Amsler St.
Torrance, CA 90505
310-320-7231
(Safety equipment)

SUMMIT RACING EQUIPMENT
PO Box 909
Akron, OH 44309-0909
330-630-3030
(Mail order performance parts)

STREET & PERFORMANCE
Route 5, #-1, Hot Rod Lane
Mena, AR 71953
501-394-5711
(Electronic Fuel injection systems, wiring harness, computers, turn-key EFI engines, Pulleys, accessory brackets)

T & D MACHINE PRODUCTS
4859 Convair Ave.
Carson City, NV 89706
702-884-2292
(Shaft and pedestal rocker arm systems,Specialty tools)

THE TURBO SHOP
1280 Kona Dr.
Compton, CA 90220
310-669-8101
(Transmissions, adapters, computer chips, wiring harnesses)

TOTAL SEAL
2225 W. Mountain View, #-6
Phoenix, AZ 85021
602-678-4977
(Piston Rings, Engine development R & D)

TRANS-DAPT
9599 W. Jefferson Blvd.
Culver City, CA 90232
310-921-0404
(Adapters, pulleys, carb spacers, Oil filter relocation kits)

ART CARR
714-962-6655
(Performance Transmissions)

ULTRADYNE RACING CAMS
P.O. Box 394
Southaven, MS 38671
601-393-3214
(Race cams)

VIBRATECH
11980 Walden Ave.
Alden, NY 14004-9790
716-937-3603
(Fluidampr vibration dampers)

VORTECH ENGINEERING
5351 Bonsai Ave., Suite-1114
Moorpark, CA 93201
805-529-9330
(Centrifugal Superchargers and related parts)

WEIAND
2316 San Fernando Rd.
Los Angeles, CA 90065
213-225-1346
(Intakes, superchargers, water pumps)

WISECO PISTON, INC.
7201 Industrial Park Blvd.
Mentor, Ohio 44060
800-321-1364
(Race Pistons)

WORLD PRODUCTS
353 Oliver St.
Troy MI 48084
810-244-9822 / 516-737-0450
("Merlin" cylinder heads, blocks, intakes)